CR

Creativity, whether lauded as th............................uted as
a driver of international policy, ...₄.......₄cs, nas been very
much part of the zeitgeist of the last few decades. Offering the first accessible, but conceptually sophisticated account of the critical geographies of
creativity, this title provides an entry point to the diverse ways in which
creativity is conceptualised as a practice, promise, force, concept and rhetoric. It proffers these critical geographies as the means to engage with the
relations and tensions between a range of forms of arts and cultural
production, the cultural economy and vernacular, mundane and everyday
creative practices.

Exploring a series of sites, *Creativity* examines theoretical and conceptual questions around the social, economic, cultural, political and
pedagogic imperatives of the geographies of creativity, using these
geographies as a lens to cohere broader interdisciplinary debates. Central
concepts, cutting-edge research and methodological debates are made
accessible with the use of inset boxes that present key ideas, case studies
and research.

The text draws together interdisciplinary perspectives on creativity,
enabling scholars and students within and without geography to understand and engage with the critical geographies of creativity, their breadth
and potential. The volume will prove essential reading for undergraduate
and post-graduate students of creativity, cultural geography, the creative
economy, cultural industries and heritage.

Harriet Hawkins is a Reader in Geography at Royal Holloway, University
of London, UK where she co-directs the Centre for GeoHumanities and
the MA Cultural Geography (Research). Her research focuses on the
geographies of art work and art worlds, and often proceeds through
creative practice-based collaborations with artists and arts organisations.

Key Ideas in Geography

SERIES EDITORS: SARAH HOLLOWAY, LOUGHBOROUGH UNIVERSITY AND GILL VALENTINE, SHEFFIELD UNIVERSITY

The *Key Ideas in Geography* series will provide strong, original, and accessible texts on important spatial concepts for academics and students working in the fields of geography, sociology and anthropology, as well as the interdisciplinary fields of urban and rural studies, development and cultural studies. Each text will locate a key idea within its traditions of thought, provide grounds for understanding its various usages and meanings, and offer critical discussion of the contribution of relevant authors and thinkers.

For a full list of titles in this series, please visit www.routledge.com/series/KIG

CREATIVITY

Harriet Hawkins

Routledge
Taylor & Francis Group

LONDON AND NEW YORK

First published 2017
by Routledge
2 Park Square, Milton Park, Abingdon, Oxon OX14 4RN

and by Routledge
711 Third Avenue, New York, NY 10017

Routledge is an imprint of the Taylor & Francis Group, an informa business

British Library Cataloguing in Publication Data
A catalogue record for this book is available from the British Library

Library of Congress Cataloging in Publication Data
Names: Hawkins, Harriet, 1980- author.
Title: Creativity / Harriet Hawkins.
Description: Abingdon, Oxon ; New York, NY : Routledge, 2017. |
    Series: Key ideas in geography | Includes bibliographical references.
Identifiers: LCCN 2016014440| ISBN 9781138813434 (hardback :
alk. paper) | ISBN 9781138813441 (pbk.: alk. paper) |
ISBN 9781315748153 (ebook)
Subjects: LCSH: Creative ability–Social aspects. | Creation
(Literary, artistic, etc.)–Social aspects. | Cultural geography.
Classification: LCC BF408 .H33 2017 | DDC 153.3/5–dc23LC record
available at https://lccn.loc.gov/2016014440

ISBN: 978-1-138-81343-4 (hbk)
ISBN: 978-1-138-81344-1 (pbk)
ISBN: 978-1-315-74815-3 (ebk)

Typeset in Joanna MT
by RefineCatch Limited, Bungay, Suffolk

For all the creative geographers who I have been lucky enough to work with, thank you for your spirit and inspiration.

Contents

Figures

Boxes

ACKNOWLEDGEMENTS

The diverse creativities in this volume are in part function of my intellectual journey across geography in search of creativity over the last few years. This includes an interdisciplinary PhD in the Geographies of Art at the University of Nottingham; a post-doc on the cultural geographies in the south-west of Britain at the University of Exeter, and a further post-doc at Aberystwyth University and University of Arizona that enabled me to work on creativity in art and science. More recently, I have been lucky enough to be surrounded by creative geographers at Royal Holloway, University of London. Here my understanding of creativity has broadened and deepened alongside colleagues and the masters and PhD students I work with, but also the undergraduates who took my course 'Creative Geographies: Spaces, Practices and Economies'. This course, run since 2012, has formed the inspiration for this book and the testing ground for much of its material. Its contents and final configuration owe much to what the members of this course responded to, ran with and found to be most challenging and interesting.

I have been lucky enough to work with a number of different creative geographers over the decade's worth of scholarship that has built up to this book, both those that work directly with and through creative practices and those whose scholarship exhibits a creativity in thought and approach. My thanks go out to all of them for their inspiration, the time they have given up and their generosity of spirit especially those who have written short texts for this book but in particular to: Nick Alfrey, Nelly Ben Hayoun, Candice Boyd, Katherine Brickell, Miriam Burke, Ruth Catlow, Ian Cook, Phil Crang, Stephen Daniels, Deborah Dixon, Mona Domosh, Felix Driver, Sarah Elwood, David Gilbert, David Harvey, Joseph

Hawkins, JP Jones, Anja Kanngieser, Innes Keighren, Annie Lovejoy, LeiLei Li, Sallie Marston, Al Pinkerton, Amanda Rogers, Libby Straughan and Nicola Thomas. Finally, special thanks are due to Brad Garrett for reminding me not to lose sight of the possibilities for creativity and everyday adventures in the midst of the day to day and to Katherine Brickell, Miriam Burke, Joseph Hawkins, Innes Keighren, Sallie Marston, Al Pinkerton and Libby Straughan, for reminding me what is really important.

1

INTRODUCTION: TOWARDS CRITICAL GEOGRAPHIES OF CREATIVITIES

'Live, Work, Create', in the early years of the new millennium this graffiti slogan appeared on buildings, underpasses and bridges across New York City, its ubiquity mirroring that of creativity more generally. For whether lauded as the oil of the twenty-first century, touted as a driver of international policy, or used to shape national identity, creativity has been very much part of the zeitgeist of the last few decades. But yet, while the stencilled slogan might seem to reflect the 'Be creative – or die' mantra of urban policy makers and planners the world over, the assertiveness of the 'Live' challenges any account of creativity that focuses solely on its economic dimensions.

Creativity is a lively field of study, but one fractured in its diversity and riven with tensions. It is also full of promise, variously understood as the saviour of the economy, as a tool of neoliberal politics and part of the diplomatic arsenal of state-craft practices, as a psychological trait and philosophical concept. It is also an embodied, material and social practice that produces both highly specialist cultural goods and is a part of everyday life, and it offers myriad possibilities for making alternative worlds. Despite all these possibilities, such an expanded sense of creativity can feel crippling to some, diluting meaning to the degree that the idea of creativity becomes useless; an anything goes term that applies to everything and so nothing. One response to these fears has been to patrol boundaries, guard definitions and restrict understandings. Yet many are aware of the challenge to retain creativity's critical and analytical precision together with its lively possibilities, to counter tendencies towards definitional promiscuity without becoming precious. This text finds a solution to such dilemmas in an exploration of the critical geographies of creativity. Across nine creative 'sites' this volume will put creativity in its place and in so doing will explore how it can make places, shape subjects, connect communities and sculpt environments. In such geographies an analytic frame is found that enables an embrace of expanded understandings of creativity, whilst retaining a sense of precision and focus that enables, in short, an understanding and even possibly the mobilization of creativity's promises.

In what ways does it make sense to conceive of creativity as having a geography? Creativity happens in lots of different places, does it matter where? Does the location of creative endeavours make any difference to their content and conduct? Further, does the creative activity affect the sites and venues at which it happens? The answer to all of these questions is yes. Creativity, in short, has a whole set of geographies, and in turn creative practices produce geographies, they make places, shape the bodies, subjectivities and minds of those conducting them, and weave together communities and evolve environments. Across a series of nine sites, from the body to the city, the landscape and environment and the nation, this text is going to explore these critical geographies of creativity, using this as a purchase point onto the richly different and often contrasting understandings of creativity. The aim at each site will be to explore tensions between different forms of creativity found at that location, demonstrating how geography matters to understanding creativity, but equally how creativity matters to the production and understanding of geographies.

It makes sense to conceive of creativity as having a geography in all sorts of different ways. Creativity is concerned with ideas, with objects, with performances, with practices, and is the affair of individuals, institutions, companies and governments. All of these have spatial dimensions. Matters of geography are involved at all stages of the creative process, whether this be the micro-geographies of the body that is thinking and practicing, or the myriad spaces in which creativity happens, from studio spaces, to the home, the gallery, opera house or street. We could think too of the planning of creative spaces, their hot-housing by governments and policy makers keen to harness creativity's economic and social possibilities, or the subversive creativities that function as critical spatial practices, challenging the status quo of planned and surveilled spaces. To query what is created, how and what it does in the world is to ask questions that are intimately bound up with the 'where' of creativity. In short, just as geographers and historians who study the production of science are aware of the importance of 'placing' practices of science, so too should we be aware of the geographies of creativity. For, as this text will argue through its nine sites, such geographies enable us to understand the intersections of the politics and economics of the creative economy with comprehensions of creativity as a social, cultural, material and embodied practice with subject and world making potentials. What is more, such geographies enable us to appreciate the potential and possibilities of creativity that sit alongside, and at times even critique, creativity's central place in contemporary economic and political spaces and processes.

Historical geographers and historians know that place matters to the production of knowledge, especially science, and creativity is no different. Unlike science, creativity was never really assumed to be a universal practice that would be the same everywhere. Indeed, creativity perhaps even sits apposed to the universal replication, routinization and standardization that science once desired. Historians and geographers of science have long asserted the need to engage with the particularities of the local conduct and content of science and have explored how profoundly it is shaped by the venues in which it is made. Yet, while scholars of science have tackled the imaginaries of the placeless places of science, creative practices have, by contrast, long been understood to be a product of the people and places in which they are found (Livingstone, 2003; De Costa Kaufmann, 2004). Whether that be a function of the inspiration offered by local scenery, the use of local materials or the evolution of local styles, an environmentally determinist streak has long dominated the histories of the production of arts and literature. But this is not to say that it is not worth spending time putting creativity in its place, to echo Livingstone's (2003) exploration of science. For too easily it seems that the places where and practices through which creativity is conducted are overlooked, but as this text will explore they matter profoundly, to the types of creativity that occur, to how it is we understand creativity and what it might do, and to what it might mean to understand and mobilise the promise of diverse creativities, rather than being disempowered by them.

Indeed, the geographies of creativity are no less significant with respect to consumption than they are concerning production. Ideas, images, objects and practices travel from person to person and place to place, from culture to culture (Livingstone, 2003). Such migrations are not, however, the same as replications. As creative practices and objects circulate, just like scientific practices and knowledge they undergo translation and transformation because people encounter them differently in different circumstances (Rogers, 2014). If creative practices need to be understood as shaped by the context of the periods and places from which they emerged, then their reception must also be temporally and spatially situated. So, if we are to understand how creativity has shaped the world we need to be attentive to how these practices have been appropriated, how they have been made and remade rather than just replicated. Furthermore, we need to work with less fixed conceptions of creativity. For, what passes for creativity is contingent on time and place, it is constantly negotiated; to 'be creative' means different things in different eras and in different places. Cultivating critical geographies of creativity will help us comprehend the diversity of understandings of creativity less as a challenge to its analytic precision. Rather, these geographies help us to explore the interesting and valuable tensions between different understandings of creativity. They also enable us to conceive of how creativity both bears the imprint of the locations of its production and consumption and is a force in the 'making' of these locations. The remainder of this introduction is going to take a closer look at the issues of diverse creativities, then explore in greater depth what the critical geographies of creativity might mean and be, and how they might help us understand the possibilities to be found in diverse creativities. It will close by introducing the nine creative sites which follow.

LIVE, WORK, CREATE: THE CHALLENGE OF DIVERSE CREATIVITIES

> something relevant may be said about creativity, provided it is realized that whatever we say it is, there is also something more and something different.
>
> (Bohm and Peet, 2000, p. 226)

'Live, Work, Create', appears for all intents and purposes to be a street artist's tag, but it is actually the philosophy-cum-slogan for New York based hip clothes and urban living brand Brooklyn Industries.[1] In many ways, Brooklyn Industries distills any number of the success stories of the contemporary creative economy, and as its name implies the company has closely tied its brand identity to geographic location. The local area and its heritage as an area of heavy industry and, more recently, garment manufacturing drives the company's marketing strategy, which includes naming products after city areas and a logo based on the Manhattan skyline. However, both the drive for a certain sort of life and the possibilities that creative products enable for lifestyle creation is distilled in the story of the brand.

The story of Brooklyn Industries intersects creative production and place with global narratives of changing geographies of making and manufacturing (Carr and Gibson, 2015; Scott, 2000). Started in the late 1990s by two artists disaffected with their day jobs, the company's origins lie in the pair's dumpster diving activities, during which they up-cycled used billboard fabrics to create bags. As the pervasive narration of their origin story details (found on the company blog, Instagram, Tumblr and YouTube), they built Brooklyn Industries 'one bag at a time not knowing where it will lead us.'[2] This incremental, speculative strategy clearly paid off; they have grown from a single sewing factory, opening in 1996 in the then run-down industrial area of Williamsburg, to in 2014 a new HQ by the water in trendy DUMBO near their recently refurbished factory block. Having initially outsourced much of their manufacturing to family-run factories in China, the global recession saw the company return to its NYC roots. Promotional YouTube videos and blog entries tell a tale of an international company now firmly rooted in place. Whilst acknowledging the challenges of manufacturing in New York, the company trades on both the manufacturing heritage of the area and their 'field to factory' ethos. This ethos sees them making their leather goods and recycled bags in their DUMBO factory and the surrounding area, and producing many of their top-selling T-shirts within 150 miles of their Brooklyn HQ. The products are sold online and through sixteen carefully curated stores known for their 'up-cycling' parties, art installations and sustainable energy practices. Aside from the brand kudos gained from being local, the company cites reasons for remaining in the city as ethics, local inspiration, and the ease of having the whole product designed and made under one roof, enabling space and time for experimentation and a quick turnover in response to emerging street-style trends. Of importance too is the ecology of creative practices that Brooklyn Industries is a part of:

fashion, graffiti, design, visual arts, music and the everyday practices of street-style, to name just a few. Key in Brookyln Industries' story is a sense that it, like so many similar companies, are not just using the city for their location and inspiration, but are also collectively contributing to the creation of the urban atmosphere. Whether this is directly through their branding and the presence of their hip boutiques and their support of arts activities, or more indirectly by just adding to the aura of their area. Indeed, Brooklyn Industries is one of the many creative companies contributing to what maintains Brooklyn's status as a 'cool' neighbour-hood (Zukin, 2010).

This is, however, not just a story of geographical work-based efficiency and place-based branding. Across many social media platforms Brooklyn Industries narrates a creative dream that folds together those three terms – live, work, create – in a tale of the satisfaction with a creative working life ethically and sustainably lived. Dovetailing with the economic facets of the story are narratives of creative labour, especially the more personal narrative of the creative practitioner, the politics and subjectiv-ities of creative practices, of creative entrepreneurialism, but also of the desire to seek a creatively fulfilling life (Gill and Pratt, 2008). Creativity, in short, is not just a way of making a living but also about making lives.

Exploring the backstory of Live, Work, Create keys us into some of the critical coordinates that frame the discussions of creativity within this text. Unfolding from the specificities of this trio of words is, of course, a story of 'work' and of the geographies of the creative economy. The creative sector has been lauded as the 'oil of the twenty-first century' and not without good reason. Creativity was one of the economic success stories of the first decade of the new millennium, one of the few economic sectors not only resilient to the global economic recession, but actually to demonstrate growth during this era. In 2008, despite a 12 % decline in global trade the creative goods sector continued to expand, doubling between 2002 and 2011 to reach a net worth of $624 billion.[3] The same report also noted that annual growth in so-called developing countries was 12.1 % higher than the worldwide average of 8.8 %, and that creative products accounted for over half of these countries total global exports. In 2014, 1 in 10 jobs in the UK and 1 in 6 graduate jobs were in the creative economy, figures that are replicated around the world.[4] But what are the creative and culture industries (CCI) exactly? What occupations and skills constitute them; what is being counted within these global and national figures? These are questions that continue to be posed by academics working in the sector. If creativity more generally can be understood in terms of a diverse set of different definitional debates, then these discussions find a microcosm in the ongoing machinations around how to define the creative/cultural industries/economy.

Within economic circles, creativity remains definitionally slippery, taking up a place within neoliberal vocabularies alongside terms such as flexibility and precarity. Marxist scholar Jamie Peck notes the presence and mobilization of creativity as a

> distinctly positive, nebulous-yet-attractive, apple-pie-like phenomenon: like its stepcousin flexibility, creativity preemptively disarms critics and opponents, whose resistance implicitly mobilizes creativity's antonymic others – rigidity, philistinism, narrow mindedness, intolerance, insensitivity, conservativism, not getting it.
>
> (2005, p. 765)

Who, as Pratt and others have argued, 'wants to be uncreative?' Given the power accorded to creativity, not only in its ability to deliver economic regeneration, but also social benefits, it is perhaps no surprise that it has also become a key feature of the policy landscape. Creativity has long had a place, albeit a shifting one, within the broader policy landscape. For many commentators there is a key transition from a policy era in the 1960s and 1970s where culture and arts were given social and cultural rationales, to a more recent era in which the economic rationales rule (Kong, 2005, 2014).

The shifting understandings of creative policy reflect the 'tortuous and contorted definitional history' of the arts, cultural and creative industries (Roodhouse, 2001, p. 505). Whilst the historic roots of the creative economy in the craft industries is often acknowledged in its contemporary form, it is seen as part of the post-Fordist knowledge economy. This economic form was brought to the fore in the late 1990s by European and American scholars and analysts who wanted to describe the sorts of people, skills and ideas that were prospering in the evolving information, knowledge, digital and/or 'weightless' economy (Amin and Thrift, 1992). Whilst often associated with immaterial production, the experience economy and the consumerist lifestyle of the creative industries are often based in a sense of material production, whose key feature is the evolution of creative products as symbolic forms (Scott, 2000, p.12). We can think, for example, about 'goods and services that have some emotional or intellectual (i.e. aesthetic or semiotic) content' (ibid). As Scott continues,

> commodified symbolic forms are products of capitalist enterprise that cater to demands for goods and services that serve as instruments of entertainment, communication, self-cultivation (however conceived), ornamentation, social positionality, and so on, and they exist in both pure distillations, as exemplified by film or music, or in combination with more utilitarian functions, as exemplified by furniture and clothing.
>
> (2000, p. 12)

One of the key areas of debate around the creative economy concerns the tensions between the economic, social and cultural roles of creative activities. When the cultural sector is folded into the creative economy, its own values and ideas are often bypassed, overlooking culture's public benefits. For, as is increasingly the case, the creative economy is often thought of as the means to: 'overcome social and economic inequalities and effect future economic growth' (Banks, 2007, p. 71). To privilege creative economies, however, is to exclude other dimensions of creativity and those individuals whose creativity is not primarily conducted through economic logics. More specifically, to focus on making money 'structures how creativity is defined, developed and employed' (Banks, 2007, p. 73). For other scholars working in this area, especially those who think about the mobility of creative policy, there is ongoing imprecision in how we characterise these industries, a lack of empirical evidence and confusion over their potential role (Kong and O'Connor, 2009). Global statistics are recognised to be very hard to create as different countries consider creative occupations in very different ways, with implications for scholarship, policy and practice.

However important the creative economy might be for economic growth and policy direction, that trio of words – Live, Work, Create – direct us firmly to understandings of creativity beyond that of the creative economy; to the relationship between creativity and living. If we must consider culture to be, as Marxist critic Raymond Williams (1958) famously wrote, 'ordinary' then we must also consider creativity to be ordinary too. To appreciate the expanded field of creativity is not just the preserve of the singular artistic genius, nor is it purely an economic process. Alongside being a practice that produces creative products, whether these are clothes and bags, art works, writing, films or advertisements, creativity is also an everyday practice; not so much working to live, as living creatively. These forms of creative living might be the vernacular creativities of hobbyists, or enthusiasts (Edensor et al., 2009a), the everyday creative practices of cooking, decorating or dressing, or the improvised creativities which are an often unacknowledged part of how we move through the world (Hallam and Ingold, 2008). Indeed, we must appreciate creativity as very much part of our everyday lives and our ways of being in the world; creativity, in other words, as a daily practice proliferating in the ordinary spaces in which we live our lives; our homes, our streets, our communities and our environments. That such everyday creativities become overlooked is a function of how urban and western-centric notions of creativity promoted in the creative economy have tended to sideline other people, other places and other forms of creativity (Edensor et al., 2009a).

Moving beyond solely economic understandings of creativity puts back into play questions concerning what it is that creativity does in the world. One answer lies in the first of that trio of graffitied words 'live'. Vernacular creativities are often about the processes of everyday life, about getting through the day – about processes of cooking, of decorating houses, of dressing; they are also moments of unconscious improvisation – making do and getting by, folding up a napkin to level a wobbly table, jamming a wooden spoon in the window to prop it open. Vernacular creativities also encompass those clearly conscious 'creative' activities such as hobbies, including knitting and model railway building, as well as Christmas light design and decoration (Edensor and Millington, 2009, 2012; Price, 2015). To explore vernacular creativities is to be returned to some of those central questions; where does creativity happen, who is creative, and, crucially for this text, what is it that creativity does?

Considering creativity as part of everyday life is also to challenge its association with the inner city and with specialist spaces of production, such as workshops, studios and galleries. Instead, creativity becomes something that happens in homes and on streets, in rural and suburban areas as much as in urban cores. There has been, for example, a growth of late in studies of creative suburbia that challenge those stereotypes of suburbia as an uncreative space. Instead it becomes realised as a space full of creative possibilities and potential for professional creative arts practitioners and amateurs alike as well as, of course, a site of those vernacular creativities associated with everyday life practices. Another group of spaces have also risen to the fore as sites of creativity, namely those interstitial urban spaces often assumed to be devoid of value. Such spaces might include wastelands and industrial ruins, SLOP – spaces left over after planning below overpasses or between buildings, for example – semi-abandoned infrastructure, or those spaces awaiting completion (Borden, 2001; Edensor, 2005a, 2005b). As studies have shown such sites form fertile loci for creative practices and play, whether this be graffiti or other forms of urban subversion or alternative forms of living.

Primarily then, that trio of graffitied words – live, work, create – signal the diversity of understandings of creativity that sit at times in tension; that creativity is not just an economic force or a political rhetoric, but is also an embodied material practice, carried out both as a form of economic labour, and in rather romanticised terms, as a labour of love. It is also a practice of living, a means of getting by in the world. Creativity, and in particular its relationship to the arts, might capture the zeitgeist of the twenty-first century, but its roots date from the late seventeenth century when creativity evolved as a term to denote the human and secular counterpart to the religiosity of 'creation' (Singer, 2010). An enduring feature of these centuries worth of discussions of creativity has been their sheer breadth and diversity. Indeed, whether looking to philosophical accounts, to discussions of creativity and business, or to critical reflections on the politics of creativity, what emerges is a sense of the variegated nature of these understandings. As Bohm (2004) notes, creativity is both a prized trait in contemporary society but also a puzzlement. Writing from within the creative economy literatures, Flew notes creativity may be big business but that it is also 'a lot of different things to a lot of different people' (2003, p. 90). This book takes in a cross-section of understandings of creativity. It does so in a way that celebrates its puzzlements, looking for richness and possibility in the tensions between diverse understandings of creativity in different times and places. It seeks, in short, to think about what creativities are and might do rather than to define what creativity is. Such a singular definition is as Singer (2010, p.24) suggests, unneeded. How then to negotiate this variegated degree of practices, how to make sense, productively, of their differences and similarities? The answer this volume offers is through the critical geographies of creativity.

THE CRITICAL GEOGRAPHIES OF CREATIVITY?

Geography as a discipline is undergoing somewhat of a creative (re)turn of late. Creativity in myriad forms, practices and understandings is now common across the discipline, whether we look to the economic, political and social geographies of the creative economy, the cultural and social geographies of creative practices – fine art, literature, film, music or vernacular creativities – or to the attention geographers have been paying to creative practices as research methods. We find studies of creative products, processes and people distributed widely across the discipline and in bringing these different ideas together we can begin to explore what such combinations bring about.

We find, for example, a concentration of work in cultural geography that takes in a cross-section of creative products – from novels, to music, films, performance and art work – as empirical objects through which to explore substantive geographic concerns. These concerns might be space, place, mobilities, landscape or the environment (Daniels, 1993; Dixon and Cresswell, 2002; Hawkins, 2013b; Pratt and San Juan, 2014; Rogers, 2014). Here the analysis of such arts practices as empirical objects offers geographers the chance to engage evolving debates around, for example, the environment, around city spaces and practices, or around how it is we understand landscape. Perhaps of late the discipline has witnessed a sense of cultural geographic sources and methods, in terms of the analysis of creative practices and products spreading far beyond the sub-discipline. Evolving too from these studies has been the uptake of creative practices by geographers themselves, for whom becoming writers, artists, curators or film-makers comes to offer a particular methodological value in engaging with geographic concerns (Hawkins, 2011; Last, 2012; Marston and De Leeuw, 2013). These studies have been diffuse, encompassing creative methods as a response to the evolving importance of embodied engagements with the environment, as well as taking in the use of creative practices as participatory methods to engage particular hard-to-reach or at-risk communities and groups.

Within economic geography we find questions of creativity concentrating around concerns with the creative economy and the cultural industries. Key here has been an exploration of the spatial patterns of these industries and the examination of their relation to other geographies of capitalism, including the evolution of the knowledge economy (Power and Scott, 2004; Scott, 2000). What has cohered much geographical interest in the creative economy has been the ongoing importance of localization to the development and sustenance of creative industries (Bathelt et al., 2004). In an era in which geography (understood as territorial fixity) was supposedly dead, when economic activity was being touted as footloose and global and as being fueled by a networked society driven by virtual working practices, apparently enabling work to happen anywhere, the patterns of localised concentration displayed by these industries made them a subject of fascination (Scott, 2000). As the creative economy has become one of the global success stories even amidst global recession, and with the exponential growth in the uptake of the creative city 'script' as a driver of urban regeneration, critical urban geographers have taken note. Thus, alongside urban-planning based scholarship there is now a vocal body of scholarship that articulates the need for critical reflection on the rhetorics of creativity (Peck 2005; Wilson and Keil, 2008). Furthermore, there is an emerging field of research that situates urban subversions, so-called tactical urbanisms, as one means by which to short-circuit and potentially to rewrite these creative scriptings of the city (Borden, 2001; Mould, 2015).

Cross-cutting attentions to creativity within economic geography and cultural geography's studies of creative practices, is an emerging field of scholarship on what has been called 'vernacular creativity' (Edensor et al. 2009a). Simply put, for geographers such vernacular creativity concerns the doings and sayings of everyday life; they open out the sites at which we think about the production and consumption of everyday life. Rather than happening in the studio or gallery, this is creativity produced and consumed in gardens, sheds or homes, or on the street; that classic locus of everyday life. Further, these forms of creativity are often resolutely not for economic gain, with value found in other attributes such as creativity's role in making places and identities.

This text seeks to work across and between these different bodies of literature through a series of three critical geographies; creativity as a socio-spatial practice; the politics of creativity; and creativity as a force in the world. The aim of these geographies is not to collapse different forms of creativity around common concerns, but rather to use them to unpack the relations and tensions amongst different understandings of creative practices and so to help embrace the promissory nature of creativity. The remainder of this introduction will explore these geographies in more detail.

Creativity as a socio-spatial practice

Perhaps the most simple of the geographies of creativity involves reflecting on creativity as a placed-practice, asking in short, where does creativity happen? The predominate geographical concern, and indeed one of the factors that drew the creative industries to geographers' attention, is their often extreme localization in an era when capital's production processes were becoming increasingly footloose, and at a point when the production chain was becoming globally distributed. Key early work on these cultural production industries queried why it was that concentrations of these industries were so important (e.g. Scott, 2000). Exploring why these localised articulations of economic activity remain so productively efficient, innovative and durable has been an enduring question of research on the geographies of the creative industries, within and beyond the discipline of geography. Indeed, far from dissolving away with intensified globalization, these patterns have become concentrated. Furthermore, they are solidified through the reproduction of creative-economy policy that focuses on industrial concentration, and in recent times has become tied into a range of public and private commitments to localism agendas.

Of course, industrial localization is not only a contemporary phenomenon. A key thinker in this regard has been Alfred Marshall (1920), an economist working at the end of the nineteenth and into the early twentieth centuries who famously wrote of the 'industrial atmosphere' of small-firm districts in the North of England, such as Sheffield, a 'cutlery town'. As well as local product specialization, Marshall noted the vertical disintegration of large firms in favour of networks of smaller firms, horizontal integration, and how in these areas, 'the mysteries of the trade become not mysteries; but are as it were in the air, and children learn many of them unconsciously' (Marshall, 1890, cited in Amin and Thrift, 2002). Reflecting on his ideas a century later Amin and Thrift (1992) explore the notion of 'Neo-Marshallian nodes in global networks'. One of their case studies is the one-product leather town of Santa Croce in South Italy (the other is the financial services industry in the city of

London). They explore the concentration and intersection of small firms in the Italian town concerned with producing leather products for export, from tanning the leather to dying it, as well as manufacturing shoes and other leather goods. They note the persistence of this industrial atmosphere, but also its contemporary situation within global networks. The theme of localization and the creative economy has persisted. Porter's wider economic work on clustering has proven important, wherein a cluster is described as 'a geographically proximate group of associated institutions in a particular field, linked by commonalities and complementarities' (2000, p. 254). This work sits alongside Marshall's notion of there being something in the air, only this has evolved to include not just knowledge and skills, but also 'buzz', whether the latter be gossip, knowledge or, as Currid and Williams (2010a) develop the formation of a creative scene, 'the economics of a good party'.

Sitting alongside, and to some extent critiquing these geographies of agglomeration is a body of literature that examines so-called 'other' geographies of creativity (Gibson, 2010). This has diversified the cluster literature in time and space to explore patterns other than those of permanent co-location. Important here have been discussions of networks, how they sit alongside clusters supporting and maintaining them, or how temporary clusters might offer an alternative to permanent agglomeration (Cole, 2008; Harvey et al., 2012). In the latter case, fairs, festivals and events, from the Milan design fair, to comic book conventions, to fashion weeks in Milan, New York, Paris have become a focus of study (Entwistle and Rocamora, 2006; Power and Jansson, 2008). Such events are often understood to convey similar benefits to permanent clustering but over a shorter period of time and sometimes cyclically, as well as conveying additional benefits in terms of network expansion, recruitment and power to build brand and reputation.

If one set of answers to the 'where' of creativity might focus on thinking through the spatial patterns of creative production, then a second set of answers might seek to be more specific, examining the changing locations of creative practice and thinking in terms of the scales at which we situate creativity. Much has been made of the economic value of the creative economy, and the possibilities for creative city scripts to revitalise urban areas around the world. Elizabeth Currid (2007a) in her work on the 'Warhol Economy', explores how fashion, music and other creative industries are as crucial to the success of New York as the more apparently lucrative and important FIRE industries (finance, insurance and real estate). A growing body of work, however, has sought to diversify beyond world cities in Europe or America, to consider patterns of creativity that might be considered marginal, in out-of-the way places or cities that are perhaps less internationally connected (Gibson, 2010). Furthermore, attention to rural and island creative economies has prompted studies of different forms of creative organization – based around dispersal rather than concentration – as well as dealing with real and perceived issues of communication and technology deficits (Gibson, 2014; Harvey et al., 2012). An important trend within these literatures has seen attention turn to suburban creativities (Bain, 2013). This work explores what it might mean for practitioners, inhabitants and policy makers if the creativity of the suburbs is recognised, reclaiming these places from their stereotypical understanding as sites of monotony.

It is not just at the scale of the city, area or region that the diverse geographies of creativity must be appreciated. Of late, we see a growing interest in sites of creativity as moving beyond specialised spaces of production (e.g. the studio, the workshop) and consumption (e.g. concert venue, museum, art gallery) to consider rather different spaces of production and consumption. So we find, for example, scholarship that explores vernacular creative production in the home, where the home space becomes a workspace, whether it is bedrooms becoming recording studios or kitchens and sheds becoming jewellery workshops or artists' studios (Bain, 2004a; McRobbie, 1998). In addition, we see a rise in homes becoming temporary gallery spaces in the course of events such as open studios, or the setting for the selling of an artistic brand online in virtual consumption spaces such as Etsy (Harvey et al., 2012; Luckman, 2015).

In line with this penetration of creative industries into the spaces and sites of everyday life, what we also find in these locations is a sensitization to those forms of creative practice that are not just about industrial production, but are also about those more everyday forms of creativity. Thus the 'wheres' of creativity become tied into both amateur forms of creative practice (e.g. hobbies and crafts) as well as everyday creative practices of decorating, gardening and cooking (Edensor and Millington, 2009; Yarwood and Shaw, 2010). Not to mention those more improvisational forms of creativity, that barely even register as such, being either part of wider practices such as thrift, sensibilities of make do and mend, or just part of going about in the world (Hallam and Ingold, 2008). As such, the 'wheres' of creative practice have come to encompass the world as a whole; no longer are the geographies of creativity confined to clusters and traditional spaces such as the studio or the gallery. Rather, creativity happens all around us, and it is the critical geographies of these processes that this book is going to explore.

Asking 'where' is creativity, where it takes place, is not just a question of the spatialities and patternings of practice, but is also to think about the situated nature of creative doings. Situated creative practices are also embodied practices (somebody, a fleshy sensing body, is doing the painting, designing, clothes making, the cooking, the DIY). Geography has recently had a turn towards the body and questions of embodiment more generally. Stemming from humanist geographers' concerns with life-worlds, feminist geographers' work on bodies and emotions, and concentrated more recently by a variety of non-representational theories, we find bodies, materialities and practices increasingly at the forefront of geographic debates (Buttimer, 1976; Colls, 2012; Thrift, 2007). Geographical studies of embodied practices have, by and large, tended to focus on practices that occur in the landscape, exploring (for example) the sensory experiences of practices of walking, running, cycling, rock climbing (Wylie, 2005; Lorimer, 2006). There is, however, a growing body of work within and without geography that explores questions of making and creative practices in terms of their embodied experiences. This includes massage and yoga, glass-blowing, taxidermy, surfboard making and various other hobbies and crafts (O'Connor, 2007; Revill, 2004; Straughan, 2015; Warren and Gibson, 2014; Yarwood and Shaw, 2010). These sit alongside wider discussions of the embodied cultures of making, crafting, repairing, and recycling that are becoming increasingly important within geography currently (Bond, DeSilvey and Ryan, 2013; Gregson et al., 2010; Edensor, 2011; Gregson et al., 2009). Creative bodies are skilled bodies, not only in terms of learning techniques, but also developing bodily intelligences, developing good eyes or ears, attuned senses that are trained over time, and come, in some cases to make actions instinctive, habitual. Geographers interested in the body of late have not just been concerned with thinking about the fleshy sensing form, but also the intersections of this with cognition, skill development and habit and instinct (Bissell, 2012; Dewsbury, 2012; Lea, 2009). As such, creative practices proffer a valuable site from which to explore these intersections of bodies, cognition and habit. For some these are important intellectual concerns, while for others political potential is paramount. The politicalities here lie in how such ideas might rework what labour is, how they might open up possibilities for being labouring bodies differently within, and even resistant to, the controlled time-spaces and rhythms of capitalist labour (Carr and Gibson, 2015; Guéry and Deleule, 2014).

Geography's recent engagements with the body have closely inter-sected with concerns with technology and questions of matter and mater-iality. In terms of creativity, this involves a recognition that creative practices often work on and with materials, indeed in many creative prac-tices the form and type of the materials and the practitioner's ability to manipulate them is crucial to the successful production. As Sennett (2009) perhaps most clearly develops with respect to craft, creative making practices can often be understood as collaborations of hand, head and matter. Furthermore, creative practices are also shaped by non-human agencies, whether this be the vital properties of materials such as stone or wood, or the agentive properties of technology to shape and transform creative subjects as well as that which is being made (Edensor, 2011; Paton, 2013; Woodward et al., 2015). In short, not only does the making of things matter, but the matter of their making is also crucial.

As well as a material and embodied practice, creativity is also a social practice. We can, of course, think of this in terms of the discussions of 'buzz' that indicate how social networks and concentrations of interac-tions become crucial to the development and sustainment of effective creative production. More than this though, it is not just that social connections and relationships sustain creativity, it is also that claims have been made for creative practices as the means to create social relations, in the words of David Gauntlett (2011) 'Making is Connecting'. These social relations might be about well-being and empowerment, so we see hobbies becoming a means by which people connect with groups of like-minded individuals (Yarwood and Shaw, 2010); they might also concern the shaping of our own subjectivities. We see this with respect to discussions around feminist craft practices and issues of identity, especially in the wake of pregnancy or at other key moments in the life course (Hackney, 2006). We also find it in the evolution of subcultural creativities, wherein practices of dress or body modification for example, become practices of identity making and display (Beazley, 2003a, 2003b; Bennett, 1999; Daskalaki and Mould, 2013; Hebdige, 1979; Shaw, 2013). Or it might be that art and other creative practices become a means of therapy, or a mode of engaging different social groups with questions such as urban super-diversity, wherein creative practices offer a means to create social encounters across the groupings of difference (Fox and Macpherson, 2015; McNally, 2015; Pollock and Sharp, 2007). Further, it could be that such social encounters catalyzed by art and creative practices become the means to build collectives that imagine social futures differently (Hawkins, 2013a). Within these collectives the futures that are imagined (and indeed the social that they involve) might not just be a human social, but might involve non-humans, attuning us to the world around us differently in the context of the environmental crises of our times (Carr and Gibson, 2015; Hawkins et al., 2015).

Political tensions

Creativity is an inherently political topic. To draw together the different understandings of creativity that geographers are concerned with is to become aware of myriad understandings of politics that are associated with these different bodies of work. Thus we find forms of politics that range from those associated with policy and governmental decisions (Harvey et al, 2012; Luckman et al., 2009), to those politics of spaces and practices associated with institutions, organizations and governance, more generally, (Bain, 2013) and the representational politics of art works or city branding (Daniels, 1993). We also need to think about creativity and the social politics of identity and community and subcultural practices (Edensor and Millington, 2012); the interventionary, tactical politics of urban subversions and some arts practices (Mould, 2014; Pinder, 2005a, 2005b), and the micro politics of individuals based in affect, experience and practice (Kanngieser, 2014). Thinking these different forms of politics together is not easy, especially as they can sometimes be based in very different epistemological contexts and have very different goals. There are benefits to be had in doing so, however, not least in building a critical framework for thinking about the variegated enrollment of creativity within a range of political causes.

Creativity has become a near ubiquitous strategy for urban growth, circling the world as cities from Amsterdam to Jakarta turn to the formulas for developing a 'creative city' provided by academics and policy gurus such as Richard Florida (2005) and Charles Landry (2006). There is, as Peck (2005, 2012) makes clear, a 'brisk business' being done in new and what some consider to be the 'fast' urban policy products – for example, consumable creative city models – being promoted by a growing number of gurus, consultants and cultural inter-mediaries (Mould, 2015). These products are being purchased by cities and towns around the world desperate not to be relegated to the rustbelt of the knowledge economy. If tool-kits for the creative city abound, promoting 'culturally inflected economic development, rebadged promo-tional strategies, and new age gentrification' (Peck, 2012, p. 42), unsur-prisingly, these mobile, neoliberal 'creativity scripts' have not been without their trenchant critics. The most concerted criticisms have come, unsurprisingly, from Marxist camps, where attention has been paid to the nature of these policies, a concern that their formulations are shallow and that they serve to produce and intensify social inequalities. Indeed, the dominance of creativity policy has been likened to a disease, so-called 'Creativity Syndrome' (Peck, 2012), a viral contagion spreading like a rash on cities around the world.

Alongside this 'putting of culture to work' as part of mobile urban policy, we find arts and creativity doing a range of political work. To think about the politics of creativity is, for example, to be very aware of the myriad forms of politics that are performed and enacted by the making and selling of creative products. Thus, as examples across this book will make very clear, bound up with creative practices are the production of places and people.

Furthermore, attending to the spectrum of political ideas and the tensions we find within creativity is to appreciate how, paired with creativity as the darling of neoliberal policy makers is a figure of creativity as central to the practices of activists and thinkers on the critical left. Thus we find David Harvey (2009) building on Henri Lefebvre to call for the rewriting our urban scripts via a new urban creative experimentalism. For Michael de Certeau (1984) creativity bubbles up everywhere, it is a means to find tactical possibilities in the everyday practices of ordinary city inhabitants (Hawkins, 2010a). Indeed, urban spaces have long been sites of artistic and creative practices that seek to intervene in the status quo, to overturn the regimes of the powerful, and to make space for others to be seen and heard. The urban exploration strategies of the Situationists in Europe in the mid-twentieth century are a good example of this. They have inspired numerous cultural practitioners since to experience and engage the city in creative ways that resists normative, capitalist-driven engagements with urban space (Loftus, 2009; Pinder, 2005a, 2005b).

Creative practices not only offer modes of intervention, artistic or otherwise, but also establish ways to live differently. So we find, for example, activist groups developing what have been understood as creative forms of action that 'invite new relations between different constituents and groups' (Kanngieser, 2014, p. 3). They do so through the creation of 'spaces where there is a questioning of the laws and norms of society and a creative desire to constitute non-capitalist, collective forms of politics, identity and citizenship' (Pickerill and Chatterton, 2006, p. 1). As Kanngieser (2014) notes these are spaces that expose the fallacies in the perception of social movements having a monopoly on social transformation and artists having a monopoly on creativity. As such, what we find is a set of practices and spaces wherein artistic practices, creativity and political activism are interwoven to create overlappings and reconfigurations of the political, artistic and creative, in doing so opening up spaces in which we might be able to live differently (Gibson-Graham, 2008; Kanngieser, 2014).

Of course none of these forms of creativity stand outside of the possibilities of so-called 'capture' by the state and capitalist systems. As Deutsche (1996) demonstrated in her scathing critique of urban public art, the avant-garde intentions of art practices can be thoroughly enrolled in the capitalist project. Urban aesthetics can all too easily become just another way of producing and reproducing the value systems and desires of an urban elite. Today, the diagnosis of a 'creativity syndrome' would suggest that the force of creativity in general might also have undergone such capture, especially as it is precisely the newly cool subversive and activist elements of creativity that contemporary capitalism is keen to draw within its ken. The melting pot for such debates has been the practices of so-called 'urban subversions', including skateboarding, parkour, yarn-bombing and urban exploration. All of these practices have been claimed at one point or another to be forms of subversive, tactical urban practice, operating as ways to inhabit urban spaces that intervene within and resist the privatization and surveillance of the city (Garrett, 2013; Mould, 2015). But yet, all of them have become steadily commercialised and sanitised, whether it be the opening of parkour parks and festivals sponsored by large corporations, or the sanitised 'knit' bombing campaigns run recently by a number of UK high-street clothing lines (Price, 2015). As Daskalaki and Mould (2013) make clear, such capture does not de facto render urban subversions unsubversive.

These examples might only exist temporarily and there is no sense in which such commercial borrowings or individual forms of 'selling out' somehow render a whole practice apolitical. Indeed, as examples throughout this book will show these are only ever processes of co-option within which possibilities for resistance and escape also reside. What comes to mind is the observation made by Guy Debord (1983) – the self-proclaimed leader of the Situationists – that trying to stand outside of capitalism is like trying not to breathe, instead he and the group experimented with positions of resistance taken up from within capitalism and with tactics that turned its own tools against it. Throughout the discussions which follow, a number of examples will demonstrate how a range of politics are possible; uncomfortable city memories are excavated and performed, urban diversity is thought differently, and interventions really do appear to intervene. These forms of creativity are perhaps not always obvious amidst those territories of arts and creative practices that are driven by commercialism and policy, but they are there, and indeed often erupt where we least expect to find them.

Creativity as a force: making subjects, knowledge and worlds

What does creativity do? What is its force in the world? As should already be clear for geographers, whether economic, social or cultural, a primary reason to attend to creativity is precisely because it does things in the world. Clearly underpinning the widespread rise of creativity during the latter decade of the twentieth and first decades of the twenty-first century has been its economic potential and the claims for its regenerative force. Hailed as central to the 'great reset' and post-crash recovery, creativity is for Richard Florida (2005) and others an economic saviour. It is the key to ongoing growth and a highly successful form of capitalism for contemporary times. The force of creativity does not, of course, lie only in its economic and regenerative value, nor only in the exchange value of the objects that result from the creative process, rather we must also question what the creative process and the consumption of creative practices and objects/experiences does for makers, participants/viewers alike.

Recent scholarship within geography has been concerned with the 'social work' that art can do in engaging us with a range of issues; from urban super-diversity to the wounds and scars of urban and community histories (McNally, 2015; Till, 2008). This builds on a longer legacy of geographers understanding and taking seriously the sense that creative objects and practices do things in the world. If we take, for example, studies of eighteenth and nineteenth century landscape paintings, for geographers what was crucial about these works was not only that their aesthetics (depictions of sanitised and romanticised ways of life) covered over and turned away from the dark side of rural life, but that these images had ideological force (Cosgrove and Daniels, 1988; Daniels, 1993). In other words, they not only 'pictured' the power structures of the ruling elite, but produced and reproduced them, covering over existing inequalities, and further entrenching them. Indeed, these paintings came to provide ideal aesthetic forms for the landscapes that were reproduced not only in England but around the world, shaping environments and lives in their image.

If the force of creative products is true for painting it is equally true for music, literature, films and other forms of creative output. Their material forms do not just reflect the conditions of situated production, but can come, in some way, to participate in the formation of places (Dixon and Cresswell, 2002; Pratt and San Juan, 2014). To date, much work in these areas has been concerned with the role of finished creative products in place-making, increasingly though we find creative processes becoming recognised for a similar value. Creative practices – whether of knitting, painting, weaving, drawing or otherwise – are viewed as making place, as forming and transforming the subjects who practice and engage with these practices, and shaping forms of knowledge production (Hawkins, 2013). Indeed, as discussions throughout this volume will explore geographers have found in creative practices not only subjects of study, but also research methods through which to engage with multiple dimensions of the world and the humans and non-humans that live within it. As such, creativity offers the means to both research and live differently.

In amongst the positivity that shapes so many of the discussions of creativity, it is important not to lose sight of its potentially dark side (Cropley et al., 2010). This might concern critical reflection on the many mobilizations of creativity as a force for uneven economic development, for contemporary capitalism's abusive labour practices, or it might involve studies of creative practices of criminality, or even of terrorism (Raunig et al., 2011). As Cropley et al. (2010) point out, it is often less about creativity being inherently bad, but rather people mobilise creativity for a range of reasons and to a huge variety of ends.

Together these three critical creative geographies run across the following nine sites of discussion that shape this book, raising four queries. Firstly, should creativity be understood through the lens of the products it produces or as primarily a process in the course of which things happen? Compare, for example, studies that analyze the creative outcomes – films, paintings, novels, and so on – as against those discussions of creativity as a process intersecting body, mind, materials and tools. Or we might think of creative research methods, where a focus on the product demands a certain level of skill, whilst an attention to what happens in the practice enables a rather more democratic approach. Secondly, creativity is often queried as either a personal or social phenomenon (Boden, 1996; Singer, 2010). In short, does creativity involve the production of something new and useful with respect to the person doing the creating, or is it socially determined? In the latter case creativity would involve producing something that is recognised as new and useful by society more generally. In her famous book exploring *The Creative Mind*, philosopher and cognitive scientist Margaret Boden identified two senses of creativity in this regard, P-Creativity is an idea that is novel 'with respect to the individual mind which had the idea', whilst H-Creativity concerns an idea that is 'novel with respect to the whole of human history' (1990, p. 32, p. 8). As she notes, 'H-Creativity is a historic category . . . [depending on] shared knowledge and shifting intellectual fashions . . . loyalties and jealousies, finances and health, religion and politics, communications and information storage, trade and technology . . . [and] even storm, fire and flood' (ibid.). A third query concerns whether creativity is common or rare? In other words, is all reality creative in some way, as process philosophers would have us believe, is creativity part of everyday cognition and therefore something that humans are capable of, or is it a rare event, something that occurs only within a small group of people? Fourthly, is creativity domain-general or domain-specific? According to the domain-general view, creativity is a general skill or trait applied to a wide variety of situations, whilst under the domain-specific idea of creativity, different types of creativity are required in different domains, in art or science for example (Singer, 2010). We might think of this in terms of creative geographies where the rise of creative methods as part of geographic research practice, requires us to explore what 'creativity' means in the context of the production of geographic knowledge. These four queries run across the nine site-based discussions that follow and are revisited in the conclusion.

Given that this is a book that seeks an analytic account of creativity by putting creativity in its place, it does so through accounts of nine creative sites; from the body to the studio, the city and the environment. These discussions, however, are not only about creativity being put in its place, they are also about how sitting at the heart of geographies of creativity is an appreciation of how creative practices make place, how they produce and reproduce the sites and places of their production and consumption, whether this be the body, the community or the environment. Each of the discussions which follow uses one particular site as a lens onto the critical geographies of creativity. These sites are by no means exhaustive, and are organised in a broadly scalar manner, from that creative site closest to us – the body, all the way up to the scale of the nation and the environment. It is hoped that such an approach makes clear how even those creative practices that produce nations and that reshape environments are also practices produced and consumed through, on and by individual and collective bodies.

The logic behind the sites chosen is manifold, but perhaps most useful for navigation purposes is the role of these sites as locations for concentrations of literature and discussion around creativity, but also as sites that most clearly bring into tension different facets of those expanded definitions of creativity that have caused so many problems. Punctuating the discussion of each site are a series of text boxes, some of these are key ideas or case study boxes, a third group present new research on creativity written by the researchers themselves. As such, space is made within the text for ideas, methods and examples that are not yet available elsewhere.

The first creative site the volume visits is the body, building on geography's recent attention to the agentive and material, practicing, sensing body to explore the rise and fall of the human creative agent. The second site continues this attention to the micro-geographies of creative practices through an attention to the specialist sites of the studio and workshop, but also noting how creative practices clearly overspill these spaces to take place in a whole suite of sites, whether this be the expansion of art beyond the studio, or the expansion of our appreciation of creative practices beyond the specialist spaces of professional practitioners. Taking up these themes, the third site is the creative home, wherein discussion examines the tensions and negotiations required in thinking through the creative home. This might be through the expansion of the home as a creative economic space made possible through digital developments, or it might be through the role of amateur and vernacular creative practices in making and unmaking our homely spaces. If the home is emerging as a key site in the creative economy, then the fourth creative site this book explores considers creative clusters as the dominant spatial forms of the creative economy. Drawing together work not only on clusters, but also more culturally inflected work on scenes and colonies, discussion examines why clusters have persisted within concerns with the geographies of creativity. It also explores some of the more recent challenges to these dominant logics. The fifth site remains engaged with some of these more social dimensions of creativity, thinking of the community as a site not only of creative production but also a site produced by creative practices. Critically engaging with the by-now-famous phrase 'making is connecting', the chapter reflects on the successes and failures of creative practices and policies as making places and shaping communities. The sixth site tells a sequence of stories of the creative city, setting these in dialogue in an attempt to cross-cut the broad literatures on the city as a creative space which so often remain separate from one another. Building on these critical discussions of the city, the seventh site jumps scales to consider creativity in the margins; exploring 'other' geographies of creativity that challenge the dominance of the cluster, but also investigating the margins as a site for diverse creative practices that escape or challenge normative creativities. The eighth site takes up these discussions in the context of the nation, exploring the persistence of the intersections of the nation and creativity, whether that be through policy or through creative practices. Here the geopolitics of creativity intersect with discussions of digital and embodied creative practices, notions of soft power as well as policy debates. The ninth and final site the book considers is that of landscape and environment. Charting episodes in the history of the relations of geographical knowledge making and creative practices it is clear that creativity has long sat central to these key disciplinary concepts. Furthermore, looking at a cross-section of projects, it becomes clear that within creative practices we find myriad possibilities for engaging with and acting within the current global environmental crisis: creative geographies as enabling researching and living differently.

The conclusion of the book will return to the three critical creative geographies and the four queries that cross-cut them to reflect on the promise of creativity and what has been achieved by putting creativity in its place. It will examine how across these nine different sites the book aimed to encourage thinking outside of disciplinary boxes, making connections across bodies of research and practice that might be based in rather different epistemological and ontological frameworks and have widely varied political commitments. The aim being not to unify the approaches and to come out with an 'ideal' definition of creativity but rather to draw the variety of work together, exploring similarities and differences for critical ends, to, in short, encourage a thinking creatively about creativity. In taking such a stance, the volume argues we are well placed to be able to appreciate diverse creativities not as critically debilitating but as sites of promise and possibility.

NOTES

1 http://www.brooklynindustries.com [last accessed 14/9/2014].
2 Lexy Funk, in video 'The Story of Brooklyn Industries', available on http://brooklynindustries.tumblr.com [last accessed 14/9/2014].
3 UNCTAD creative economy report 2010, [ONLINE] Available from: http://unctad.org/en/pages/PublicationArchive.aspx?publicationid=946 [Accessed 25/7/2015].
4 Creative Industries Economic Estimates, January 2015, UK, Department for Culture, Media and Sport, [ONLINE] Available from: https://www.gov.uk/government/uploads/system/uploads/attachment_data/file/394668/Creative_Industries_Economic_Estimates_-_January_2015.pdf [Accessed 25/7/2015].

2

BODY: EMBODIED CREATIVITIES
AND CREATIVE LABOURS

factory work exhausts the nervous system to the uttermost; at the same time, it does away with the many-side play of the muscles, and confiscates every atom of freedom, both in bodily and in intellectual activity. Even the lightening of the labour becomes an instrument of torture, since the machine does not free the worker from the work, but rather deprives the work itself of all content.

(Marx, 1976, p. 548)

by means of movements, lines, colours, sounds, or forms expressed in words, so as to transmit that feeling that others experience that same feeling- that is the activity of art . . . that one man consciously and by means of certain external signs, hands onto others feelings he has lived through, and that others are affected by these feelings and also experience them.

(Tolstoy quoted in Singer, 2010, p. 91)

Myth one, for Karl Marx, the factory labourer was an alienated labourer, thoroughly affected body and mind (as the quote above makes clear) by the particular sensory-mental condition cultivated by repetitive production-line work. In some quarters, this figure of the disenchanted factory worker has formed a critical backstop for the emergence of the oft romanticised figure of the creative labourer and the creative body. This is a creative labourer who is mentally and physically fulfilled by their work, a labourer who loves what they do and thus will put up with any hardship in pursuit of their goals.

Myth two, for Tolstoy, the artist was in possession of unique communicative abilities, able to transfer their thoughts and feelings to an audience. This essentially Romantic view finds an artist's feelings embodied in their art, and an uncomplicated transference of these feelings into how an audience feels towards the world. Such ideas sit at the heart of the dominant myth of the creative subject, that of the figure of the singular, transcendent creative genius. This myth is based in a belief in the creative practitioner as possessive of some sort of unique trait, often one which cannot be developed through skilled practice alone. For some, influenced by Freud, the artist is a regressive figure, finding solace in child's play and phantasies that they incorporate into their work, offering a sort of wish fulfilment for audience and artist alike. For others artists are possessive of a specialised ability to get at the individuality of things, bringing us into 'immediate communion with things and with ourselves' (Singer, 2010, p. 98). Singer notes, 'for reasons of survival and practical accommodation, we scarcely see the actualities in our life, he [Bergson] says 'in most cases, we confine ourselves to reading the labels a fixed to them' (ibid.). Creative art thus exists as a partial remedy of that condition.

The breadth of scholarship this chapter explores takes place across the backdrop of the much mythologised figure of the creative labouring body. Whether the idealised craft worker, a labouring subject fulfilled through their physically and intellectually active labour or the self-exploiting passion worker, the creative subject is usually hard working, often badly paid but always driven by the love for what they do. Taking leave from the two myths outlined above, this chapter will explore two principal ways of thinking about the creative body. The first is the figure of the creative subject understood phenomenologically. This takes form in a set of studies shaped by an attentiveness to the experiences of the creative subject, their skill sets, their manipulation of materials, their use of tools. This is an account of the micro-geographies of creativity and involves an attunement to the inter-relationship between creating bodies and worlds. A second way of thinking about creative bodies might take up some of those more political elements of Marx's alienated labourer to explore the political economies of creative labouring bodies. This is to attend to the forms of labour that creativity demands, to critique any illusions of romance or work driven by passion to engage with the conditions and effects of creative labour and its forms. Oftentimes, such discussions appear quite different, and indeed the two bodies of literature are often distinct from one another, yet as the discussion here will hopefully show, both sets of ideas share in common a centralisation of the messy, fleshy creative subject, their cognitive and physical practices, and the affect and emotions that adhere within their creativity. The chapter thus proceeds as follows; firstly it will explore the rise and fall of the agentive, autonomous creative subject; this will include considering the embodied and material nature of many creative practices. It will then move to discuss the figure of the creative labouring body, reflecting on the ideology of creative work, both the pleasures and challenges of the aesthetic labours of creative practitioners. The chapter will close by exploring the figure of the creative body in contemporary geography, examining how creative practices have offered geographers a key to understanding their embodied experiences of the world.

THE RISE AND FALL OF THE AGENTIVE CREATIVE SUBJECT

There are an increasing number of studies that track myriad histories of the evolution of the idea and practice of creativity (e.g. Maitland, 1976; Negus and Pickering, 2004; Singer, 2010; Sternberg, 1988; Tatarkiewicz, 2012). Before the eighteenth century, creativity was assumed by the general population to be something done by the gods. Gradually, however, we see a movement from the religiosity of creation myths to the human-based secular understandings of creativity that are now more common. In short, to look at the history of creativity is to explore an intertwined sense of divine, natural and human creativity. A retelling of the history of creativity as an idea is, of course, beyond the scope of this text; instead the discussion that follows focuses on some of the key shifts in understandings of humans as creative agents. Across time and space there is an interplay of divine creativity and human agency, and of human embodiment and non-human forces. The result is a challenge to creativity as a uniquely human power enacted by a transcendental genius.

Divine creativities

If the figure of the creative practitioner and their embodied skills, often in the form of the craft worker, has become central to discussions of contemporary creativity then this has not always been the case. For a long time, the role of a divine creator was greater than any human creativity. The creation of the cosmos, of humans and animals, as well as the creation of various art forms and institutions has been explained in myriad ways across time and space. Common to many, however, has been the role of the divine creator. Across history we see a series of specificities regarding the definition of creation and its processes (see accounts in Pope, 2005; Singer, 2010). In Pharaonic Egypt (a location associated with great creativity), for example, the concept of individual creativity as the product of a singular artist did not exist. Instead, the rules and techniques of craft were passed down from generation to generation and practiced until they were mastered. In both Egypt and Mesopotamia the idols of the gods were made according to strict formulas, with the creator understood to be the servant of the divine. Artists were thus anonymous and a creator's goal was to serve the wishes of the patron and ultimately the gods. Astrology was also a significant feature of early understandings of creativity (Singer, 2010, p. 23); in other words, what humans do is 'written in the stars' not invented by them. When human lives and actions are determined by gods or stars, then humans are resolved of responsibility for their actions, and creativity as we understand it today is near impossible (Singer, 2010).

The Ancient Greeks had no term 'create' or 'creator'; instead they used the term 'to make'. Art – in Greek *techne* – was the making of things not found in nature and included no 'creativity', as such (Singer, 2010, p.107). For the Greeks, the ideas of creativity and creator implied a freedom of action, whereas their concept of art involved subjugation to laws and rules. Indeed, if it had involved elements of creativity then the work would have been in the Greek's view, bad; art was about the discovery of the laws of perfect nature, not the invention of things. *Techne* thus levelled all arts and crafts, understanding them as skilled practical behaviours. The role of the painter, for example, was seen as that of an imitator not a creator, for the artist through skill achieved the duplication of a divinely inspired blueprint, a duplicate inspired by celestial exemplar (Singer, 2010, p. 109). Plato, in *The Republic* asked 'we will say of a painter that he makes something?'; he answered 'certainly not, he merely imitates'. As sixth-century Roman official and literary writer Cassiodorus wrote 'things made and created differ, for we can make, we cannot create' (cited in Tatarkiewicz, 2012, p.247). Moreover, not only is the artist an imitator rather than a creator, he is also governed by rules (ibid.). The great exception to this view of the arts was poetry; poets were seen to make new things and to do so freely, and they alone were primary creators. This split between creation and making was formalised when the Latin *creatio* 'creation' came to designate God's act as distinguished from the Latin *facere* (to make), with which it had previously been associated.

Many of these ideas around creation and divinity were scattered and diluted into European ideas of creativity, at the base of which was the act of creativity being that which was accomplished by God. Medieval uses of 'creativity' thus refer to the biblical act of creation in Genesis Book 1, verses 1–3. Furthermore, there was an orthodoxy that God created things *ex nihilo* and that creativity was a past fact, rather than an ongoing activity. The principal association of creation as a divine act was to prove persistent. Indeed, we can trace this into more contemporary object-orientated understandings of creativity that stress the finished work, as distinct from those process-based approaches which stress the activity of creation and the processes of making (Ingold, 2010a, 2010b; Singer, 2010).

It took a long time for a human agency to become a part of the meaning of creativity. During the Renaissance purely human creations were viewed with suspicion and even as harmful; human powers of creation were only valid when they were tinged with a divine aura (Singer, 2010). Towards the end of this period, key artistic figures of the sixteenth and seventeenth centuries publically challenged the idea of creativity as either the product of divine inspiration, or as a form of imitation of a divinely created nature. Art Historian Vasari, for example, notes that nature is conquered by art, Raphael shapes a painting in accordance with his own ideas, and Da Vinci wrote that he employs shapes that do not exist in nature. The case was, however, far from clear-cut, with many still insisting that artists remained principally copyists or imitators, second creators (Tatarkiewicz, 2012).

Such resistance to ideas of human creativity had a triple source. Creation was reserved for creation *ex nihlo* – from nothing – and thus was inaccessible to man. Further, creation was a mysterious act and the Enlightenment did not admit mysteries. Finally, artists of the age were attached to their rules, and creativity seemed irreconcilable with rules. As such, the link between creativity and divinity proved persistent, and was not only a function of religiosity so much as wider intellectual and art world conditions (Singer, 2010). It is not until the eighteenth century and the Romantic era that a strong relationship between the capacity of the human mind and creativity is forged.

The Romantic Imagination: Art's claiming of creativity

By the nineteenth century and certainly into the twentieth century, art had claimed creativity. Such that when in the twentieth-century creativity in sciences and nature began to be detected this took the form of the transference to science and nature of concepts proper to art. To understand this claiming of creativity we can turn to the creative imagination of the Romantic era, an idea born from the Enlightenment and evolving throughout the eighteenth century that raised artistic creativity to the fore.

The history of the creative imagination from 1660 to 1920 was that of a human drama that unfolded by stages (Engell, 2013). It began in Wordsworth's terms as a 'mild creative breeze' that 'rustled through and animated a forest of individuals, stirring each to participate in it, and to add his own force until it turned into a Zeitgeist' (ibid.). Eventually, the creative imagination was to become the basis of a new theory of the mind, replacing what could be understood in simplistic terms as a passive/receptive theory of the mind. As was clear

> to those who empirically examined the creative productions of the mind (whether art, science or philosophy) . . . such productions were characterized not merely by a passive association of old ideas in new ways. They were marked essentially by a fusion, a relational character, so that a new element or idea was created from the old.
>
> (Wheeler, 1983, p. 44)

Thus arts and literature, creative acts, become shifted from their classical understanding as imitative to their modern usage. For Engell (2013) the creative imagination was a product of the Enlightenment that led to Romanticism, rather than a product of Romanticism. It did so by turning the focus away from art as the imitation of nature and turned interest to the subjective inner world. Art becomes thus conceived of as the expression of the human mind, through symbolism in which the productive nature of the mind manifests itself.

Within Romantic working, poetry in particular, originality was highly valued and was seen to come from two key sources – from within creative practitioners or external to them. The first of these two ideas is creation – or the creation of something from nothing; creators bring their works into existence from an intangible force that works within or through them. There is a certain sense of a moment of inspiration in the moment of creation, creative originality being less about work or skill and rather about an unconscious force that results in few borrowed elements (Millen, 2010, p. 96). Escaping the challenges of industrialisation and the burden of history, this concept was popular in the Romantic era. Creation was, however, the gift of the select few, it was a gift that could be cultivated, but it could not be learned in *toto*. Understood in these terms, creativity bears a sense of the era's striving for individuality within a growing population, an early desire to stand out from the crowd. But it is not wholly about the human creator, for creation is also a supernatural force for which the creator is merely a channel. This energy could not be allowed to run amuck, however, but order and skill could come to tame and contain these creative energies. The Romantic creative processes thus centres on divine power rather than acquired skill.

The second concept of creative originality common to the Romantics was that of invention, wherein creativity arose through the rearrangement of existing parts; emerging not from nothing it involves reworking existing material into novel constructions (Millen, 2010). Invention, in contrast to creation is associated with the hard work and skill that shapes the poem or painting. The building blocks of invention might be the aesthetic traditions or socio-political context, prior readings or viewings (Millen, 2010, p. 100). Invention still, however, does not only come from hard work and wide knowledge, but might also involve an unconscious quality; some sort of flow of organic originality from within the poet to bestow new value upon pre-existing materials (ibid., p. 101). It is said that the result of these Romantic inheritances is to find our ideas of creativity lying somewhere between these two poles (Millen, 2010; Singer, 2010). Indeed, the current understanding of intellectual property and copyright which have shaped much of the discussion about the creative economy suggest that there is much of value in these ideas. Further, as Singer makes clear, Arts for Arts sake emerged as a reaction to many of these Romantic sensibilities (2010, p. 116).

Through the Romantic creative imagination, the Homeric catalogue of polar opposites became united; man was reconciled with nature, the subjective united with the objective, the internal mind engaged with the external world, time with eternity, matter with spirit, the finite with the infinite, the conscious with the unconscious, and so on (Engell,

2013). In short, imagination becomes the process to understand and view both world and the self; moreover it became the means by which to imagine and recreate the world anew. The creative imagination thus promised to the arts a crowning role in philosophical thought, in knowledge and power and even in religion (Engell, 2013). The increasing confidence in the creative imagination from around 1740 onward Engell argues, 'led poets and critics to trust and to believe in it, to sense they had a mission not only to fabricate a new world view, a reappraisal of man and nature, but even more to swaddle this thought and energy around human feelings' (2013, p. 8). As he goes on, this concept of the creative imagination 'enlarged the humanities and increased the expectations put on secular art ... and the promise and burden of these expectations continue today' (ibid.). From the Romantic era we inherit a whole series of ideas about creativity and the creative imagination, about the figure of the creator, and questions of skill, hard work and divine intervention. One of the ongoing perspectives that the divine view of creativity brings is an object-orientated understanding of creativity that focuses on the finished project, rather than thinking about the creative process itself. In recent times, however, discussions of creativity have come to be dominated by an attention to process, and it is to such discussions that we now turn.

Creativity as an embodied, material practice

What does it feel like to have the skills to sculpt clay, to shape wood, to meld metal?

Geography, in recent years, has undergone a bodily turn. In the face of objective, distanced views of the world from above, often typified in the image of the map, geography has come to embrace embodied ways of being in, and knowing the world. Following a range of other disciplines, it has come to a broader recognition that bodily experiences, affects, sensations and know-how are far from 'remaindered, shrugged-off as left-over traces of un-reasonable noise' of living, but rather should be recognised for the 'dynamic contributions to the (re)production of everyday life' they offer (Woodward and Lea, 2010, p. 5). Within geography, scholars have explored practices of running, walking, yoga and climbing, foregrounding the embodied experience of these activities. Attention has been given to both physical experiences (hard, soft, stretchy, cold) as well as the emotions (happiness, elation, exhaustion, fear) that these practices give rise to. Drawing on the phenomenology of Maurice Merleau-Ponty, these scholars have emphasised the importance of these embodied experiences of being in the world for knowing both it and ourselves. No longer are landscapes something to be looked at or thought about, rather they are spaces to be moved by and through. Embodied subjects – the runner, the walker, the geographer – are no longer able to remain apart from the world in and through which they live, rather they are thoroughly entangled with it; seeing, smelling, hearing and touching. It is through our senses that we come to know the world, but it is also through them that we are created as human subjects living in the world.

Collectively, geography's studies of the body have been shaped less by a headlong rush towards the senses and sensualities of lived experiences, but have rather been sculpted through a series of sober and more considered studies. In these we witness the braiding together of the external bodily senses – in particular sight and touch – with reconfigured sensory and cognitive topologies that are better equipped to account for the complexities of the body's sensory system and the variegated modalities of thought. For what such accounts of embodied practices make clear is the sensitivity needed towards the differentiation between 'sensations' (information routed via distributed nerves and sense-system clusters, Paterson, 2009, p.779) and 'sensuous dispositions', which draw attention not only to the 'immediacy of conscious sensation and cutaneous contact', but also the 'historically sedimented bodily dispositions and patterns of haptic experience that become habituated over time' (ibid.). Amongst their evolving attention to the body, geographers have yet to turn their attention in any concerted manner to the creative body and the embodied nature of creative experiences (although see Straughan, 2015). Other disciplines have, however, spent rather more time exploring and engaging creative bodies.

Where we perhaps see this most clearly has been in the rise of literatures around craft (Adamson, 2007, 2009). The combination of body – principally hand – and mind that craft-making enables has been at the root of the twenty-first century revaluation of 'craftsmanship' (Sennett, 2009; Thrift, 2007). Offering a form of homo faber that is a 'means of composition and of channeling which involves bringing together discipline, and concentration, understanding and inspiration in order to bring out potential . . . working both for its own sake and as part of a community of ability' (Thrift, 2007, p. 15). Craft is, in short, understood as a combination of the embodied skill of the in-corporated mind ('discipline', 'concentration', 'understanding'), and of material and affectual relations between human and non-human materials, tools and environments ('composition', 'inspiration', and 'channeling'). The master-potter Bernard Leach, on asking a distinguished Japanese potter how people were to make and recognise good pottery, was given the answer 'with their bodies' (1940). Such ideas have been detailed of late through the emerging suite of ethnographies of creative doing. Erin O'Connor's sociological accounts of spending several years learning to blow glass are a good case in point. Her ethnographic work details her bodily experiences as she overcomes awkwardness, develops skill, and learns to manipulate her tools (see Box 2.1).

Box 2.1 EMBODIED, MATERIAL GLASS-BLOWING PRACTICES

The glass-blowing studio becomes a site of the embodied and material practices of creativity in the ethnographic accounts of sociologist Erin O'Connor (2005, 2006, 2007). Her accounts build a dialogue between maker, tool and material, emphasising the need to shift from solely cognitive readings of practice towards corporeal readings that take account of the place of cognition within embodied practice. Her suite of published studies focus on her own processes of learning to blow glass, exploring apprenticeships and the process of achieving proficient practical knowledge, as well as developing phenomenological perspectives on the body's interactions with the material world. She is also concerned with questions of studio talk about tools and how to use them; 'tools are one of the many resources necessary for artistic creativity' (2006, p. 177).

Through ethnographic accounts of working with tools, O'Connor explores how these implements, shears in her case, expand the practitioner's phenomenological body and dispose practitioners towards the matter to be worked upon (2006, p.178). She talks through the process of learning to blow glass, from her first visit to the studio where she watched the gathering of glass from the furnace using pipes and was introduced to other tools such as tweezers and shears, as well as less obvious tools like wet newspaper. As O'Connor describes, it is through these tools the novice comes in contact with the glass, comes to learn embodied techniques of glass blowing including how to respond to the material through the tool. She describes the process of becoming proficient with the tools and building a relationship with the materiality of the glass. She recounts her evolving relationship with the tools; 'I had taken on the shears as an extension of my hand, my grip melded to the handles; I no longer struggled to take them in hand as I had before' (2006, p. 189). In order to understand this process she draws on Polanyi's work on tacit knowledge

> we may regard this as the transformation of the tool . . . into a sentient extension of our body . . . in this sense we can say that when we make a thing function as the proximal term of tacit knowing, we incorporate it in our body- or extend our body to include it.
>
> (Polanyi, 1966, p. 16)

The tool here becomes a feature of tacit knowledge, knowledge known in the body, an object of subsidiary awareness through which attention is directed to the object of focal awareness, the glass. What O'Connor describes is the evolution of proficiency as a movement to see tools as 'merely an object-in-use, to go further, such that tools become synonymous with the body, the glass becoming foregrounded in practice' (2006, p. 190). In her accounts she develops a powerful figure of the maker as a listener to material and the tool as helping to shape the maker, as such, enabling them to listen. This is a question she says of 'following the glass', foregrounding the materiality of the process and drawing out the agency of materials.

Such discussions echo Martin Heidegger's discussions of the craft of the cabinet-maker, in which he writes

> the learning [of a cabinetmaker's apprentice] is not mere practice, to gain facility in the use of tools. Nor does he merely

Continued

gather knowledge about the customary forms of the things he is to build. If he is to become a true cabinet maker, he makes himself answer and respond above all to the different kinds of wood and to the shapes slumbering within wood- to wood as it enters into man's dwelling with all the hidden riches of its nature. In fact, this relatedness to wood is what maintains the whole craft. Without that relatedness, the craft will never be anything by empty busy-work, any occupation with it will be determined exclusively by business concerns.

(Heidegger, 1977, p. 14)

The materials and tools do not totally disappear in the process, but rather become a focus of the maker's attention, become the force that shapes the process of making and the resulting object produced.

As O'Connor's accounts make clear, however, this is not just an account of human-non-human relations; she was also concerned with the human social worlds created through these material interactions. She explores how exchanging ideas about tools between novices and professional glass-blowers contributes to the formation and solidification of glass-blowing social worlds. O'Connor (2006) recounts arriving at the studio one day with a new blow pipe and how it prompted a series of exchanges between her, her instructor, and other students about exchange of knowledge. This contributed to the formation of a community of practice. Tool talk, often accompanied by holding tools in hands, weighing them in palms, not only expresses the glass blower's embodied relation with material, but also contributes to the solidity of the group in the studio. Tool talk offers a 'disclosure of the embodied empathy among glassblowers as regards their maker-material relation' (2006, p. 191) and is an example of how 'people create group lives with hands as well as tongues' (ibid.).

Of course, the embodied experiences of blowing glass, making pots or doing taxidermy cannot just be understood through the immediate experiences of the sensing body, rather, as O'Connor's discussion develops, such a sensing body intersects with tools and technologies, it is also a body that is variously skilled, that is more or less trained, and more or less aware of conventions of practice, for whom the practice is more or less habitual. Geographer George Revill (2004) writing on his experiences of playing French folk music and learning to dance, observes the importance of habit and technique. He notes how over time certain notes and even pieces of music can be played out of habit, without thinking. Such ideas of the 'skilled' body and the 'habit' body have yet to be much studied within geography, and both ideas seem, at first glance, to challenge the idea of the creative spirit. For both figurations have long philosophical trajectories, beginning from what was for Descartes and later Kant a negative attribution associated with habit as mechanistic automatism, inhibiting the freedom of the mind. To take the habit body, Bissell (2012) and Dewsbury (2012) note that 'habit is an indispensible trophe for considering the constitution of bodies, cleaving open crucial questions concerning the forces that compose matter and thought' (Bissell, 2012, p.121). As they point out, and as other literatures that bear on habit make clear, habit frames for us modulations of repetition and difference, obduracy and transformation, enabling us to question the dualisms that split organic from inorganic, voluntary from involuntary agencies, freedom from determinacy, inside from outside and activity from passivity' (Bissell, 2012; Dewsbury, 2012). Challenges lie in thinking through 'habit as refinement rather than as automatic, shifting what was once enabled as bodily technique divorced from the mind to being composed, on closer engagement, through a 'calculating capacity'. In such a capacity, habit can be held in reserve until it is required by a situation, or in a reflective pause between idea-as-will and action-as-habit, the reflection on habit in short, refines action (Bissell, 2012).

Other research has moved us away from habits as merely refinement achieved through repetition to an awareness of volatile habit bodies, and an appreciation of how becoming skilled might also mean cultivating a mode of experience that is composed by the cognitive weighing up of what is already known. In other words, as Lea (2009) writing of the skilled practice of Thai Yoga Massage notes, 'the eruptive moment that disturbs the automatism of habit and challenges the refinement of skills, is sourced from the affective materialities of the leaky body being worked on; such that a 'becoming skilled' takes into account changing contexts and their integral importance to learning'. Thus skill becomes on the one hand increasingly refined, whilst on the other demands from the experienced practitioner a stance towards the world that involves an attentiveness to the situation and forming a more improvised, immersive, finding of 'one's way through things, coming through one's body in order to understand what one is capable of' (May, 2005, p.111 quoted in Lea, 2009, p. 467). There are lessons then to be learnt for thinking about the creative body and the skilled body and the evolution of creative practices that might be more attentive to these micro-geographies of subjectivity and practices. For these are creative geographies that take place in and through neurons, through muscle fibres and through intimate encounters between sensory and habituated bodies, thinking minds, and volatile, agentive materials – clay, wood, or metal – that do not always behave in the way expected.

Of course, there is more to creative embodied practices than just bodily doings. One of the critiques of much geographical scholarship on the phenomenological body has been its focus on sensory life worlds of experience, rather than appreciating the situated nature of that body within a wider realm of experience, within a social context and within a sense of history and memory. Appreciating the habits bodies have picked up, as well as those they have cultivated through practice, helps attune us to a creative body as other than the immediate moment of creation. After all, to play a musical instrument, to dance, to sculpt a surfboard, to knit, requires learned, practiced skills. There are, however, social, cultural and political-economic factors that also bear on these embodied practices. Revill, for example, in his study of playing and dancing to French folk music, notes how a certain bodily economy, a form of technique based on small precise movements rather than large expansive ones, was seen as particularly French. As such, the bodily forms and deportment of dance become associated with certain traditions and national identities. Of course, creative practices also take place within institutions too; institutions that teach the practices, examine and maintain standards, as well as those that support their performance. Sofie Narbed's discussion of contemporary dance in Ecuador, demonstrates how all these forces can come to bear on the dancing body (Box 2.2). Focusing on surfboard making Warren and Gibson's (2014) book-length study of surfboard makers situates their material-embodied practices of board shaping within local and global surfing cultures, within craft traditions and commercial demand. Through the lenses of skill and technique, they query the kinds of bodies able to shape surfboards over a long period of time and examine how the physicality of makers bodies are, in turn, also shaped by the prolonged production of these boards. They detail the contraction of muscles and tendons, the pollution of lungs and the gradual onset of aches and pains that challenge the surfboard makers and can eventually mean they have to leave aside their craft to care for their bodies.

Box 2.2 CREATIVITY AND THE DANCING BODY

Sofie Narbed

The darkness turns to a dimly lit red. It picks out a figure, large in layers of material, moving with her back to the audience, bare feet stamping back and forth as she dances to a small, decorated altar. With her torso tilted forward, her movement is fitful, agitated, her steps punctuated by quick turns that send her brightly coloured skirts and shawl swirling around her body. She is a mass of material in motion. As she steps, back and forth, back and forth, she hides her face in the crook of her arm, before lifting it to shake a fist angrily at the heavens. Suddenly the driving drums that have been accompanying her movement stop and the figure halts. Coming out of her feverish motion, she stands erect and turns slowly to look at the audience. The light, a cold white now, picks out her face for the first time – a mask in shiny hard patent black framed with black fur and trimmed with gold. She clutches her skirts in gloved hands, clasped tightly in front of her body. Her head makes minute disarticulated movements, tilting curiously from one side to the other as she fixes the audience in her gaze. In a moment of stillness, she starts, letting the skirts fall abruptly from her fingers.

(Fieldnotes from a performance of *Mama Negra* by the artist Wilson Pico as part of his work *Fervorosos Pasos*, Quito, May 2014)

There are many things that contribute to the emergence of a dancing body for the stage, including its history of training and

---

Continued

technique, personal explorations and improvisations, the spaces in which it practices, and the audiences to which it performs. This is a constantly negotiated and inherently creative process in which the dancing body intermeshes affective, sensuous, embodied experience with the social, cultural and political discourses that shape, and are shaped by, its practice. This dynamic can be seen in the performance of *Mama Negra*, which forms part of the Ecuadorian artist Wilson Pico's work *Fervorosos Pasos* or *Impassioned Steps* in which he interprets, through the practice of contemporary dance, a series of characters from popular festivals in the country (Figure 2.1). The work is the product of the project *Cuerpo Festivo* or *Festive Body*, a two-year investigation that Pico undertook with his daughter the anthropologist Amaranta Pico into twelve indigenous, Afro-Ecuadorian and *mestizo* festivals, with a particular focus on the role of the dancing body within them. The *Mama Negra* or *Black Mother* is one of the principal figures in a festival of the central Andean city of Latacunga held in celebration of the *Virgen de las Mercedes*, who is said to have saved the town from destruction after an eruption of the nearby volcano Cotopaxi in the eighteenth century. The festival also coincides with the sowing period, a time for blessing the earth, in the Andean agricultural calendar. The origin of the character of the *Mama Negra* herself is largely lost to the past, narrations including her conceptualisation as Jesus' wet nurse, seeking to please the Virgin with her attentive care; a vengeful figure appearing to neglectful devotees; and as a commemoration of the liberation of enslaved Africans in Ecuador. Debates aside, however, she is a recognised symbol of fertility, and processes through the streets on horseback, nursing baby Jesus and spraying donkey's milk at the crowd. The character – large, female and black – is played by a man in blackface and is a complex and contentious manifestation of the tenacious class, race and gender dynamics that have shaped and continue to shape Ecuadorian society (see Weismantel, 2003 for discussion). In recent decades the festival has also been adopted as part of the city's independence celebrations and now exists in two related versions, one more touristic than the other. Like many

Figure 2.1 Fervorosos Pasos (Photo: Andrea Cuesta)

traditions, the festival today is a multifaceted articulation of diverse and blurred heritages that reflect local, national and transnational concerns, Pico's creations standing as one other expression of a character and festival that is in constant evolution.

Continued

In the reinvention of the *Mama Negra* for the contemporary dance stage, creativity works in dialogue with both the micro-geographies of the dancing body and the wider geographies within which it moves. At the level of the body, the imaginaries, histories, and practices embodied by the *Mama Negra* become the basis for the detailed investigation and experimentation of movement and character. Pico describes this as a process through which the body is 'turned festive', shedding and rearticulating learnt techniques and languages through improvisation to seek the emergence of the character anew. In this process, the artist critically responds to a context in which traditional practice is commonly 'folklorised' and stylised for the stage by exploring the character through ideas of the grotesque, seeking to turn a familiar figure strange – the movement of the hand that greets the public is broken and disjointed, the head cocked, the shoulders restless. Pico's *Mama Negra* trots unstably on raised heels, feet bent as if on her parade horse, her body gliding jauntily along the floor. She moves, at turns in devotion and irreverence, both to the familiar brass sounds of the festival's *banda de pueblo*[1] and to soaring violin or the earthy drone of a didgeridoo. In Ecuador, such exploration also forms part of a wider postcolonial geography of re-appropri-ation that seeks to creatively engage with traditional and popular practices in a medium where both national and international ideas of contemporary dance frequently privilege particular kinds of practice.

> So much teaching is wasted because, as it doesn't come from an official entity, it tends not to be valued as it should be. As a result of cultural colonialism, particular techniques in dance and theatre are imitated as if they were the only possible options and people copy productions that the international media extol as the canon of validity, leaving to one side our own resources, ideas and approaches. Popular festivals are a rich source that, with respect and great care, should be better known and recognised by the performing arts of the country.
> (Wilson Pico in Pico and Pico, 2011, p. 32)

For Pico, then, attention to popular festival and ritual thus becomes one way of reclaiming and decolonising the dancing body and its practice within international networks of contemporary dance that can work to shape, order, value, and hierarchise practice in particular ways.

Institutions, however, occupy multiple and complex roles in dance production and while at times they might act as arbiters of what constitutes 'correct' or 'desirable' practice, they can also serve to facilitate particular kinds of productions in relation to specific social or political contexts. The project *Cuerpo Festivo*, for example, was supported by state funds through the Ministry of Culture, which has placed particular emphasis on strengthening national identity and interculturalism through the arts in Ecuador. In a country with a richly diverse population, and stark ethnic and racial inequalities, government discourses frequently centre on ideas of 'unity in diversity' in the construction of a plurinational state (see Senplades, 2013). While the work *Mama Negra* is a thoroughly independent production, then, state support for the creative investigation and representation of traditional practices through such projects could perhaps be seen as part of ongoing concerns for the recuperation and reinterpretation of collective imaginaries aiming at the 'cohesion of the Ecuadorian nation' and the creation of a 'new contemporary Ecuadorian identity' (Ministerio de Cultura del Ecuador, 2011). Here creativity in the dancing body is thus not only about individual experimentation and exploration but also about ideas of a collective heritage and identity to which various groups and discourses might lay claim. In this process, movement emerges both through the dancing body and the spaces and institutions that dance with and through it in its practice.

Acknowledgements: Many thanks to Wilson Pico for conversations about his work. This research was supported by the Economic and Social Research Council [grant number ES/I026525/1]. All translations are the author's own.

Attending to the creative process: decentralising human creative agency

Perhaps one of the foremost thinkers to draw attention to creativity as a process is the twentieth-century German philosopher Martin Heidegger. Simply put, for Heidegger human creativity is an act of revelation; painting, drawing, writing, or sculpting reveals something about the truth of our existence (Mitchell, 2010; Thompson, 2000). This bringing forth of truth sits alongside another form of creation that Heidegger identifies, physis, the bursting forth out of itself that happens in nature. The creative impulse for humans is thus a means to unveil what is ordinarily hidden from everyday awareness (see Bolt, 2011 for a summary of Heidegger's ideas in relation to art).

For Heidegger, however, creation is not the act of the genius artist. Rather, he argues that modern subjectivism has misinterpreted creation 'taking it as the sovereign subject's performance of genius' (Heidegger, 1977, p. 200). To create, he notes, is 'to let something emerge as a thing that has been brought forth. The work's becoming a work is a way in which truth becomes and happens' (Bolt, 2011, p. 104; Heidegger, 1977, p. 185). The act of creation therefore is to let something emerge as a thing, rather than to make a thing. If so, a query Heidegger's work often raises concerns the role of the artist in the process of creation (Bolt, 2011). In great art, Heidegger suggests, 'the artist remains inconsequential as compared to the work, almost like a passageway that destroys itself in the creative process for the work to emerge' (1977, p.166). Heidegger thus suggests that the artist is just a conduit through which art emerges.

Where Heidegger's thinking has become of most influence in contemporary thinking about creativity has been as a means to rethink the relations creative makers have with materials and tools of practice. This rethinking has moved us away from mastery to explore these relations in terms of concernful dealings. For Heidegger such dealings with tools, materials and ideas enable something never-before conceived to be revealed: in short, understandings of the world grow out of practice. For, in place of the performance of the genius artist, and in contrast to post-modern understandings of art as a cultural construct, or art as created through discourses on art, Heidegger requires us to keep focused on what is revealed or brought forth into appearance through the creative process in practice.

Heidegger's foregrounding of creative process has been an important starting point for many interested in creative practices, including anthropologist Tim Ingold (2013). Ingold is concerned to reverse the tendency to read creativity 'backwards', starting from an outcome in the form of a novel object and tracing it through a sequence of antecedent conditions to an unprecedented idea in the mind of an agent (2010a, p.16; 2013). Building on the distinction he makes between objects and things, Ingold argues that the role of the artist is to 'join with and follow the forces and flows of material that bring the form of the work into being'. Creativity, in short, lies in the forward movement that gives rise to things. To read things forward, Ingold argues persuasively, is to focus on improvisation. To improvise is to follow the ways of the world as they unfold, rather than to connect up a series of points already traversed. Such a world that is always in the making, in a constant state of becoming, suggests new understandings of creativity. Less creativity as abrupt and sudden innovations and rather creativity thought in terms of adaption and response to the possibilities and barriers presented by an ever-changing world.

Pondering the creative process Ingold (2010b) alights on what he calls the 'textility' of making. Textility, for Ingold, emphasises materials and forces in developing an account of the negotiation between material and human action. Thus the processes of creativity, of making, draws together embodied action, material properties, knowledge and know-how. One of Ingold's principal sources is the painter Paul Klee. Klee writing in 1920 notes that 'art does not reproduce the visible but makes visible'. As Ingold (2010b) explains, art does not replicate finished forms, but joins with forces that bring form into being. The essential relationship following Klee, and also Deleuze and Guattari (1988) is thus 'not between matter and form, but between materials and forces' (Ingold 2010b). It is about how materials of all sorts, energised by cosmic forces mix and blend with one another in the generation of things (Ingold, 2010b; 2013).

Such an understanding overturns what is called the hylomorphic model that suggests, as Aristotle reasons, to create anything you need to bring together form (morphe) and matter (hyle) (see account in Ingold, 2013). This hylomorphic model of creation was deeply embedded in western thought. Form was imposed by an agent with a design in mind, while matter, rendered passive and inert became that which was imposed upon. Such ideas continue to structure how we think about creativity. Ingold (2013) looks in his discussion of making to overthrow this model.

He uses ideas drawn from Deleuze and Guattari to replace the ontology that gives primacy to states of matter and their form in production, to instead value processes of formation and the flows and transformations of material. Taking on what Klee says of art – form giving is life – Ingold argues this more generally for makers and creative practitioners. They are, he notes, 'wanderers, wayfarers, whose skill lies in their ability to find the grain of the world's becoming and to follow its course while bending it to their evolving purpose' (2013, p.92). To read creativity forward in this way requires that we embrace improvisation, to improvise is to follow the ways of the world as they open up, rather than to recover a chain of connections from an end point to a starting point on a route already travelled (ibid.).

There are a number of implications of such a reading of creativity forward, such a dwelling in/on process. These include a growth in studies of the micro-geographies of creative practices. This entails often auto-ethnographic exploration of the processes and materials of creative prac-tices, whether this be working stone, glass or wood or wider processes of crafting (Edensor, 2011; Ingold, 2013; O'Connor, 2005; Paton, 2013). The relationship between the creative worker and the material on which they work is far from clear and straightforward. Indeed, concern-full dealings with both the tools and the vibrant animated matter with which creative practitioners collaborate have become a crucial part of thinking about the micro-geographies of creative practices (Carr and Gibson, 2015; DeSilvey, 2006; Paton, 2013). As such, to understand the force of creative practices in the world is to understand the intersections of skilled and unskilled bodies and animated materials.

Geographic concerns with 'vibrant' matter, in other words with the forces and animations of matter – for example, stone, wood, wool, metal – have begun to usefully inflect discussions of making (Bennett, 2010; Ingold, 2013; Luckman, 2015). Here we see concerns with the animate nature of materials and their agency, such that the relationship between material and transcendental creative maker is disrupted, no longer do we revere human agency in manipulating materials. Rather, materials come to have agency, force and character that can help the maker, be worked with if the maker is skilled and has a feel for materials, but can also go against and thwart a creative practitioner, introducing surprise and contin-gency into the process and the finished 'product' (Bond et al., 2013; DeSilvey, 2007b; Edensor, 2011; Ingold, 2013).

The idea of animate or vibrant matter (often as theorised by Jane Bennett [2010]) has come to play an important role in geographical discussions of the agency of materials. Bennett is concerned with 'the capacity of things – edibles, commodities, storms, metals – not only to impede or block the will and designs of humans, but also to act as quasi-agents or forces with trajectories, propensities, or tendencies of their own (2010, p. ix). These forces and propensities are mobilised in the discussion of the stone-metal flesh dialogue in shaping and carving stone, or the glass-metal-flesh dialogue in shaping glass (see Box 2:1). As Paton notes of quarrying stone,

> the interplay of person and matter is not about the imposition of one material onto another, but a growing exchanging of material properties that form unique yet constellated relations. To make something with a material is a creative relationship, an open-sourced reciprocating highway of sensual orders.
>
> (2013, p. 1077)

In common, these accounts suggest a relational account of the process of creativity. Ingold describes it thus,

> the gradual unfolding of that field of forces set up through the active and sensuous engagement of practitioner and material. This field is neither internal to the material nor internal to the practitioners (hence external to the material); rather, it cuts across the emergent interface between them.
>
> (2001, p. 342)

Such accounts don't just offer the vibrant matter of materials and tools, but also ensure that we are aware of our own vibrant bodily matter as human makers; making processes don't just shape the creative product, they also shape the maker. Whether this be the new neural pathways created as habits form, the sculpting of muscles and tendons through repetitive manual labour, as well as the debilitating effects of such labour as backs are bent, muscles tighten and strain and wrists and joints ache from hours of repetitive manual labour.

It is not just in the making of things that we find materiality entering into discussions of creative practices. Also important for these concerns is the role of materialities in practices such as parkour, skateboarding or urban exploration (Garrett, 2013; Mould, 2009). These skilled urban practitioners feel the materiality of the urban environment through their bodies and their boards, working with it to develop the moves they want to execute, learning the different surfaces, textures, and angles and what they do or don't enable in terms of their creative practices (Borden, 2001).

What such accounts draw to the fore is a distributed sense of creativity. To appreciate the importance of process, the vibrancy of matter, the importance of tools is to refuse to grant sole agency to the human maker. In a series of different ways, all these approaches deny the transcendental human creative genius, refuse the sense of creativity as solely an interior force and demand that we understand it as an embodied and material practice that is historically and socially contingent.

Creativity and improvisation: a way of being in the world

Taking an expanded sense of creativity seriously, as the introduction to this volume argues we need to do, requires that we not only appreciate what we might call specialist forms of creativity – artists, craft practitioners, and so on – but also that we appreciate creativity in ordinary and mundane ways, and, moreover, creativity that might not even be a conscious act. In a very expanded sense of creativity, Hallam and Ingold (2008) argue that creativity is something,

> that living beings undergo as they make their ways through the world ... this process is going on all the time in the circulations and fluxes of material that surrounds us and indeed of which we are all made- of the earth we stand on, the water that allows it to bear fruit, the air we breathe.
>
> (2007, p. 2)

Creativity then is indistinguishable from living, creative improvisation and is 'the forward movement of keeping life going, that involves a good measure of creative improvisation'. As such, creativity becomes an improvisational quality that is present across all forms of cultural activity and requires people to adapt, often unconsciously to particular circumstances (Edensor et al., 2009b, p. 9). Rather than foreground creativity as about the production of novelty, this is a creativity that is found 'even and especially in the maintenance of an established tradition . . . [for] traditions have to be worked to be sustained' (ibid., p. 5–6). Reflecting on these ideas Edensor et al. (2009b) note that creativity thus comes with a particular temporality that is 'located in the adaptation of pre-existing ideas and intentions to a fluid present, and is thus constantly evident in the circulation and fluxes of materials that surround us and indeed of which we are made' (p. 8).

Thinking about creativity as improvisation requires an appreciation that creativity emerges through the experience and practice of doing; it is part of the unfolding of life. As Crouch (2009, p. 133) notes 'in our lives we construct, handle, make sense, cope, respond and anticipate amid a complex collision of influences, unbidden occurrences and desires, only partly planned'. Thus our way in life is 'continually altered and responds to the performance of others' (ibid., p.4). As Crouch summarises, 'creativity relates to life's emergent potential to shift and change as becoming, rather than as stasis or fixity' (p. 133). As such, far from creativity associated with novelty and innovation, those ideas for which it has gained so much kudos, this is a creativity that foregrounds the routinised and habitual everyday practices in mundane realms where a slower pace dominates and where things are down to chance and contingency as much as, if not more so than, planned action. As Paul Harrison (2000) explains, the everyday is not always about dull compulsion and repetition but also 'in the everyday enactment of the world there is always immanent potential for new possibilities of life' (p. 498). Quotidian practice is open-ended, fluid, generative, a sensuous experience that is 'constantly attaching, weaving, disconnecting: constantly mutating and creating' (ibid., p. 502).

It is hard though to conceive of these improvisational creativities outside of a wider assemblage of forms of more conscious and planned forms of creative living. Writing, for example, of the cultural improvisations of allotment gardening, Crouch argues for an attention to how 'creative energies inhere in everyday living, in the hidden and possibly unbidden, quiet excitement of the performance of mundane life' (2009, p. 141). He describes a creativity in assembling and recycling pots, pans, baths, sheds made of assemblages of things, all 'fused with vegetable and flowers, weeds, puddles and bare earth' (p. 135). This 'soft collision' is joined by creativity expressed through the patterning of the ground and the intimate movements of the body amongst vegetation, earth, insects and air (p. 137). As his account and other discussions of gardening explore, such activities are not just those of organic cultivation, often for consumption, but are also creative acts of self-discovery and innovation. Such actions also cultivate generosity based on the collective enjoyment of growing and the sharing of practices, materials and crops (Adams and Hardman, 2013; Crouch, 2009; Hallam and Ingold, 2014; Milbourne, 2012). For Hallam and Ingold, such mundane creativities are slow and routinised, part of a liberation of creativity rather than its repressions. As Crouch's account develops, however, via Deleuze and Guattari, in the slow becomings can lie an intensity in the sense of feeling and connectivities that are cultivated (p. 141). Intensities of feeling that sit alongside the creativities of human and non-human kinds.

If the discussion in this chapter has focused largely on human and divine creativities, it is important not to overlook the other forms of creativity that exist. Of course, it is not just humans and gods who have been considered creative, but also a range of non-humans. As discussed earlier, Heidegger finds a form of bursting forth in the natural world that he views as akin to an artistic creativity. It is also clear that animals are capable of some of those improvised forms of creativity discussed above, as well as forms of aesthetic practice – song, territory creation – that are part of their evolutionary process and often related to mate selection (Dixon et al., 2011a). Further, process philosophers, principally Alfred North Whitehead, situate a form of creativity as central to their metaphysics. For Whitehead, the focus should be on dynamic becoming rather than a static being; further, creativity is the condition of the possibility of being and thus cannot be a mode of it (Rose, 2002). Indeed, for Whitehead 'a creative advance into novelty is the most fundamental feature of the universe' (Garland, 1969, p. 361). Philosophers are unable to agree on exactly what creativity is and means within Whitehead's work as he does not offer a full-length treatise on this important concept. What is clear is that creativity is more than domain general in Whitehead's account, it is fundamental to understanding the becoming of the world, as Singer puts it simply 'all reality is creative in some way or another' (Singer, 2010, p. 10).

Thus far, this account has explored the rise and fall of the agentive creative subject, exploring key moments in history but also modes of contemporary study- embodiment, vibrant matter – that challenge the transcendental creative genius. The chapter will now build on these figures of the embodied creative practitioner to explore the figure of the creative body that is offered in the political economy discussions of the creative labourer.

THE CREATIVE LABOURER: CREATIVITY CREEP

> Recasting work as art, and recasting art as work, it is hard to talk about creativity without involving a profession of some kind.
>
> (Rothman, 2014)

The creative labourer is not a new figure. In the guise of the artisan, the creative labourer has long been identified with struggles against industrialisation and capitalism, and with the making of the English working class. Research into the nineteenth-century craft industries identified them as a labour aristocracy, valued for their skill and political consciousness. While the twentieth-century embrace of Arts and Crafts, especially in England and the US, saw craft take up a radical position against mass manufacture. Richard Sennett (2009) recently revived the figure of the craftsman-artisan as the antithesis of demoralisation and degradation. He builds an account of contemporary craftsmanship that finds in such handed practices a source of ethical living. In contrast to the alienated factory labourer, the mythical figure of the creative labourer is un-alienated from their work fulfilled by a practical use of hands and an engagement with materials.

This section will reflect on competing figures of the creative labourer. It will explore how their bodies and subjectivities are made present (and also rendered absent) across creativity discourses and within the interdisciplinary literatures. Discussion will consider the trend-setting liberal bohos, the affective and immaterial labour of the creative precariat and the passion workers, appreciating both those characteristics celebrated in these myths, but also exploring them as ideologies that dangerously hide from view the dark side of creative labour. Not only are such figures interesting in their own right, but they are of importance in the context of contemporary capitalism's widespread adoption of the figure of the creative worker (Gill and Pratt, 2008; Lazzarato, 2012; Hardt and Negri, 2004). Valuing the creative worker's high levels of self-motivation and exploitation, love for their job and willingness to withstand high levels of precarity, the creative labourer has become the ideal worker sought by capitalism the world over. As Richard Florida (2005) and others call for the support and growth of the 'creative class', it is important that we appreciate what constitutes such creative work, both its pleasures but also its often hidden demands.

The bohemian and the creative precariat

The idea of bohemianism first came to describe the lifestyle of Parisian artists in the 1830s, living eccentric and edgy lives concentrated in lower-rent parts of the city such as Montmartre. Bohemian communities eventually grew up in a range of areas, including Greenwich Village and Soho in New York, Wicker Park in Chicago, often in what were initially the most rundown areas of the city, only for these areas to become gentrified by these communities and become desirable (see Chapter 7). The imaginary of the bohemian is a powerful one, both in terms of the romance of the creative, free-spirited and often politically progressive lifestyle, but also in terms of the possibilities it offers for urban regeneration. More recently Richard Florida (2005) has adopted the term and its associated imaginary as the foundation of his 'boho index', an index that ranks cities based on the proportion of the population working in the creative economy, together with the proportion of the population that are gay. As Florida notes 'creative, innovative and entrepreneurial activities tend to flourish in the same kinds of places that attract gays and others outside the norm'.

While bohemianism was often associated with a certain creative life-style, it was also associated with financial insecurity, all part of the myth. As the image of the creative, bohemian worker becomes a form of worker increasingly sought under capitalist relations (by workers and industry leaders alike), it is important to address the darker side of this bohemian myth. Under the creative industries paradigm, creative labour has become seen as 'intrinsically self-actualizing and a meritocracy . . . of workers as a 'creative class' marked by (relative) affluence, diversity and the pursuit of an experience-orientated elite-consumerist lifestyle' (Arvidsson et al., 2010, p.296). For Florida, his creative class is a group of 'exceptionally talented individuals' whose key role is to 'produce new ideas', and who are motivated by the search for 'abundant high quality experiences, an openness to diversity of all kinds, and above all else the opportunity to validate their identities as creative people' (Florida, 2005, p. 36). As this group become increasingly popular and desirable for city planners and governments to attract and for companies to employ, the lived experiences of these workers and the conditions under which they labour have become a pressing issue. For the political left it is a question of rethinking resistance, as well as a 'need to reassess and redefine the concept of exploitation under current global conditions' (Kanngieser, 2012) of which precariousness is key. As Berardi writes, 'precariousness is no longer a marginal and provisional characteristic, but it is the general form of the labour relation' (2009, p.31).

Creative labour has often been associated with the 'productive capacities of the soul itself . . . intelligence, sensibility, creativity and language' (Berardi, 2009, p. 192). Moreover, for these workers labour is valued as 'the most interesting part of his or her life', such that they will happily not escape work but will lengthen the working day 'out of personal choice and will' (ibid., p.79). As such, resistance to labour often needs to be redefined, but further the conditions of labouring and their effects need to be considered. As workers work out of interest and care for their jobs, they are subject to a range of biopolitical effects, from sleep disorders to unhealthy work-life balances, to enhanced stress and the effects of having identity and work closely bound together (Kanngieser, 2012). Furthermore, the precarious, often informal nature of these ways of working – long hours, insecurity, lack of set career path, often intern-based working, low pay, contract or project work – actually becomes glamorised. As Neff et al. (2005, p. 331) argue '"Hot" industries and "cool" jobs not only normalize, they glamorize risk, and the entrepreneurial investment required . . . leads to a structural disinvestment to exit'.

A growing number of empirical studies have identified traits observable across studies of the fashion, art, television and film industries. Such traits include a divided workforce, bifurcating between a small controlling elite and an oversupply of workers with generic skills whose employment is poorly paid and precarious. This creative precariat often engage in self-employment, freelance or short-term contract work, will often work as interns, and have to rely on social skills in the construction and maintenance of networks to enable them to find work (Neff et al. 2005). Indeed, some discussions of creative labour suggest there is very little that is actually creative about creative workers. Instead of the image of self-realised, cool, alienated labour that has come to characterise the creative economy, many do work that more closely resembles non-creative sectors of knowledge industries such as call centres. As Gill and Pratt make clear, however, one of the most consistent findings on research on work within the creative sector is that it is experienced by most who are involved with it as profoundly satisfying and intensely pleasurable (at least some of the time). A vocabulary of love is repeatedly evidenced in such studies, with work imbued with the features of the Romantic tradition of the artist, suffused with positive emotional qualities ... Research speaks of deep attachment, affective bindings, and to the idea of self-expression and self-actualization through work.

(2008, p. 15)

We have not therefore rid ourselves of the sense of the creative, romantic aura around creative work, rather our sense has grown more 'social, practical and mercantile', as we demand a 'spiritual wage' in exchange for the production of things (Rothman, 2014). The discussion that follows will explore some elements of creative labour in more detail.

The passion worker

Doing what you love – DWYL – has become the unofficial work mantra of the twenty-first century (Duffy, 2015; Tokumitsu, 2014) and is often associated with the creative industries and, in particular, the social and digital media sectors. Accompanying such aspirational labour – where pleasure, autonomy and income seemingly co-exist – is a sense of an industry assuring individuals that 'their labour serves the self and not the market place' (Tokumitsu, 2014). The result enables the industry to pay workers very little, if anything, for long hours and highly precarious working conditions. DWYL is the latest in a line of terms developed to describe these forms of labour from passionate labour (Postigo 2009) to venture labour (Neff, 2012) and even Playbour (Kucklich, 2005).

Indeed, a number of empirical studies are starting to explore the contradiction of the 'ideology of creative labour' and the apparent reality. In a study of the work of Milan's fashion workers (see Box 2.3), Arvidsson et al. (2010) note how 'perceptions of work as creative and self-actualizing contrast with a reality marked by strong hierarchy imposed hyper-flexibility, little autonomy and, in general, few possibilities for self-actualization' (2010, p. 297). They argue, however, that this ideology should not simply be dismissed but understood to have power as a material and concrete reality. In this case, they are interested in how the ideology of creativity serves an important function in the construction of the subjectivity of creative workers, shaping motivations, self-image and the notions of the value of their work. As such, they develop Gill and Pratt's (2008, p. 18–20) argument that when considering creative labour, scholars should attend to the meanings cultural workers give to their activities, rather than layer these meanings and practices with our theories. Arvidsson et al. (2010) go further still to note how the ideology of creative labour also serves to add value in the brand-centric creative sector. In this sector, connotations of coolness and creativity provide a means to recruit talent and labour, as well as enable symbolic production to support and legitimate otherwise immeasurable values. In other words, the ideology of creative labour produces and maintains the very idea of creativity by which the value of the creative industries product is sustained.

To appreciate the power of the ideology of creative labour should not be however to dismiss the lived realities of passion work and its often associated 'corrosion of creativity' (McRobbie, 1998). As studies of musicians, fashion workers and other creative professions demonstrate, often little time is devoted to creative practices, with much more time spent (up to 90%) on activities that include financing, distribution, public relations, branding and marketing. For Hracs and Leslie (2013) this equates to a Do-It-Yourself culture of music production, enabled by a combination of accessible digital production processes and online distribution that is radically different to the vertically integrated industrial system of record production (see also Leyshon, 2014). Whilst this appears to be shaping a more democratic creative sector with lower barriers to entry, it also intensifies some of the negative aspects of creative labour. Considering 'aesthetic labour' offers one way to understand some of the challenges creative workers face. Aesthetic labour, often discussed in relation to the interactive service industries, recognises the demands placed on workers when their embodied capacities and attributes are called on as they cultivate and perform a certain workplace persona (Crang, 1994, Warhurst and Nickson, 2001). This intersects with wider notions of emotional and affective labour; labour that requires the management of feeling, or demands emotional traits from the workers for successful working practice (Hochschild, 1983). In a study of musicians in Toronto, Hracs and Leslie (2013) note the changing nature of music work has seen a rise in live performance and a growth in the use of social media and other techniques to cultivate an online presence. As they note, live performances require a management not only of look, but also of body language. Further, as individual musicians rather than record companies manage their own marketing, the cultivation and maintenance of an online presence, a 'digital body', is an exhausting and labour-intensive process that falls to the musicians themselves. Not only must the 'digital body' be managed, but as Hracs and Leslie note there is a pressure to make connections with fans, to stand out from the crowd by way of building online relationships with potential audiences. Such processes are rarely quick, often involve management of self and image as well as an investment of time outside the working day, ensuring that creative practitioners feel like they are always 'on', whilst at the same time experiencing a 'corrosion of creativity' as they spend less and less time on the 'creative' elements of their practice.

Box 2.3 MILAN FASHION INTERNS

Milan is a global centre for fashion production with one of the world's largest agglomerations of fashion firms. The city together counts some 12,000 companies involved in fashion production (Power and Jansson, 2008 p.6) and nearly a third of Italy's 65 garment districts are located around the city. From 1996 to 2005 the Italian fashion industry grew and there was a shift in the site of value added, from the material production of garments to the immaterial productions of design, events and communication. Thus there has been a shift in the structure of fashion employment from material garment production to immaterial production. As Arvidsson et al. (2010) note, what has emerged is a brand-centric system which moves control of the production of trend and fashion from the street level to a fashion industry largely led by the large brands and their communication strategies. This moved a business model away from one geared towards following the dynamics of consumer tastes towards one based on brand presence and consistency. This included cultivating the growing market for accessories, sunglasses, wallets, key rings, belts, and so on, helping the value of Italian fashion brands grow by 50 to 100 % between 2004 to 2008.

In the course of their ethnography with low-rank fashion workers in Milan, Arvidsson et al. (2010) noted the concentration of workers engaged in this brand-based activity. Such brand work is generally underpaid, precarious and involves long hours, only a third of their sample had a contract, and only 9% were permanently employed. There was a lot of project-based work, lasting less than a year and much flexible self-employment, 66% of the sample had relied on parents or family money for economic support in the previous two years, and two-thirds claimed they were unable to raise a family on their salary. Internship was rife, regular employment when gained, led to long working hours, and insecurity of contracts was often used as a deliberate policy. Bosses expected obedience from workers during the day and often at night, working hours were flexible, extending when needed, with no fixed hours meaning no end to work. 66% of those on project-based contracts worked outside of regulation hours. The working environment was also marked by strong hierarchies, and fixed divisions between workers and bosses and intense competition between workers. Such that while people might have large social networks, they often have few friends as these social networks, intensely important to career development, are made up of colleagues who are always competitors. The picture painted is of a bifurcated world, where the everyday drudgery of long hours of underpaid work underpins the glamorous world of fashion events, parties and illicit drugs. Further, there is a lack of mobility from the lower ranks to upper ranks of management, this structural problem means little turnover at the top and young talent is often lost abroad.

Despite these conditions, Arvidsson et al. (2010) record that for many of their interviewees their work was mostly deemed to be satisfying, perceived as autonomous and flexible, and also as providing opportunities for creativity and learning experiences. Yet on exploring the forms of creativity that could be exercised in the workplace, it seemed that creative decisions were taken at the top end, in part to ensure brand consistency. At the lower end work seemed to revolve around menial supportive

Continued

tasks. They conclude that creativity for these workers does not so much describe an actual reality as a way of giving sense to and legitimising a labour process that is, in fact, marked by high levels of fragmentation and insecurity (2010, p. 305). They also note how creativity had bled out of the actual labour practices to become associated with a wider creative lifestyle. The job became associated with belonging to a particular creative scene, enabling a common world of parties, socialisation and occasional celebrity moments, shared consumption interests and lifestyle habits. The work then fuels a sense of the 'possibility to imagine oneself as belonging' as the lifestyle is lived vicariously. Of course, the workers at the bottom of the chain don't get to attend the higher-end networking events; access to such events can only be offered as a reward or incentive by bosses.

From Arvidsson et al.'s (2010) account, and others like it, fashion work, like creative industries work more generally, emerges as passion work. As such, is it (it is) valued in non-monetary terms, separating the financial reward for hours worked from the identity value and cultural capital bestowed by these forms of labour. In this coincidence of the production of value and the production of subjectivity, creative workers are part of the wider shift towards the biopolitical governance of labour. Here, the sector's provision of forms of subjectivity to their workers becomes a way to shape and govern the valorisation process and ensure the jobs are valued despite often bad working conditions.

As this discussion of creative labour has suggested forms of creative work are complex and varied. Given that the creative economy is one of the great economic hopes of the era, offering a diffuse model for economies around the globe as well as for 'ideal' forms of labour, it is important that the embodied practices and daily experiences of creative labour are not written out of these global discussions. If the creativity economy is to remain important the world over then to appreciate the nature of creative labour (its benefits and challenges for workers and employers alike) is a political imperative and a crucial part of understanding the dynamics and power of the creative economy more generally.

CREATIVE GEOGRAPHIES – THE BODY – SENSING WORLDS

The body has long been a key geographic site, and in this final section I want to shift focus from thinking about the bodies of creative workers to explore briefly how creative practices, more broadly, have helped geographers to grasp and understand relations between bodies and worlds. Further, I want to explore how a turn to creative practices as research methods enables geographers to think through their engagements with the world more generally. Whilst these discussions will run throughout the text (see, for example, reflections in Chapter 7 on the city, Chapter 8 on the margins, and Chapter 10 on landscape and environment in particular this discussion will aim to lay the groundwork for these more specific examples. It will begin by way of a consideration of humanistic geographies of the 1960s (Buttimer, 1976, Driver and Martins, 2005) and how these embodied foundations saw geographers turn to creative practices as a means to map and remap bodies and explore their relationship with spaces. It will explore how this foundation also laid the groundwork for geographers' current turn to creative methods, from drawing to writing and performance practices, as a means to know and engage place and community; how in short, these particular engagements of bodies and creative practices enables a doing of research differently. Running throughout will be a sense of the challenges of embodied methods, including issues that can emerge around writing practices (Dewsbury, 2010). In other words, what does it mean to write these sensing bodies, both deploying creative writing as a means to do so, but also querying what sorts of writing skills are akin to such ways of sensing and engaging the world.

Life worlds and body-scapes: humanistic geographies – and their legacy

> We shall not have a humanistic geography worthy of the claim until we have some of our most talented and sensitive scholars deeply engaged in the creation of the literature of the humanities. Geography will deserve to be called an art only when a substantial number of geographers become artists'.
>
> (Meinig, 1983, p.325)

The relationship between humanistic geographies and the sensing of the world is a long one, and one within which creative practices have taken up an important place. Heavily influenced by phenomenology's evolving concern with our bodily engagements with our life worlds, humanistic geography of the 1960s and 1970s sought a more subjective, embodied way of engaging with the world. For many, a natural place to turn to was creative practices and the arts and humanities disciplines that analysed them. For humanistic geographers, there was a move to find in art fewer facts, as had been the trait with previous engagements, and rather more what has been termed, 'nuggets of experience', that not only provided particular types of information about places, but further offered ammunition in the evolving project of humanistic geography (Tuan, 1977, p. 274). Writing in 1983, Wreford-Watson notes that arts practice stood for all the 'empirical traits' that quantitative geography hoped to escape – personal experience vs. categorical abstractions, uniqueness vs. generalization, the qualitative rather than the quantitative, and a celebration of the expressionistic and experiential (Watson, 1983, p. 386). Turning to the deployment of arts practices today, especially in the context of our sensing of space and place, many of these traits still hold true, albeit replacing the phenomenological lens dominant in humanistic concerns with one shaped by the tenants of post-phenomenology, principally pre-subjective concerns with affect, force and matter.

The body, once peripheral to geographic concerns, has become 'very much part of its conceptual and methodological core' (McCormack, 2008a, p. 1823). Indeed, geography has become engaged in a vast interdisciplinary project that maps and remaps bodies as 'volatile combinations of flesh, fluids, organs, skeletal structure and dreams, desires, ideas, social conventions and habits'. This is a theoretically ambitious project that encompasses phenomenological life worlds, as well as the volatile bodies and vital materialisms of feminist scholars and bio-philosophers (Grosz, 1994; Paterson, 2009; Thrift, 2007). Key questions remain, however, around 'how and in what ways the body is geographical', and as McCormack (2008a, p. 1822) notes, 'there is much we still do not know about how the relation between bodies and geographies might be understood, experienced or experimented with'. There are two key ways in which arts and creative practices more generally have intersected with geography's bodily concerns. Firstly, creative practices have become a core site for the mapping and remapping of our bodies and our bodily boundaries. Secondly, creative practices become the means to answer questions about how 'cultural geographies are also corporeal geographies.' That is, in what ways do bodies participate in the processes through which we make sense of the world both individually and collectively (McCormack, 2008a, p. 1826)? It is through bodies that we live and know space, and in diverse creative practices geographers have found the resources to study the form and dynamics of these spatial doings of bodies in more detail. Arts practices and dance practices, in particular, have come to offer geographers ways to explore the making and remaking of the spaces in which we live. Creative geographical practices enable us to think about both the spatialities and spaces of our bodies, but to think 'with and through the spaces and encounters of which these bodies are generative' (ibid.).

Mapping bodies

Creative practices, whether thinking about or doing arts practices, dance or performance have come to offer geographers a 'lab' for sensory exploration. As such the picture, the installation, the performance and the dancing body all form spaces for encounter in which the human body and the cartographies of its spaces and senses are made available for study (Hawkins, 2011; 2013). One of the primary foci of these studies has been the sensory cartographies of our spaces closest in. As Cosgrove (2008, p. 5) wrote in *Geography and Vision*, 'the conventional classification of five discrete senses, while intuitively appealing in that they relate to distinct organs and locations in the body, seems to be heading towards the scientific oblivion of the four Aristotelian elements'. This is an insight with important implications for how we not only know space, but also how we think about the spaces of the body.

For geographers, the movements, experiences, learning practices and cultural conventions of dancing bodies have become a key means through which a number of corporeal questions have been addressed (McCormack, 2008a, 2014). In particular, dancing bodies have formed a crucible for thinking through the relationships between representation and non-representation. Here they have aided in the reworking of the former as practice, being part of geographers' broader concerns with practice-based thinking, embodiment and a consideration of distributed agency, that underpins the far from singular projects of non-representational thinking (Thrift, 2007).

One of the primary sites of critique for geographical scholars thinking about art has been the concern with sight. Indeed, geography is often thought of as a visual discipline (Driver, 2003), but recent years have seen a rise of interest in the other senses for both scholarship and methodological experiment. Across these studies we see the benefits of artistic practices as offering the means to enable us to explore the dynamics of a remapping of bodily senses, and a recalibration of the sensory hierarchy, engaging the relations of both the external senses as well as the internal senses (Hawkins, 2011; Hawkins and Straughan, 2014). Vision, long the dominant sense within geography, is gradually coming to be toppled from its leading place in the hierarchy of the senses. This is not so simple, however, as merely down-casting the eye, and relocating other senses atop the hierarchy. Rather, core to research has been to open up variegated modalities of vision, principally by situating sight within a fleshy, multi-sensory body. We might think, for example, of geographical studies that have examined the 'invisible architectures' of sound installation, the place of smell within exhibitionary geographies, or the intersections of touch and vision in the haptic senses and with hearing (Cameron and Rogalsky, 2006; Hawkins, 2011; Hawkins and Straughan, 2014). It is not just sensory cartographies that arts practices have enabled an exploration of, but also the materialities of bodies and their boundaries. Digital media art and art-science practices enable audiences to explore different bodily spaces and scales, whether this be the 'plasticity' of the human brain, or bodily tissues on a nano-scale (Hawkins and Straughan, 2014). Exploring the 'nano-art' practices of Paul Thomas, Hawkins and Straughan (2014) examine how the sonic, visual and tactile dimensions of his installation Midas, disrupts any sense of the human body as a discrete fleshy container, instead mingling bodily insides and outsides in a way that overturns bodily boundaries and renders bodies vulnerable and constantly emergent, rather than fixed and static. As such, creative practices have become a common means by which geographers have opened out explorations of the senses, substances and boundaries of our bodies. Creative practices become in McCormack's (2014) words productive of the kinds of 'thinking spaces', of experimental sites and situations that make possible engagements with the affective, precognitive and non-representational aspects of the world.

Researching through the body

If the creative practices of others offer geographers the chance to study the spaces and sites of the body, then enrolling our bodies within the doing of creative practices has become an important means for thinking about the body as a tool *through* which research is done (Crang, 2010a). Geography was long argued to be methodologically emaciated, especially in regard to its own bodily project, '[w]e simply do not have the methodological resources and skills to undertake research that takes the sensuous, embodied, creativeness of social practice seriously' (Latham, 2003, p. 1996). Geographers were accused of displaying an 'unwillingness to experiment with techniques that go beyond the now canonical cultural methods: in-depth interviews, focus groups, participant observation of some form or other' (Paterson, 2009, p. 766). Indeed, as Crang observes, there had long been a significant gap in geographer's learning through the bodily sensations and responses that occur as an inevitable part of the embodied experiences of the researcher; catchily, he suggests that 'touchy feely' techniques have perhaps not been touchy feely enough (2003).

Recent years have seen the growth of research methods for studying the senses within which creative practices have had a key part. For some, visual methods have played a key role in helping to record such explorations, such that visual culture techniques from photography to videography have grown up to accompany auto-ethnographic accounts (Garrett, 2011; Hawkins, 2015a; Simpson, 2011b). For others, particular value has been found in practicing the arts whether this be dance, drama, music, or indeed experiencing of art works – for exploring other, often embodied ways of knowing places. For such scholars, creative practices such as these are of value for their positing of the body 'directly in the field as a recording machine itself . . . knowing that writing these nervous energies, amplitudes and thresholds down is feasible, as such jottings become legitimate data for dissemination and analysis' (Dewsbury, 2010). Indeed, what geography has seen of late is a whole series of experimental creative methods of research, or 'research creation' many of which situate the sensing body front and centre. Such practices challenge geographers to operate at the edges of their accepted skill sets, to develop new techniques and practices, to be experimental and to be open in not only how they conduct research, but also how they conceive of and present research too (Hawkins, 2013b; 2015a). Such creative geographical research methods are not only confined to cultural geography (P Crang 2010), where we might expect to find them, but interestingly have moved throughout the discipline, such that a range of geographical scholars are now turning to arts practices, either alone or in collaboration to develop their research practices.

Box 2.4: DANCING RESEARCH CREATION

Geographer Derek McCormack has spent a number of years exploring moving bodies by participating in a range of movement practices, from contemporary art and dance, to dance therapies and other therapeutic movement practices. In his book *Refrains for Moving Bodies* (2014), he presents seven experimental engagements with these movement practices that enable an examination of what it is that such creative experiments might offer for our understandings of bodies, space and affect. In line with other non-representational theorists McCormack is concerned with 'how bodies and spaces co-produce one another through practices, gestures, movements and events' (p. 2). He is especially interested in how we can learn to be attuned to the affective qualities of spaces, how we can explore, in other words, qualities of emotion, mood and feeling. This might be the vibe of a space, a sense of collective excitement or the sense we all sometimes get in a space of fear, a sense of calm or liveliness in the countryside, or tension, or stress in our friends. Rather than feeling restricted by disciplinary habits or research and methods, McCormack asks 'what sites, techniques and concepts provide opportunities for expanding horizons of potential' (p. 191). His answer is to turn to what he calls 'research creation', a process that takes 'thought as a laboratory for creative practice and creative practices as a platform for thought' (Rubidge, 2009, p. 3, cited in McCormack (2014, p. 187); see also McCormack, 2008a).

In McCormack's case his practice-based focus is dance and movement practices. Variously experienced at these practices, McCormack makes it clear that we don't necessarily need to be

experts to be able to benefit from what we can learn through participating in these creative practices. Whether dealing with performance art, dance or movement therapy, each of the creative practices
McCormack explores is engaged as a site for thinking-feeling. He
writes through these accounts of experience in the first person,
offering deeply felt personal experiences of participating in and
watching movement practices. His text captures what it was like to
be doing these movements, whether pleasurable, uncomfortable,
physically easy or difficult, shaped by rhythms or engaging with
others. McCormack describes these as ethico-aesthetic practices,
each understood less as 'overcoded by the traditions from which it
emerges, but as an ecology of practices with the potential to
generate modest variations in moving, feeling, thinking' (p. 115).
Whilst not determined by their histories, McCormack is concerned
to detail the cultural-historical specificity of those practices he
studies. These specificities can be very important to understand
these practices. For example, one of his case studies is the avant-
garde Laban movement study. One of the founders of contemporary dance practice, Rudolf Laban, numbered amongst his
collaborations and fans of his corporal experimentations the Nazi's
and the management consultant Frank Lawrence. In exploring the
history of these practices, however briefly, McCormack reminds us
that creative practices have sometimes troubling pasts that we
need to engage as we experiment with their potential.

McCormack's personal accounts of his experience not only
bring us as his readers into corridors, halls and studios with him
and his collaborators, but they bring us as geographers face-to-
face with spaces and techniques of research practice for which we
are quite often ill-equipped. Such encounters raise questions of
our skills as geographers. McCormack's account offers us much
material to reflect on how these experiments unfold in practice,
the stutterings and hesitancies of things that are uncomfortably
felt and awkwardly enacted. The points where bodies and minds
do not know are less moments of failure to be uncomfortably
passed over, but are instead passages in the process of research

Continued

creation. The process of practicing is more important than producing a finished product.

As well as writing about these movement experiences,
McCormack also experiments with what it might mean to diagram
these practices. In responding to a movement practice called Five
Rhythms, McCormack offers us a combination of text and image.
Each image is a mechanically drawn abstract created from various
thicknesses and lengths of lines, some free form, others more
regimented. Each one is accompanied by a written description;
'venture from home on the thread of a tune, moving along
sonorous gestural, motor lines'. Each passage goes onto suggest
one of the five rhythms, one notes, 'think of angles, jagged edges,
karate', another, 'beginning to move out of chaos into lyrical, into
lighter notes, lifting sounds .... a play of ducking, weaving,
chasing through lines that always seem to be opening onto others'
(p. 88). McCormack never explicitly spells out the relationship,
rather leaves the images and descriptions to tell their own story
for us.

McCormack closes his text by offering us an imaginary of
geography. This imaginary is more or less disciplinary. It is an
imaginary that seeks the spaces in which to experiment with 'the
creation of new modes of encounter between artists and scholars
from a variety of disciplines' (Rubidge, 2009, p. 3, cited in
McCormack (2014)). This is a geography whose power lies in its
presentation of a series of openings onto the possibilities of
creative experiments, or better still, propositions – lures to feeling
(Manning, 2008) – that draw us towards creative experiments.

CREATIVE BODIES

Throughout this chapter we have encountered a range of different creative
bodies, from the idealised mythical bodies of the transcendental artistic
genius or the satisfied creative labourer, to the embodied practitioner and
their struggles with tools and skills, and their sensing of process. The
discussion has explored how creative bodies encounter and engage with
the vibrant matter upon which they work, and how they are themselves,

worked on by such matter. Such appreciations of messy, fleshy, sensing bodies have been brought to bear on the ideas of the creative labourer that are so common in contemporary capitalism. Bohemia might seem like a wonderful place, and Richard Florida might be urging all cities to increase their 'boho' index, but yet the ideologies of creative labour – the passion worker – are just that, ideologies that tend to hide the dark side of these labour practices. The precarity of creative workers, the emotional and aesthetic forms of many of their labours that demand performance and management of physical and emotional bodies, as well as the corrosion of the creative elements of their work, suggest the need for some careful thought about those myths of the satisfied creative worker. What is more, as geographers turn towards creative embodied practice as a new site for some of their own research methods, we might consider not only how creativity gives us a new appreciation and new mapping of the body, but also how it requires that as scholars we use and engage with our bodies in new and challenging ways.

NOTE

1   *Banda de pueblo* is a traditional town band of brass and percussion whose sound is typically associated with 'official' national music. They are common in festivals, processions, and other public events throughout the Ecuadorian Andes.

3

STUDIOS, GALLERIES AND BEYOND

The sites of creative production and consumption are often much myth-ologised. Indeed, specialised spaces such as the artistic studio, the hallowed halls of nineteenth-century European museums, or the once avant-garde white-cube modernist art gallery, are the site of many powerful imaginaries. The twentieth-century Pop-Artist Andy Warhol, for example, described his studio as an 'office space', while for the eighteenth-century landscape painter J.M.W. Turner, the studio was more a state of mind than an actual place (Daniels, 2011b; Postle, 2009). Other accounts have emphasised the studio as an archive space, or its place in the formation of a practitioner's creative identity; perhaps most famously Virginia Woolf's emphasis on female creative practitioners needing a 'room of one's own' in order to develop their work. Despite such mythologisation, the actual spaces in which creative practitioners make work have, until recently, largely been overlooked. Within geography, attention has been turned to studios over the last few years as part of wider creative networks. Bain, for example, describing arts practice as 'embedded within the culturally constructed context of the art world and located within the place-based culture of the studio, the home, the neighbourhood, the community, the city, the nation' (2004b, p.425). Spaces of the gallery, and other locations of creative consumption, whether it be concert hall or the theatre, have often been better studied. Indeed, there is a rich repertoire of studies exploring how such expensive and privileged spaces of creative consumption have been parts of displays of power and national identity, or how the politics of such displays served to curtail public behaviours creating a civilised populace. Recent years have seen the growth of studies that explore less well-known gallery spaces, reflecting more provincial or overlooked stories of their politics of display (Neate, 2012).

If work on specialised spaces of production and consumption is growing apace, then this scholarship has been accompanied by research studying how such specialised spaces are being challenged. Such challenges come from many directions, perhaps most significantly from evolving traits in creative practices and from shifts in the creative economy enabled by evolving digital spaces for production and consumption and the growth of the experience economy. So, for example, mid-twentieth-century art was shaped by an important shift that saw the focus of production and consumption shift from studios and galleries to sites and situations. In other words, art became made outside specialised studio spaces, with artists working in communities and landscapes. And, in the place of galleries, art became increasingly consumed in the city streets and in the environments that were its subject and medium. To take another example, as digital tools and technologies became more common the organisation of industries such as the music industry underwent a series of shifts in the spaces and practices of their production and consumption. Once expensive and exclusive practices such as recording music became accessible to the degree that bedroom recording studios have become increasingly common. Furthermore, the internet and websites such as MySpace, SoundCloud, Kickstarter, Instagram and others, made once-specialist sectors such as promotion, funding and access to markets/audiences, accessible not only to signed, professional musicians, but a whole range of amateurs and pro-amateurs.

This chapter will explore this trend for the movement of creative production and consumption beyond specialised spaces. It will begin within the studio spaces, both investigating the micro-geographies of creative production and exploring the situation of these spaces within wider art worlds. Bridging discussion of creative production and consumption, discussion will then consider four specific examples of challenges to such specialised production and consumption spaces offered by changing artistic practices; the evolution of digital platforms and software, as well as ideas of the experience economy. The chapter will close by turning to sites of creative consumption. Beginning from the gallery space, discussion will explore wider literatures on audiencing before reflecting on the relationship between creativity, the digital dilemma and the experience economy. In particular, it will explore tensions within the growth of digital consumption spaces that have opened up previously specialist spheres to wider producers and consumers. At the same time, however, such apparent democratisation has seen practices such as curation exported from specialist gallery spaces to the wider creative and experience economy in order to help producers stand out from the crowd and help consumers find the best of the creative sector. This is not therefore a straightforward story of the erosion of specialist creative spaces and practices, rather it is a story cross-cut with tensions and hybrid practices that both challenge and preserve specialist spaces and practices of creative production and consumption, as well as adopt and adapt their practices for development in the wider, perhaps, democratised creative sector.

THE STUDIO AND THE MICRO-GEOGRAPHIES OF CREATIVE PRODUCTION

Studios have long been privileged sites for exploring the material and intellectual transformations that constitute the making of an art-work or other creative form. In recent years they have come also to be understood as sites for the transformations of artistic subjectivity and identity brought about by the processes of art making. What valuably emerges from a number of studio studies is a reinforcement of the complexity of the creative labours of the creative practitioner discussed in the previous chapter. While today the studio might be conceptualised as a valuable site for creativity, encompassing processes of transformation, performance, production and transmission (Daniels, 2011b), historically there was a distinction between the workshop, where physical making was carried out, and the studio, which was a space for reflection. In Renaissance Italy the space for the production of art was a *bottega*, a workshop; in contrast, the studio was a private space for reflection and study. The word 'studio' itself comes from the Italian *Studiolo*, from the latin *stadium* meaning a space for reflection and study (Daniels, 2010; Sjohom, 2012, p. 41). Consciously or not, such a distinction shapes many discussions of the studio as a space for material making and transformation in the form of ongoing experimentation and refinement; as a site for the archiving of practice and process and as a site for the more immaterial dimensions of creative practices. Whilst not denying the materiality of processes such as reading, note-making and reflecting, the studio is clearly a site for processes such as thinking, reflecting and conceptualisation; processes that intersect with material practices of drawing, making or sculpting materials.

The ethnographies of studio-working that have emerged in recent years have gone a good way towards dispelling myths of the studio as a romantic refuge or a space of frenzied material productivity. Erin O'Connor's auto-ethnographic accounts (detailed in Box 2.1) explore the embodied practices of glass blowing and query skills, learning and technology and the role of sociality in creative practice. Whilst Sjöholm (2012, 2014) offers an account of the coming together of the studio as a space full of materiality, but also as a space of and for immaterial processes, a space for ideas, thought, knowledge, reflection, memory and intention. In the course of these discussions the studio emerges as a material and immaterial archive, a 'space for ongoing as well as finished projects, colour, paintings, scraps, scribbles, prototypes, ideas, chaos, order language' (2014, p. 505–506). In thinking through this entwining of the material and the immaterial processes in the studio, Sjöholm explores the artist as collector of objects and as archivist often of their own work. She examines processes of categorizing materials and things as not only a process of sorting and thinking about these objects, preventing drowning in these materials, but also as a process of transformation whereby 'art digests reality and comes into being' (Harrison, cited in Sjöholm, 2014, p. 505). The archive is also a place where careers are stored and developed, a site where artists look both forward and back in their practices.

In geographic accounts of the studio what is often emphasised is the coming together of the studio as a place for making work and a place for construction of self and identity of the practitioner within and without the art world (Buran, 2004; Bain, 2004b; Sjöholm, 2012, 2014). Indeed, Alison Bain (2004b, 2007, 2013) has developed important work in recent years that explores studios as sites for the production of self and identity as well as creative works. Interestingly, Bain's work, together with other studies, often emphasises the role of gender in the desire for, acquisition and maintenance of studio spaces within and without the home. Studies of new mothers have emphasised how the home as site of craft production offered both a means to regain identity and the ability to be economically productive during motherhood (Fisher, 1997; Grace et al., 2009; Luckman 2013). However, often such home-based creative practices involve the improvisation of studio spaces rather than the demarcation of contained times and spaces for creativity. Fisher's (1997) empirical research on the domestic rural creativities of new mothers finds a generalised condition of porosity. She describes how 'production flows into home space' (often literally in terms of dust, fumes, noise etc.) with home often 'tumbling into productive space', in terms of childcare responsibilities or the inability for makers to get distinct times/spaces in which to work. Getting and making space to work becomes problematic, and spaces often have to be multi-functional. This can require improvisational creativities to rig up ingenious means to enable the efficient shifting from one type of production to another, drawing and redrawing boundaries in quick temporal succession. As Fisher (1997) notes, kitchen tables used to prep and eat breakfast become workspaces, then a few hours later a space for food preparation and consumption again. For many of those she talked to, Fisher noted the desirability of creating specialised and separable spaces for production and reproduction, even if this meant curtains and bits of string, rather than separate rooms. Such temporary and flimsy barriers served both a symbolic function but also a practical one, enabling space to be made by allowing communication to be retained, disturbances heard and questions answered. Further, as Gray's studio of amateur dramatics and domestic space makes clear, creative practices rarely stay put, but rather profuse the home with materials, ideas, resources and inspiration (Box 3.1). While it might be positive to see these processes of living and creating these ambiguous spaces and sets of identities as forms of improvised creativity, for many creative practitioners it is either the negotiated spaces and times of home working and ad-hoc studio spaces, or nothing. Interestingly, oftentimes complex negotiations of home space were preferred over renting external studio spaces. For while non-home-based work spaces often seemed desirable – appearing more serious, for example – they were also less flexible, less human and often untenable due to costs or incompatibility with caring and domestic responsibilities.

Box 3.1 DOMESTIC SPACES OF CREATIVITY: THE AMATEUR DRAMATIST

Cara Gray

Figure 3.1 'Set-in-a-Box'

Above is a photograph of a 'Set-in-a-Box', found in the home of an amateur dramatist. This material object taught me about the imagination, flexibility and resourcefulness of the amateur dramatist. It introduced me to the 'nature of amateur dramatics': where scripts aren't sacred because of the challenges of assembling a full cast; where a water bucket and a pipe can be engineered in such a way to create the illusion of a running tap on stage; where craft is

Continued

nurtured by the self-sufficient *making of* theatre. Danny Miller (2005) asserts that 'if we learn to listen to these things we have access to an authentic other voice' (p. 2), 'these things' being the materialities of everyday life. Through *portraits* of domestic spaces and the materials that inhabit them, Miller affirms the need to explore the biography *of* and the relationships *that* move between persons and 'these things'.

On one of my early trips to Letchworth Garden City I was lucky enough to be invited around for lunch by Pat and John, both long-standing members of The Settlement Players. The house, an Arts and Crafts style roughcast building, encapsulated everything that I had read about Letchworth's architectural aesthetic and philosophy. As a cultural geographer looking at amateur dramatics, domestic spaces intrigue me, especially with the potential links I see to current cultural geographical ideas emerging within the growing field of work on creativity. Geographers' have been investigating the materialities and sociabilities of creative practice as well as placing analytical focus on the spatialities and sites of creativity (Bain, 2004b; Daniels, 2010; Sjöholm, 2012).

Alison Bain (2004a, 2004b) examines the artists' studio as a situated space for the construction and maintenance of artistic identity. Bain exposes the process by which a space is compromised, and the resulting necessity of 'making do', by considering the fashioning of a creative space amongst possible everyday inconveniences and vice versa (for example, locating a workspace within a domestic environment). This helped me think about the reciprocity between creative and domestic spaces, particularly given the manner in which this relationship is inevitably heightened in amateur practices.

In the kitchen, I was instantly greeted with their dramatic identities: washed bottles were being collected for props and pieces of 'Theatre Club' admin were scattered around the counter. John took some folders from the bookshelf for us to look at; bursting full of hand-drawn set designs, notes on lighting, cross sections and detailed sketches of particular parts of the stage. These sketches triggered stories of his successes and challenges associated with past performances. The stories would not have been so easily told without 'these things'.

Notated after-thoughts: 'Concerns' 'Solutions' 'Requirements' scattered each page; the *making of* theatre that I was witnessing through these pages was a carefully crafted process. These ring-bound portfolios were working collections of ideas and inspirations; some successful, some that never went further than being pencilled thoughts on a piece of paper. All were safely archived in amongst books and photo albums, where photographs of a husband and wife on stage lay unsegregated from other snapshots of personal memories.

Jenny Sjöholm (2012) suggests that an artist's studio can give insights into their working process as well as their life, through the fact that this physical space is filled with 'material and memorial archives of [their] work, biographies, intention and thoughts' (p. 12). Looking at the artists' studio as an archive introduces an interesting way for us to trace the geographies of artistic production. The same can be said when looking at the *studios* of the amateur dramatist. In this instance the kitchen, a site typically associated with everyday domestic doings, becomes a site where one can trace the materialities of theatre.

The 'Set-in-a-Box', when opened, revealed a scaled-down model of the theatre at The Settlement (home of The Settlement Players) and was explained to me as a tool used in preliminary set designs. These can cost hundreds of pounds and so John handmade this beautifully crafted box himself. It seemed to me that the sheer nature of the amateur acted as a catalyst for the nurturing of craft. John's craftsmanship was reminiscent of the Arts and Crafts' philosophy of creating beautiful things without relying on industry, and indeed the way in which this philosophy helped to shape and materialise the town.

With Social and Cultural Geography's recent interest in creative processes, and the relationship between the materialities and practitioner, this self-sufficient *making of* theatre is something that I think is valuable to research. Returning to Miller (2005) it is important for us to 'wonder at, the world of small things and intimate relationships that fill out our lives' (p. 7).

It is not only female creative practitioners whose caring responsibilities requires them to find a spatial fix to their need for home-studio spaces, but male artists too can struggle to negotiate the spaces and responsibilities of home and work. In her studies of Canadian male heterosexual visual artists, Bain (2005, 2007) details how these artists often struggle to balance visions of hegemonic masculinity in the arts and their own desires as artists and fathers. Often these struggles for legitimacy play out through a spatial fix to help manage their artistic identities, a fix that often sees them set up carefully delineated spaces for work, either within or away from the home. Indeed, as she notes, these spaces and times were often jealously guarded, aping artistic myths where 'separateness' had become understood as the essential quality of any artist. As such, myths of marginality, of 'outsider status' and of the need to have creative freedom remained potent and had a strong hold on the artists themselves. Many of the men that Bain worked with struggled to reconcile their roles as husbands and fathers with the image of the lone creative genius, eschewing family life. As she notes in her sample size (80), male artists were less likely than their female counterparts to work at home while their children were young, and for those that did they had 'near complete spatial reign or had doors that closed and locked to retreat behind' (p. 253). As she explores 'spatial control is a significant strategy used by men who have combined artist practice with fatherhood' (ibid.), enabling them to negotiate both their desires to care and their roles as careers and the hegemonic notions of masculinity in the visual arts.

In their complex geographies and presencing of practice, explorations of the studio space draw out a range of dimensions. These dimensions include the micro-geographies of artistic practices, the materialised 'doings', the embodied practices and the sayings and unsaid elements of the creative process, the collections of objects and papers, and the processes of inspiration. It is, of course, not just in the studio that creative practices happen. Despite its role as a locus where artistic processes are made present, recent studies have demonstrated how the studio is not, and never has been, comprehendible as a discreet, atomised place of production (Daniels, 2011b; Sjöholm, 2012). Rather, it is a mobile – 'materials and implements on the move' – highly connected space, linked into wider networks, spaces and places of artistic production and consumption in which it plays a functional role. The studio has thus always been a part of wider artistic networks. These have included, the home, the field, the lab, as well as the position of the individual artist's studio within larger 'creative clusters', and the role of collections of studios in urban regeneration, and the place of the studio as at once also a site of display and a painted scene (Bain, 2003; Daniels, 2011b; Dixon, 2008; Harvey et al., 2012; Hawkins, 2013; Ryan, 2013). An increasingly rich body of work explores spaces of artistic creation beyond the studio, whether this be the landscape or the city space. As Daniels (1993) and Ryan (2013) explore with respect to eighteenth-century painting and Imperial photographic practice respectively, some artists create outside, taking their studios with them in the production of work through en-plain air or in the field encounters. Oftentimes artistic and writing practices can involve shuttling between spaces of inspiration and studio spaces, often using devices such as notebooks and sketchbooks or, increasingly, mobile phones or iPads as a means to engage and record thoughts and images 'in the field' (Brace and Johns-Putra, 2010). In short, the studio becomes situated as a space of practice in relation to a whole series of other sites of collection, documentation, rumination, development and information. Indeed Sjöholm (2014) describes studios as 'relational workspaces' linked firmly to other geographies of creative practice. To think about these 'other' geographies of creative practice this discussion is going to turn to Becker's concept of networked 'art worlds'.

ART WORLDS

In his famous account of the 'Art World', sociologist Howard Becker (1984) details multiple connected art worlds. His examination presents art production as a collective action that results from the artist operating amidst highly networked groups of people, 'whose cooperative activity, organized via their joint knowledge of conventional means of doing things, produce(s) the kind of art works that art world is noted for' (1982, p.x). Becker (1984) thus offers us an account of art making that is not just the preserve of the singular genius, but rather sees creative practice as situated within and enabled by a complex network of relations between creative producers and those who support the production and consumption of their work. Writing about the 'extensive division of labor' (p.13) and the 'elaborate cooperation' (p.28) among many people, he notes the role of groups such as critics, audiences and others who support the arts, whether this be wealthy patrons, or commercial interests (i.e.

club owners, dealers etc.) or the interests of states and other political entities. He is also careful to note the importance of those who make and sell the tools of the trade, musical instruments, paint or canvas, and so on.

Becker's work was important at the time of writing and since for piercing the veneer of those ideas of creative genius and the myth of the singular force and vision of the artist that has been in circulation since the Renaissance. For Becker, art works are what the art world 'ratifies as art' (p. 156). For alongside the sets of interacting people and enabling tools and technologies, art worlds also produce and are produced by sets of conventions and normative standards of taste, which can and cannot be pushed and extended in various different ways. While, for some, Becker's emphasis on the role of the social conditions in the production of art was too extreme, requiring rather more attention to the power of the works themselves, there is no doubting the importance of his accounts to understandings of creative production. Indeed, Becker's work is an important foundation point from which to establish the complex geographies of creative making that recognise the need to move beyond the artistic genius acting alone.

Geographical accounts of artistic production have long taken seriously the need to understand art works by entwining the geographies of their production and consumption with the geographies within the work (Daniels, 1993; Leyshon et al., 1998). As such, geographical accounts of creative production have often noted the complex and manifold nature of the geographies of artistic production. Some studies take a more cultural geographical perspective, wherein the content of the work is entwined with the situated nature and social and material conditions of its making. This might include, for example, the inspiration provided by landscape (Brace and Johns-Putra, 2010), the role of an urban zeitgeist (Rycroft, 2011) or the importance of a group of like-minded people in an artists' colony (Lübbren, 2001). For others, a more sociological and economic perspective is required, wherein concerns with things like resource demands are centralised. This might include, the need for fabrics or sewing materials in a fashion cluster (Crewe, 1996) or the need for a range of services to help record a documentary or produce a film (Bassett et al., 2002). What this adds up to is a set of concerns with how creative practices are 'embedded within the culturally constructed context of the art world and is [are] located within the place-based culture of the studio, the home, the neighbourhood, the community, the city and the nation'

(Bain, 2004a, p.425). Further, as Nash makes clear (talking specifically about images) while we can take account of artists' and curators' intents, it is also important to consider 'the relationships between the initial time and place of the image's production, the location and places figured within the works, and the endlessly variable arenas through which they circulate' (Nash, 1996, p.125). What such discussions centralise is that to consider creative production is to consider both situatedness and mobilities and to reflect on material and social factors across a whole range of spaces and at a whole range of scales.

A range of geographic studies have recently taken the mobilities of the art world, rather than its spatial fixities as a point of analytic departure. Such literatures produce accounts of creative practices as constituted through networks and mobilities – of people, art works, ideas and styles. This is a point that Morris (2005) makes very clear in his account of the cultural geographies of Abstract Expressionism, noting the international networks of painters, dealers and critics. Exploring the circulation of people, ideas and objects, largely driven by Europe's twentieth-century geopolitical issues, he noted the formation of an art that was neither of the USA nor of Europe, but rather was an Atlantic art. In another example of the importance of networks and mobilities to the production of art, Amanda Rogers explores the theatre production, Madam Butterfly. She is interested to offer a rather more situated account of the role of trans-local circulations of creative practice. Exploring the writing, casting and staging of the play in New York, London and Singapore, Rogers tells spatial stories of artistic production that link localised practices of writing, casting, rehearsing, advertising and performing into global production and performance networks. As she argues 'even when creative practices are situated, they operate through networks and flows that link locations together (2011, p. 663). Exploring the creative work that the production does, but also the creative work that goes into producing the play, she highlights how creativity is produced trans-locally, through 'physical and imaginative connections to other locations' (2011, p. 665). Unlike other accounts, Rogers is less interested in how the play in question might represent these places and mobilities, but rather how it produces and is produced by them, with local-local connections 'provide[ing] routes along which people, ideas and finance can travel, shaping artistic production and reception' (p. 665). In these studies, to place artistic production is not to evolve accounts that figure only located works and situational imperatives. For, and following contemporary accounts of place as well as empirical work on creative practices, to appreciate placed creative production is to embrace both the role of networks and mobilities that produce creative outputs, but also those networks and mobilities that are produced by them.

To explore these questions of the geographies of creative production and consumption, further discussion is now going to turn to explore four examples of the changing geographies of specialist creative spaces and their associated practices.

RETHINKING SPECIALIST SPACES OF PRODUCTION 1: SITE-SPECIFIC ART, FROM STUDIO TO SITUATION

Creative production has clearly always been a geographically located practice, but for many creative practitioners the relationship between their work and the place at which it is produced and consumed has become a conceptual concern – becoming termed situated practice (Kwon, 2001). Situated practice denotes a movement, broadly captured by the phrase 'from studio to situation', that engages the shifting geographies of the production and importantly also the consumption of creative practices. In other words, art is understood to move not only beyond the studio as a primary location of production, but also beyond the gallery as a primary site of display. Indeed, there is arguably a reconfiguration of the place of the gallery along with other specialised spaces of creative consumption (i.e. the concert hall, the theatre, the cinema) that requires that we complicate the geographies of sites of creative consumption and their relationship to sites of production (Hawkins, 2011, 2013; Neate, 2012).

What we could perhaps call a geographical turn in art (Hawkins, 2013a, 2013b) has witnessed the production and analysis of much contemporary art under the terms of 'site', 'situation', 'location', 'context' and 'place' (Pearson, 2010; Rendell, 2006, 2010). This geographical turn in the art world represents, for the art world, a qualitatively different relationship with 'site'. As Krauss (1994, p. 37) puts it, 'the specificity of the site [/ space] is not the subject of the work but . . . its medium'. There is far from a consensus within the art world, however, as to how to deal with this relationship between art and site. Looking across a range of studies is to witness site understood to denote everything from a physical space, wherein the artist is engaging with the physical attributes of a space, to site understood in terms of social and humanistic explorations of place and the local, to a sense of 'site' as primarily a social site based around a community, or, site as the discursive spaces of intellectual debate, or even the psychic spaces of self (Jackson, 2011; Kwon, 2001; Pearson, 2010; Rugg, 2010). Often, this variety is elided into a sweep of change that has, perhaps unhelpfully, claimed to see a movement from a physical site to art engaging with a 'discursively determined site that is delineated as a field of knowledge, intellectual exchange or cultural debate' (Kwon, 2001, p. 93). A key theme within this work has been community or socially engaged art. This huge and diverse field, variously termed relational aesthetics (Bourriaud, 2002) dialogic art practices (Kester, 2004) or the more established new genre public art (Lacy, 1995) covers a range of practices. These might include communication events organised in urban-based art gallery spaces, the orchestration of street protest and the community building practices of event-based rural arts projects. The result of this diffusion of geographically engaged art has been the breakdown of general categorisations of art as 'site-determined, site-referenced, site-conscious, site-responsive' (Pearson, 2010, p. 8). In the place of general categories what evolves is an enhanced engagement with the specificity of each instance of work, integral to which is a rather geographical querying of 'what and how site means' and 'is' (Pearson, 2007; Rendell, 2010).

What this means in practice is that we find art produced in dialogue with the street, the community and domestic space as well as various natural spaces and landscapes. Given that geography's recent disciplinary interest can be largely attributed to such forms of art, there are numerous examples to choose from. As geographers have explored, the importance of the site might relate to its physical attributes, such as Andy Goldsworthy or Robert Smithson's use of natural material and geomorphological conditions in their land art; its socio-political features such as its histories and the need to overcome trauma (Mackenzie, 2006a; Till, 2012) or ongoing issues of urban super-diversity (McNally, 2015) or historical traditions and myth (Hawkins and Lovejoy, 2009; Pearson, 2007). Disciplinary interest can largely be attributed to the understanding of arts practices as forms of world making. Such world making might evolve through offering critical reflections on places and intervening within their politics, or it might involve imagining and bringing about new futures for places and communities. Specific examples of these works will not be discussed here as they feature throughout the text, whether it be the participatory drawing projects of local communities discussed in Chapter 6, or the knitted birds discussed in Chapter 10. If even specialist practices are moving beyond specialised spaces in their production and consumption of works, what is interesting to see is how this has been enabled for other creative practices through the evolution of the digital.

RETHINKING SPECIALIST CREATIVE SPACES 2: DIGITAL CREATIVITIES

Capitalism is shaped by 'ceaseless experimentation' (Thrift, 2007, p, 379), a seemingly inexhaustible ability for innovation and adaptability in the face of new developments, both positive and those that might be perceived of as a crisis. The development of digital forms of creative production and consumption has proceeded at a pace, whether we are talking about bedroom studios, iPads being used as creative tools, or the role of now-mundane technologies such as mobile phones in the development of creative communities or vernacular, participatory forms of creativity. Leyshon's book *Reformatted* offers one of the first in-depth accounts of how digital developments have shaped part of the creative sector. He explores how the music sector has been remade in terms of both production and consumption as networks between producers and consumers and within different parts of the industrial system have been reshaped.

The music industry has often been explored as a series of networks, with as Hennion (1989, p. 402) argues progressive attempts being made to 'extend what was first localised in the studio'. Such networks constitute, as Attali (1984) argued, functions of composition, representation and repetition; in other words, creation, the capture of the sound, and its reproduction and circulation around the world. At the core of these networks sits the recording company and the recording studio, with the traditional route for musicians being seen as getting signed. Leyshon develops a telling historical geography of the recording studio in the UK, and the role of socio-technological change in the evolution of these spaces that were once the backbone of the music industry. In Leyshon's account software is central, and in particular shifts in software during the late 1970s and the mid-1990s onward. In the early twenty-first century there were around 300 'economically significant' recording studios in the UK (Leyshon, 2014, p. 117). Studios, as Leyshon recounts, offer dedicated spaces for the recording of music, time is sold in these spaces often in the context of 'projects', an album for example. It is not only spaces that these studios provide, but also dedicated technology to enable recording to happen, as well as the skilled labour able to operate these technologies. However, if recording studios were once highly privileged sites accessible only to those with significant resources, then the advent of digital recording technologies and software has widened the normally narrow routes to reproducibility that for a long time controlled the music industry (Leyshon, 2014, p. 120).

Recording studios long faced significant challenges in a market characterised as a oligopsony, that is, one with few buyers (Leyshon, 2014, p. 121). With a narrow market, competition was intense, ensuring studios needed to both keep costs down but also invest in the latest equipment to ensure they could meet clients' needs. Whether focused on recording new material or on post production the challenges studios faced were intense. Whilst much attention has focused on recent growth of at-home recording studios and the challenges these posed to the industry, as Leyshon notes, these are not just recent developments. Whilst until the 1940s recording studios were specialist spaces, with technologies ensuring the lock-in of expertise at these sites, since the Second World War, he argues, recording technologies have lowered the costs and barriers to entry; thus digital recording and software is just the latest stage in this process. Leyshon describes the laboratory like atmosphere of the purpose-built EMI studios at Abbey Road in North West London. Part of large vertically integrated organisations, technologies evolved that were distinct to each organisation and studio group, whether it be EMI in London or Warner or Columbia in the US. In this context, studios were not really seen as money-making locations on their own terms but as part of a general resource of the organisation and were thus often not really used to capacity. This changed, Leyshon suggests, in the 1970s when studios began to be run as cost-centres, and opened up to non-label artists and producers. Further developments in technology enabled flexibility of recording sessions, enabling producers to move between rooms but also studios with their work. The result was an increasingly competitive sector that was further complicated with the shift to software-based digital recording systems, which could be run on PCs or Macs. Such a shift not only required new investment on the part of the studios, investment that required ongoing upgrades, it also reduced the need for time and studio space (2014, p. 134). What is more, whilst still expensive (£20,000) for a professional system, entry level equipment was cheap but still of a high enough quality to enable home recording. As such, artists, whilst not necessarily recording at home, would often do pre-studio work in their home studios, a feature that was later to become dominant as equipment further reduced in price.

The story of the recording studios then is one of increasing challenges to the specialist spaces of music production, which sees digital developments over a number of years challenging a studio system once focused on the vertical integration of large companies, to develop instead a more distributed studio system, which sees home and professional studios both playing an important role in the production of music.

RETHINKING SPECIALIST SPACES OF CREATIVITY 3: CRAFT WORK AND THE DISPLAY OF CREATIVE PRACTICE

If once the spaces of creative production were private spaces, hidden parts of the commodity chain of creative products, recent trends have intensified processes in which we see creative practices becoming a public part of how creative products are sold. As such, creative practices are sold not only as 'experiences' but through the commodification of the making process, such that displaying the skill in making the product becomes the means through which it is sold. This is perhaps nowhere more apparent than in the growth of studios as sales spaces, especially common in those creative clusters that collapse spaces of production and consumption (see discussion in Chapter 5). Here, the working processes of creative practitioners, their skills, become a central part of how their products are sold. To buy something from one of these spaces is not only to buy the woven fabric, silver earrings or pottery bowl, but also to consume the experience of the weaver at work, the materialities and embodied experience of being in their studio space, of seeing the raw materials from which your product was made, the technologies and skill involved, and maybe even commissioning or customising the product. This process is as much a marketisation of the maker, their practices and creative studio spaces as it is a selling of the created object. For some makers, such a display of labour is a way to stand out from the crowd; for others, such a marketisation of their process is a means to justify the often high prices of their products compared to mass-produced items. This takes its toll, however, as the

Chapter 2 noted, as there is considerable emotional and physical labour involved in having your spaces and self always on show, of always being ready to talk to your customers, to display your practice, to perform the role of creative practitioner.

Interestingly, a collection of scholarship has emerged that unites these discussions of the marketisation of creative labour processes with a set of reflections on how labour practices that might have once been seen as service work have become understood as craft work. A series of studies have revisited a cross-section of service work and found within it less alienated labour but rather forms of craft labour and passion work (Ocejo, 2012, 2014a, 2015). Such an aestheticisation of service work brings the practices of the studio and the workshop into the wider spaces of consumption, with implications both for the creative workers and the consumers.

Whereas once craft skills belonged to the spaces of the workshop and studio, now they are being found on the high street. Certain jobs are thus emerging as 'cool' producing flexibility, enjoyable work and the chance for people to display their cultural capital (Lloyd, 2006; Neff et al., 2005). Exploring everything from cocktail bars to bakeries, florists and high end butchers – often named craft, artisan or local – and located in places such as Chelsea Market in New York, or Borough Market in London, Ocejo (2014b) and others have taken a closer look at these kinds of labour. Studies have made sense of this increasingly segmented service economy through ideas of craft consumption and production, focusing on skilled techniques and high-quality raw materials (Campbell, 2005; Ocejo, 2012, 2014a, 2014b; Watson and Shove, 2008). As such work, such as specialised cocktail bartenders and craft butchers, is understood less as service labour but as demonstrating forms of trade-craft involving questions of taste, of skill, of functional aesthetics and knowledge (Ocejo, 2012, 2014b). Furthermore, service work is rehabilitated, perhaps dangerously so, away from being repetitive and mundane, with little space for individuality or creativity, towards offering the kinds of emotional fulfilment associated with creative labour. Sometimes these service jobs are becoming seen as creative careers; sometimes they are popular options to help subsidise core creative careers in visual art, music or performance for example.

As Ocejo (2014a, 2014b) notes, however, not all segments of all industries and occupations can transform themselves from manual labour service work into creative, aesthetically driven professions. There is potential in some service industry jobs to encourage independent and innovative thinking, allowing the 'window of creativity to open for service work' (Ocejo, 2012, p.179). Taking the example of butchery, Ocejo explores what he calls the 'meat philosophy' of upscale shops where culinary capital is created through the display of skills of butchery, their creation of rare cuts and promotion of wholesome animal-rearing practices. Butchers thus promote a philosophy of quality; they are 'taste-makers, and cultural producers, rather than just material producers performing manual labour' (p. 117). Through ethnographic work, he explores how the butcher's shop and the butcher's practice become sites for the production of new elite tastes in meat through the technical skills of butcher and the social skills of service work. The butchers here perform a functional aesthetics, presenting meat to the consumer as appetizing but also as 'cool', while ensuring it can be prepared easily and in a manner that optimises taste and stays true to the philosophical foundation of the shop.

Thus in a society where creativity contributes significantly to contemporary urban economies, it is not just that service workers provide for the creative class, rather service work has come to be reinterpreted under the creative economy as a respectable creative career. Night-time scenes, for example, are an integral part of what makes a creative city attractive, and have long provided opportunities for creative workers to subsidise their creative careers. As Ocejo (2012) explores, cocktail making cannot just subsidise other creative careers but can be appreciated as a form of craft production, combining the knowledge and skills of creative drink-making with the skills and practices of serving the drinks,

> as bartending becomes infused with production, service and consumption aspects of the job, with aesthetic considerations based on legitimacies, cultural practices and a desire to create a unique sensory experience for customers.
>
> (Ocejo, 2012, p.184)

It is not just the making and serving of drinks, or the cutting of meat that might be seen as the creative part of this process, but also as Vadi notes (see Box 3.2), the creation of an atmosphere of a whole restaurant develops other parts of the service sector as also creative in nature. It is important to note that this is not to suggest a wholesale reformatting of the entire service industry, indeed this is a niche occupation and for many who make and serve cocktails or butcher meat the skill-set is very different. Further, it is interesting to note how while there might be some sort of sense of creative engagement with one's job, these jobs often remain low paid and precarious, they are, by and large, non-unionised and the career structure is far from clear. They often involve irregular hours, may not be suitable for people at different stages of their life course, and often rely on tips to ensure that a living wage is earned. There is a risk that to bring craft to service work is to romanticise such labour, 'creativising' occupations as a way to justify unacceptable working conditions.

It is also interesting that craft would be the choice of term here, for while respected in the nineteenth century, craft in the twentieth century was allocated a subordinate position in relation to other forms of creative practices. While Sennett (2009) may have located in craft-skilled ethical practice, for others it was a marginal, or minor practice associated with women and hobbyists and with unproductive economic formations (Banks and Hesmondhalgh, 2009; Hughes, 2012). When craft and design skills are brought into economic-and-regeneration focused discussions, what oftentimes becomes clear is a mismatch in scale and intent. As Hughes' study demonstrates, successful micro-businesses that often employ one person and sometimes part time are disparaged as 'lacking ambition and vision', and decried for their lack of growth and concern with modern management methods (Hughes, 2012). Such comments belie the fundamental mismatch and misunderstanding that remains within creative industries policies of the motivations that might drive people to develop craft practices and businesses in the first place (Luckman, 2015).

Box 3.2 CREATING RESTAURANT EXPERIENCES

Priya Vadi

Restaurants are creative spaces in myriad ways; in terms of the food they serve, their décor, the way they serve food and the

---

Continued

atmosphere they create. This case study focuses on a small chain of Indian London-based restaurants called Dishoom. Dishoom differs from the other Indian restaurants in London, in the sense that it pays homage to the nostalgia and romanticism of Bombay cafés that are declining in their numbers in Mumbai, India. Dishoom also reflects the wider Indian diaspora through the celebration of festivals and cultures such as Holi and Eid, by holding special events. Perhaps one of its most memorable events is the Chowpatty Beach pop-up at Southbank in 2011, where a Mumbai beach environment was recreated:

> . . . Shamil came up with the idea of doing Dishoom on acid. That's what it was like. It was glaring bright colours, like the bicycles outside. We had walls constructed of the Times of India newspapers, the lampshades were made of glass jars, we had traditional *gola* machines we were making real *golas* [shaved ice with syrup] . . . We did things like the naughty coconut, so literally pierced a hole in threw a load of rum and people just drank it . . . the *golas* were probably the most national part of it really because we did naughty *golas* which you wouldn't normally get in Bombay. We spiked it a little bit with alcohol it made it fun people enjoyed it.
> (Interview with manager, Covent Garden, January 2013)

This case highlights that authenticity and tradition is maintained through the serving of *golas* using the traditional machines (Figure 3.2), but there is also a divergence from authenticity, like the *golas* with alcohol. What is presented at Dishoom is a modern take on the traditional. Here the foods remain traditional as one would typically find in Bombay cafés, like the *Lamb boti kebab* and Chicken berry Britannia. However, the creative outlet in terms of cuisine is through the cocktails, like the *Paan sour* and *Chaijito*. The *Chaijito* takes the basis of the mojito, but has additional Indian ingredients such as ginger, Dishoom chai syrup, and coriander.

Figure 3.2 Gola machine at Covent Garden from Dishoom Chowpatty
Beach pop-up

The most striking part of Dishoom's restaurants is the décor; it is
a utopia of the Bombay café, inspired by a range of Bombay cafés
in India, such as Britannia, Bademiya, and Leopold's, rather than
copying one in particular. Dishoom's website states,

> 'Creating a pastiche of an Irani café – perhaps an attempt at
> literal authenticity – would probably be a misguided errand.
> We're not in Bombay, we're not Irani immigrants and this
> isn't the 1930s. However, we can still pay loving homage to
> this rich period narrative in a modern space in London. The

Continued

marble tables, bentwood chairs, ceiling fans, ageing mirrors and monochromatic palette are the backdrop which starts to tell the story'.

(Dishoom, 2011)

These spaces tell stories in terms of creativity and curation, by creating an ambiance that is playful, vibrant, and modern. What is important to note from the quote about the décor, is that the geography is co-opted, spatially and temporally reinforcing that authenticity is not the point. The point instead is to create an environment which co-opts the essence of Bombay cafés in India but squarely placing the restaurant in London: for example the King's Cross restaurant is named 'Dishoom Godown', *godown* meaning warehouse. A back story was created and paired with the restaurant, as is with the Shoreditch and Covent Garden restaurants, and published on the Dishoom blog (see www.dishoom.com/blog). The Covent Garden restaurant is referred to as 'the slightly more showy cousin' (Dishoom, 2012), compared to the eccentric Shoreditch restaurant. The Dishoom Godown tells the story of an Irani man arriving at the Victoria Terminus in Bombay in 1928 who starts selling Irani chai and baked goods from a stall at the corner of the *godown*, later expanding and dominating the *godown*, eventually becoming an established café and is regarded as an institution (Dishoom, 2014). The story is reminiscent, a sort of collective story of the Bombay cafes in India as described by Conlon (1995). The geographies of London and Bombay are reflected at all three restaurants through the décor: the Shoreditch restaurant is reflective of the hipster effect of the area, with visible infrastructure, portraying a slightly dishevelled look; the Covent Garden restaurant is more glamorous, where objects are placed with some sense of order; Dishoom Godown portrays an ambiance of the railway as an institution during the time of Indian Independence, portrayed through a series of portraits of those involved in the Independence movement like Gandhi.

Dishoom exemplifies the case of creativity in a restaurant setting through a number of facets: food and drink; décor;

narrative; and a celebration of nostalgia and diaspora. It is through paying homage to, rather than recreating or replicating, and diverting from authenticity that has allowed Dishoom to be a creative environment.

RETHINKING SPECIALIST SPACES OF CREATIVITY 4: VIRTUALITY, CO-CREATION AND THE RISE OF THE AMATEUR CREATIVE SUBJECT

Creative producers face a range of challenges in their desire to stand out from the increasingly crowded digital marketplace. One common strategy has been to distribute the creative process to the consumer. Once more, it is often digital developments that have intensified such practices of co-creation, expanding the creative process from the singular figure of the artist to include the customer in making creative decisions. Such co-creation exists in the individual design of products and experiences across a range of sectors from fashion and video games to advertising and online content provision, as well as contemporary art (Banks and Deuze, 2009).

If the co-creation of content or the sharing of creative labour was once a marginal economic condition, it is increasingly recognised as central to many forms of economic production, not just those associated with the creative industries. This is especially the case in the digital economy where consumers, fans and audiences are 'the drivers of wealth production within the new digital economy, their engagement and participation is actively being pursued, if still imperfectly understood by media companies' (Green and Jenkins, 2009, p. 213; Banks and Deuze, 2009). The result has been the development of models of 'decentralised creativity', where co-creative activities see producers and consumers in constantly shifting roles, challenging and reshaping understandings of media work. Questions have been raised as to whether this means that consumers become industry workers, and whether the labour market for professional producers becomes compromised? (Banks and Deuze, 2009, p. 420). In both cases the winners are generally seen to be the firms and companies whose content and sites are being engaged with.

This shift in labour from the producers and companies to the consumer indicates for some a significant change, not only in the understanding of the spaces and practices of creative production, but also in the relationship between the domains of work and leisure, the professional and the amateur (Banks and Deuze, 2009, p.421). These shifts are, as scholars remind us, not to be thought about in any straightforward way as liberating, democratizing or exploitative (Banks and Deuze, 2009, p. 421). For some, this harnessing of user-created content by big media business such as Google, Sony, and so on, involves the extraction of surplus value from the unpaid labour of the consumer. Further, co-creation can be understood as a form of outsourcing that contributes to the precarious employment conditions of professional creative workers. As Deuze and Banks note, however, 'the success of media production may increasingly rely on effectively combining and coordinating the various forms of expertise possessed by both professional media works and creative consumer citizenship, not displacing one with the other' (2009, p. 421). The story as they develop, and as others also acknowledge is complex, there is no simple sense in which creative specialist spaces and practices are being challenged and these amateur creatives are being exploited, but equally no easy way in which these are merely pleasurable activities (Terranova, 2000). These relations are neither a production of capitalism, nor developed as a response to the needs of capitalism, but are the 'results of a complex history where the relation between labour and capital is mutually constituted, entangled and crucially forged during the critique of Fordism (Terranova, 2004, p. 94).

GALLERY AND MUSEUM SPACES

Entering a major museum or gallery space you often sense a change in atmosphere, the architecture choreographs behaviours, movement slows down, voices become hushed, and reverent attention is given to what is on display. Whether it be the ornate architecture spaces of nineteenth-century institutions, like the Louvre in Paris or the National Gallery in London, the minimalism of white-cube art gallery spaces popular in the twentieth century, or the more recent spectacular surroundings of 'star-chitect' designed spaces, gallery and museums spaces are designed to affect their audiences. A series of challenges to specialist spaces of creative consumption exist as much as they do to creative production. Further, trying to split discussions into those concerning creative production and those concerning consumption is, in some respects, to fight a losing battle. As studios become spaces of consumption and as creative practices move beyond studio spaces to other spaces of display as well as onto the high street creating hybrid spaces and practices, it is clear that the other hybrid being created is that of creative production and consumption. This last section of the chapter explores the geographies of display and creative consumption more generally, beginning as with the previous discussion from the specialist spaces of creative consumption and moving gradually to consider those traits that challenge these spaces and practices.

A politics of display

Studies of galleries and museum spaces and their contents have most often been studied as part of the 'politics of display'. Large museums and galleries usually had their roots as state or monarchy sponsored projects. As has been shown in a range of studies, such projects offered multiple benefits; offering a way to promote a nation and brand its identity; a symbolic show of wealth and power and a means to civilise their populace (Bennett, 1999; Duncan, 1995). In a now classic study, Duncan demonstrates how the buildings, contents and practices of museum and gallery going in the nineteenth and twentieth centuries engaged their viewers in the performance of ritual scenarios, which replicated and affirmed ideas, values and social identities.

While the late twentieth century was dubbed the 'Museum Age' given the numbers of new institutions that were opening, and the increasing democratisation of these institutions, the history of these spaces is centuries old. Once highly exclusive, the history of many institutions lies in the opening up of and public display of princely or private collections to the general public. While museums and galleries had previously been considered as collections of things or as works of architecture, Duncan (1995) pioneered a new approach when she considered these as spaces of ritual, moreover ritual that produces and reproduces identities. Such discussions have evolved in two directions, firstly to consider these spaces as performing national and other identities, and secondly to consider the effect of these rituals on their audiences. Often focused historically rather than in the contemporary context, such discussions have foregrounded how museum spaces constitute civilising rituals, helping not only to form national identity, but also to constitute a citizenry. Of course, such discussions might encompass not only permanent institutional spaces, but also the temporary spaces of festivals, especially large events such as London's Great Exhibition of 1851.

While historical studies are based more on a theoretical audience rather than the experience of an actual audience, more contemporary work on specialised spaces of creative consumption has been interested to investigate the experiences of an actual audience but also the practices of those who create and control museum spaces – the curators. Interestingly, in both cases we see discussions of audiencing and discussions of curation as once specialised processes increasingly moving beyond the specialised spaces of creative consumption.

Work on audiencing within geography is just emerging, as scholars are reflecting on the experiences of art works within and beyond gallery spaces. Whilst some are interested in the embodied experience of art works (Hawkins, 2010b; Warren, 2013), others are concerned with the effects of art works on their participants, exploring how these works shape behaviours and attitudes (Hawkins et al., 2015; Ingram, 2014). If the latter tend to develop social science methods to engage with audiences in community and participatory arts projects, then the former develop embodied experiences of arts practices. Just as the phenomenological body has been important to studies of creative making, so it has also been a vital part of discussions of audience experiences. Artists working in the 1960s deliberately set out to engineer bodily experiences of art works for their audiences, and in their own deployment of the theory of Maurice Merleau-Ponty led others of the era and since to phenomenological accounts of audience experiences. Within geography, Hawkins (2010b) takes up phenomenology and post-phenomenology to reflect on the embodied experiences of a piece of installation art. Experimenting with poetic ways of writing this experience, she explores how these were arts practices designed to both be explored through all the senses as well as experiences that configured reflections on sensory experience, whether it be sight, sound or touch. Like other studies of audiencing, what is put at stake in this discussion is an account of experience as one based in sight alone, with the viewer being radically challenged as a multi-sensory being in which the objective eye is thoroughly emplaced within a sensing, emotive, messy, fleshy body. Warren's (2013) study of the audiences of an outdoor sculpture park notes their embodied experiences as they encounter these works amidst their semi-natural surroundings, and explores how they are inspired to reflect on the pasts and futures of these works, their making and decay. Of course, considerations of both embodied and symbolic questions of audiencing as well as didactic concerns should not be restricted to what are often considered elite creative spaces such as galleries. Rather, whether we look at accounts of cinema spectatorship, at engagements with experiences of nightclubbing or festival going, we find a similar focus on audiences as embodied agents, who appreciate both affect and experience as well as a work's narrative and symbolic content.

It is not just audiences who inhabit the specialist spaces of the gallery and museum, but also the curators who select and organise the objects within an increasing wide range of gallery spaces. If Duncan's (1995) approach focused on the experiences of audiences, then a more recent approach, the 'relational museum', like Becker's art worlds, understands museum practices and curatorial practices, in particular, as created through webs of relationships between people who make, exchange, collect and display. Intersecting the makers' histories and practices, the personal stories of the collectors and the institutional histories of the museums, this work develops a multifaceted understanding of spaces of display that unites creative production and consumption. The results are myriad stories of colonial relations between travellers, anthropologists, local people and collectors, together with the shifting interests of the cities and institutions as intellectual centres that drove collection strategies. People and things equally power these stories that often follow objects through these networks. If visions of curation practices are becoming increasingly connected, understanding how this specialised practice sits in networks, and how it expands beyond the practices of the museum and gallery spaces, we ought also think about how, like creative production practices, curation, too, has been adopted as a set of practices that work beyond the gallery space.

Curation has of late seen an expansion in the spaces, practices and objects it is associated with; so much so that Balzar (2014) diagnosed a contemporary condition of 'curationism', wherein the practices of curation spill beyond the rarefied spaces of galleries and museums to become iconic of contemporary society. Balzar is not the only one to recognise the rise of curation as a practice, attention to 'big data; and the so-called 'internet age' including social media has spawned a series of business-driven perspectives on curation (Anderson, 2015; Rosenbaum, 2011). For Wright (2014), 'curation is creation', coming to mean the 'intentional identification, collection (sometimes) repackaging, augmenting, updating and sharing of content'. This content might concern a range of activities, individuals and projects, including restaurants, fashion and music, as well as personal details. Such ideas, as Taylor Brydges explores in Box 3.3, are importantly not divorced from more traditional curation practices. Instead, they deploy the same words and practices and curators become, just like their gallery based peers, intermediaries between the producers of creative goods and the consumers of these goods. As the final section in this chapter will explore, such intermediaries have a potentially important role to play.

Box 3.3 CURATION BEYOND THE GALLERY SPACE?

Taylor Brydges

Curation, long associated with the art world, has in recent years gone beyond the museum and entered the mainstream. According to *The New York Times*, 'the word "curate", lofty and once rarely spoken outside exhibition corridors or British parishes, has become a fashionable code word among the aesthetically minded, who seem to paste it onto any activity that involves culling and selecting . . . now, among designers, disc jockeys, club promoters, bloggers and thrift-store owners, curate is code for "I have a discerning eye and great taste". Joosse and Hracs (2015) understand the intermediary function of curation, 'to involve the interpreting, translating and shaping of the marketplace through the practice of sorting, organizing, evaluating and ascribing value(s) to specific products'. These new actors increasingly filter supply for cultural products such as food, fashion, music and furniture, by offering advice and inspiration for consumers in physical and virtual space.

Case study: Curation in the fashion industry

One field where curators are increasingly prevalent is the fashion industry. As the fashion industry continues to undergo considerable restructuring, both established and up-and-coming actors are struggling to reposition themselves as dominant actors in the sector. One result of these changes is that curation has become a strategy to compete and 'stand out in the crowd' (Hracs et al., 2013).

Curation appears to be taking different forms across spaces. For some actors, such as fashion buyers and independent retailers, curation has been employed as a tool to create value through unique retail experiences in physical space, while fashion bloggers' content is curated online on a diverse range of social media platforms for readers (or 'followers'). Fashion bloggers provide a particularly interesting case for exploring the impact of curation in online spaces.

Before the era of social media, power in the fashion industry was often concentrated in the hands of the few; whether that be a key designer or industry magazine. New digital technologies and Web 2.0 have allowed emerging actors, such as fashion bloggers, to sidestep traditional established intermediaries, and become curators in their own right. Increasingly, fashion bloggers are replacing traditional journalists and other fashion veterans in fashion show front rows, thanks to their ability to communicate to and mobilise their thousands of online followers. Social media provides these new curators with a direct line to consumers across many different platforms – from their personal style blogs to Instagram, Twitter and Facebook – as they tell their followers what they are wearing, how to style it, and then where to go out and buy it. The intimate and conversational tone of many bloggers serves to establish trust between the blogger and his or her readers, who often begin to feel a personal connection to the blogger.

For some of the most successful fashion bloggers, their curatorial ability ranks them amongst some of the most powerful brands in the fashion industry; for example, in September 2014, *Women's Wear Daily* singled out Chiara Ferragni (better known as The Blonde Salad) as the first blogger who has successfully turned a personal style blog into a global brand and business. Ferragni, who has over 3 million Instagram followers and nearly a million

Continued

Facebook likes, is projected to make $8 million USD in revenue this year alone, from her blog, advertising, brand collaborations and footwear collection. In comparison, American *Vogue*, an institution in the fashion industry founded in 1892, only has approximately 200,000 more followers than Ferragni.

Popular American fashion blogger Leandra Medine (or The Man Repeller) a cultural phenomenon in her own right, boasts a smaller but still impressive 725,000 Instagram followers and 215,000 followers on Twitter. In addition to writing a memoir at 23, Medine has also had design collaborations with brands such as Nina Ricci and Superga. A growing number of fashion bloggers, like Ferragni and Medine, are making the transition to lifestyle blogs, offering curated advice on a broader range of subjects, from food to fitness and travel, which suggests their reach may grow beyond that of traditional fashion actors.

The future of curation?

While fashion bloggers provide an interesting exploratory case study into the field of curation, in the ongoing debates regarding the usefulness of curation as a concept for geographers, several questions remain. For example, who can be a curator? At what point can someone be deemed a curator? Does being a curator require professional qualifications, or is it a distinction that can be self-appointed? What motivates curators; money, status, identity, reputation? And, what can be curated? Is the archival function typically ascribed to museum or art gallery curation a necessary precondition, or can the label be affixed to any thoughtful collection of goods? A third question central to geographers is where can curation take place? Do processes of curation differ in physical space, as opposed to virtual or temporary spaces?

If discussion thus far has explored how the practices of creative consumption move beyond specialist spaces, to end discussion turns to consider how creative consumption practices have come to permeate everyday spaces and how in so doing these creative practices have become refigured. The internet has recently emerged as space that reshapes creative market places, whether through the advent of digital download and player services, or whether through its potential as a democratic space of creative display; a means by which audiences and market places can be reached by those shut out from conventional creative spaces of display and sales. The growth of websites such as Etsy, MySpace and general social media are enabling routes to market and accessible spaces of display for emerging and established artists, musicians and craft practitioners. Not only can pro-amateurs reach people with their creative practices, but professionals, such as musicians, can have success without necessarily having to engage with the traditional structures of the creative industries. What is more, software and internet-ready digital formats such as iTunes and streaming services such as Spotify and iMusic, as well as cinema services such as Netflix, and Amazon Prime and more specialist players such as the British Film Institute, have shifted the spaces and practices of creative consumption. Not to mention pirate-download services that enable pre-release and cinema films to be downloaded and watched within consumers' own homes long before they might otherwise be available. The experience is further enhanced by the increasingly advanced home cinema systems available, but also complicated by the ways in which digital downloads and the growth of the mobile phone as a content consumption device have reduced the quality of playback many consumers demand. The result is a complex and ever-changing terrain of creative consumption.

Exploring the music industry is to witness some significant shifts. Not least, as digital formats challenged classic record sales, the music industry shifted to favour live performance as a primary source of income, whilst at the same time recognizing the increasing ubiquity of our music consumption through phones as the latest development of portable music players. As Leyshon notes, with digital formats challenging record sales the industry began to reconsider live performance

> the conventional argument in rock analysis has been that live concerts exist courtesy of the record industry; their function is to promote records, to which they are subordinate (and for which purpose they are subsidised). But this argument no longer seems valid.
>
> (Frith, 1997:5)

Indeed, as Leyshon demonstrates, revenue in the UK music industry has been greater from live than recorded music since 2008, and since 2010 globally. As David (2009) argues, the 'shift back' to performance is to the benefit of music production, if to the detriment of the capital intensive part of the music industry – record sales. Increasingly, in this saturated marketplace and where consumers are increasingly unwilling to pay for content, what has become valued and considered worth paying money for is 'embodied, affect-rich, place-based experience of performance' (Leyshon, 2014, p. 158).

Just as digital networks enabled others to produce music outside of the music industry, so they also allow musicians to find routes to market and to live performance that circumvent record companies. Fund-raising websites such as Kickstarter are playing an increasingly important role in enabling emerging musicians to crowdsource money for recording albums and for concerts. Such digital democracy, however, has led to what Hracs et al. (2013) describe as the 'dilemma of democratization' wherein, the lowering of barriers to entry has enabled people to enter the market place, which has in turn increased competition and made it harder for practitioners to extract a livelihood from their creativity. As a result, creative practitioners have to work harder and harder to stand out from the crowd. One of the ways creative producers have been mobilizing the digital to help them distinguish themselves has been to develop their online profiles, whether this be Etsy profiles for craft practitioners, Myspace music profiles for the music industry as explored by Hracs and Leslie (2013), as well as more general social media profiles developed and maintained through websites and apps such as Twitter and Instagram. Aesthetic labour in the digital age, however, requires the constant main-tenance of profile and identity online, a stressful never-ending task that erodes the distinction between work and life, and prevents creative workers from ever turning off, as they come to sell themselves as much as their products and skills. As such, one of the unintended (and often understudied) facets of the erosion of specialist work spaces is the continued merging of work and life to the detriment of a creative worker's quality of life. As such, to consider these changing spaces of creative production and consumption is also to reflect on the politics of such expanded spaces and the erosion of differences, that can be both enabling but also disabling.

CONCLUSION: BEYOND SPECIALIST SPACES

This chapter has focused on the paired contemporary traits of both the erosion and maintenance of specialist spaces of creative production and consumption. On the one hand, we find a fetishisation of the spaces and practices of creativity, whether this be growing support for studios, the continual increase in museums and iconic institutions, as well as the expansion of imaginaries of creative labour to encompass occupations previously thought of in terms of service work. On the other hand, traits from within and without the creative sector see the expansion and erosion of these specialist spaces and practices. The art world is shifting arts practices outside of the studio space and beyond the walls of the museum. Furthermore, wider cultural traits such as the growth of digital spaces and the evolution of the experience economy demand ever-more creative approaches to consumption. This can be paired with trends that see specialist spaces and practices circumvented to enable production to occur outside of specialist space and routes to market to be democratised. With such erosion of boundaries, however, comes a series of dilemmas. In order to stand out from the crowd, creative practitioners exploit both themselves and their spaces to marketise their products further and deploying strategies seek to both enhance specialisation, whilst also enrolling the consumer into the production process. Further, as consumers seek to negotiate these more democratic online cultural offerings, flooded with more and more options and data, specialist practices like curation offer a solution, helping them navigate the creative consumption place. Throughout this discussion, one of the spaces that came up both as a site of creative production and consumption was the home, and it is to a more in-depth look at the creative home and the negotiations that take place there, that this discussion now turns.

4

HOME: NEGOTIATING CREATIVITIES

In front of a live audience, two men stand in the midst of a fake home set, one dressed in a suit and tie, the other in a flannel shirt and jeans. The man in the suit, Tim – the tool man – Taylor, turns to Al (in the flannel) and pulls over a wooden door in a stand. He explains they are going to insert a security lock. Al hands over a drill with a 1.5 inch bit, and Tim exclaims, 'that's a girly drill'; he explains, 'you want a job done, you gotta do it right, what do we want?' he asks, and the audience laughs and choruses in response 'we want more power' – the show's catch phrase. Off-stage a scantily dressed woman enters with a very large drill, clearly substantially too big for the task of drilling a hole in the door, Tim brandishes it above his head, grunting like an ape, the studio audience applauds and laughs.

A woman sits in front of a computer screen late at night, answering customer queries, filling orders, photographing her hand-made jewellery. She gets up to attend to a crying baby or a five-year-old with a bad dream. She resumes her work, packing up orders to take to the post-office the next day, uploading images to her website of new items she finished making that afternoon, checking her profile, replying to Twitter posts and posting images of the new products on Instagram and Facebook. The next day she will finish organizing her home studio and photographing it to offer her customers an insight into the environment where their objects are created.

The home has gained a lot of geographic attention of late (Brickell, 2012; Blunt and Dowling, 2006). A key focus of these discussions has been the practices and politics of the making and un-making of home. Cross-cutting such discussions are a series of tensions and intersections between the home as a site for economic production, for social reproduction and for the production of subjectivities. Recently attention has turned to home 'un-making', considering how the acts of individuals, communities and nations – whether violent, exclusionary or as a result of forces like the digital – might transform (permanently or temporarily) a homely space into an unhomely one. In the midst, however, of the debates about the practices and politics of the making and un-making of home, little attention, as yet, has been paid to the home as a site of creativity, and moreover to how creative practices might have an important role to play in these processes of making and un-making.

To explore creativity in the home is, as this chapter will explore, to situate it as part of interlinked economic and social reproduction, the latter encompassing 'the social relations, objects and instruments that enable the maintenance of everyday life within capitalism' (Marston, 2000, p. 233). It is to explore the home as a site of myriad creativities, from the economically orientated small businesses such as the many millions of women who use Etsy to help their hand-made goods reach market places around the world, to the practices of daily home-making and DIY that enrol men and women alike in the making and remaking of home and of self through creative practices. Of particular interest is how it is we might talk about creativity with respect to the 'small-scale social, physical, cultural and emotional infrastructure of the household where labour power is reproduced on a daily basis' (ibid.). The household is thus a site of social reproduction, but also one where capitalist consumption practices are entrained, a site of micro-level social processes that bring together social reproduction, biological reproduction and economic production.

Over the last few centuries the role and place of creativity in the home has shifted. The evolving nature and increasing mechanisation of housekeeping and household tasks during the twentieth century resulted in a decline in potential for creativity in the home. With industrialisation it was noted, for example, that 'work in terms of housekeeping has been simplified and minimized through labour and time saving appliances and consigned goods' (cited in Hackney, 2006, p.23). What was often saved or eliminated, however, were those types of reproductive labour that were seen as the most creative and fulfilling. Mundane aspects of housekeeping remained but those areas seen to once produce great satisfaction such as culinary crafts, sewing, knitting, weaving tatting, and so on, were minimised through efficiency savings. Interestingly, these same practices have recently been subject to hobbification, turning what was once a part of the 'work' of running a household into a leisure pursuit (Hughes, 2012). Historic practice of thrift – so making do, making up, mending, or weaving cloth for clothes, knitting and making clothes – have now been linked to gentrification, such practices (costly as they are of time and also potentially of money) becoming the preserve of those rich in both (see, for example, the discussion in Box 4.1; Strasser, 2000).

Amidst reflections on the rise of creativity as part of the twentieth century's expansion of leisure time, it is also important not to overlook the role of the home as a site for expanding forms of often economically incentivised creative practices (Hracs and Leslie, 2013; Leyshon, 2014; Luckman, 2015). Of course, creative labour has long happened at home, but now technological developments are enabling home workers to reach markets in ways not previously possible, helping what was once dismissible as 'pin money' to become an increasingly important source of income. As such, participation in the creative economy whilst at home is part of a wider intensification of working lives, partly enabled by the affordances of technology that are restructuring the relationships between domestic and working spaces.

Working through a series of case studies, this chapter will examine the home as a site of both formal creative economic production (Leyshon, 2009; McRobbie, 1998) and of vernacular and amateur creative practices. This encompasses home-making practices, such as building dwellings, cooking, decorating and gardening, as well as more obviously 'creative' practices of crafting and hobbies including DIY and amateur practices such as drama or arts (Delyser, 2014; Crouch, 2011; Yarwood and Shaw, 2010). It will explore how these different forms of creative practices are closely intertwined, sometimes indistinguishable and, importantly, how the creative processes and objects produced (economically orientated or otherwise) are often productive not only in the sense of home, but also of self.

Whilst recognizing the inter-linkages of economic and non-economic forms of creativity, this chapter will necessarily develop two key sections, firstly exploring the home as a site of economic production. Discussion will consider the negotiation of creative practices within the time-spaces of the home, and it will explore the role of the digital in enabling the growth of the home-based creative economy, examining the pros and cons of such digital penetration. The second section of the chapter will turn to explore vernacular creative practices and creative practices of home-making, reflecting on how such home-making practices are also important practices in the making and shaping of identities.

NEGOTIATING THE HOME AS A SPACE FOR ECONOMIC PRODUCTION

Feminist geographers have long focused on the site of the home as a site of production, both in terms of social reproduction, but also a myriad of micro-economic forms of productivity (Marston, 2000). We might think about this in terms of livelihood, 'this means recasting and interweaving the processes and complex geographies of production and reproduction. Most importantly it recasts these processes as practical and meaningful, rather than abstract and logical, in ways that challenge the boundaries of home and work' (Whatmore, 1991, p.147, cited in Marston, 2000). For creative practitioners the home has long been recognised as a key site of creative labour, especially for women (Bain, 2004b; McRobbie, 1998). The growth of digital technologies, however, has expanded and enabled home-work for a range of different creative practitioners (Hracs and Leslie, 2013; Leyshon, 2014). This includes technologies that enable home-working, such as the growth of the digital recording studio in bedrooms or garages (Watson, 2009), as well as the new routes to market provided by online sites such as Etsy.

Home in the general parlance is equated with not working, so to work from home often requires creative makers to establish and negotiate work spaces, with sometimes exhausting consequences when work and leisure collapse, as well as managing others who assume that to work at home is to be less serious, less effective. To assert the importance of the home as a site for creative production is therefore to participate in feminist challenges to the invisibility of women's domestic labour which long enabled home-based working to be dismissed as invalid and unimportant when compared to work done outside of the home. Creative practices and their economic validity become processes of transformation of subjects and the spaces of the home. These are far from simple transformations; indeed as we shall see there are ongoing confusions of home and work as makers struggle to negotiate the making of home, of self and the creative outputs they produce.

The processes by which domestic spaces of nurture and reproduction can evolve into sites of creative production and economic venture are far from simple linear ones. Such powerful combinations of the intimate and the economic often involve complex negotiations of time and space. Many creative practitioners, when talking of the spaces in which they work, reflect on the ongoing challenges of getting space for creative practices in the home. As discussion in Chapter 4 on the studio demonstrates, making space for creative practices within domestic spaces is an often complex process that brings to the fore ongoing debates about the responsibilities of social reproduction, about gender roles in the labours of caring and home-making, and about the sometimes tense and fraught relationships between home and work. Such debates have resurfaced as digital and online resources intensify the possibilities of creative home-working as well as the challenges of the interpenetration of the home and economically productive work.

While debates around the interpenetration of home and the creative economy are long standing (McRobbie, 1998) they have been shaped in recent years by emerging discussions of online spaces. This not only reshapes the spaces of the home and challenges home/work boundaries, but also reformulates the idea of work through the creation of highly fluid work-spaces via the affordances of digital technologies and online platforms. A top online site for such home-based creative work is Etsy, known as 'e-bay for the hand-made', with 15 million members and $63 million in sales each month. In 2013 this sales volume represented the movement of 4,215,169 items worldwide. Women make up over 90% of Etsy's sellers, many of whom set up shops online to enable the negotiation of a family friendly work-life (Pace et al., 2013). A common narrative around the Etsy-seller is the economically active woman who on giving up her job to have children sets up a micro-enterprise to combine contributing to household income and ongoing participation in world of work with caring for children. According to Kalin, one of the male internet start-up team who conceived of Etsy, the sellers are 'mostly stay at home moms and college students looking to supplement their income rather than make a full-time living' (cited in Luckman, 2013, p. 251).

Studies increasingly show, however, that for some 'Etsy mums' activity on this site is a top income generator.

As a global shop-front with high visibility, Etsy challenges the conservative vision of women's craft activity as invisible, as mute and as undertaken in the home and thus outside the auspices of the larger commercial economy (Luckman, 2015). Sites such as Etsy are of interest for their 'ability to render international marketing and industrial networks accessible to sole traders and micro-enterprises', Further, there is no gate-keeping around conditions of access to Etsy, anyone can sell and anyone can buy. This both enables unprecedented access to international markets from people's home-studios/workshops, but also makes for more democratic taste-making, rather than relying on the owners and curators of shops, boutiques and galleries as intermediaries. As a result the gap between producer and consumer is smaller and the roles often blur; Etsy sellers are often Etsy buyers, but also given the hand-made nature of the products a certain amount of customisation is available.

As such, Etsy challenges the construction of women's domestic making as purely amateur creative production and /or focused largely on the acquisition of pin money. Indeed, studies have situated these forms of craft entrepreneurialism as more than an extension of the 1950s housewife's sense of 'making do' into the monetised market place. As Luckman (2015) notes, Etsy: 'complicates how we think about craft based creativity in terms of amateur versus professional division'. Indeed, Etsy has become a key site for reimagining the fluid spectrum of 'pro-am' activity, demonstrating a kind of pro-am entrepreneurism enabled by the social and economic expansion of the internet (Luckman, 2013). Economically, these are serious small business operations, 'no longer relegated to a corner of shared family space or the table in between family meals, these women are frequently photographed in their home studio, with home rooms set aside to conduct their creative business' (Luckman, 2013, p.262). In personal terms, 'being able to work from home and/or flexibly has been embraced by many women in particular as an important compromise between paid work and unpaid domestic responsibilities' (ibid.). Frequently overlooked in the literature are the other values that are gained for women able to access markets via Etsy. Clearly economic benefits are often privileged, but what Etsy sellers, and others who have turned to craft and creative economy businesses often note is the self-fulfilment that these practices bring about. At points in life-course, such as having children or retiring, questions of identity and well-being can come to the fore. Oftentimes involvement in craft businesses of creative enterprises is seen as an important part of engaging with changed life circumstances in a positive way (Grace et al., 2009; Hughes, 2012).

For all the positive attributes of the forms of creative home-working practices enabled by Etsy and other sites, we should also be aware of the digital dilemma as it plays out in this context, becoming a force for home un-making. Etsy selling is often equated with doing what you love (Luckman, 2013) and whilst initially attractive, as such home-making practices allow for flexibility in the conduct of income-generating work alongside unpaid domestic responsibilities, online sales practices can acerbate the tendency for home-working to blur home and work to the detriment of the practitioner. Furthermore, research suggests that the growth in such paid employment in the home often does little to shift gendered divisions in household labour. In line with studies of creative labour more generally, discussions of Etsy often examine the demands of subjective and affective labour that are placed on these creative subjects (Hracs and Leslie, 2013). Standing out from the crowd in busy online marketplaces often requires that Etsy sellers do some significant selling of self, both in terms of building their virtual profile, often based around shots of themselves and their home-working life-styles and spaces, but also by building personal relations through communications with their customers. Communications from home, whether in terms of email, or the virtual shop-fronts of Etsy often emphasises the person, selling the image of the practitioner as much as the product (Pace et al., 2013). With the consumer buying into both an ethical economy (where the provenance of the item is clear and gained directly from the maker), but also buying into the stories of that person and the biographies of their craft and business practices. The creation and maintenance of such 'virtual bodies' through social media can be exhausting, demoralising and require a significant amount of time and energy. Creative work is work that anyway expands beyond the walls of the office or the workshop and into society at large and, of course, this is even worse if the main work place is the home (Lazzarato, 2012). Creative work done from the home then can often mean to always be 'on'; times and spaces and work and leisure become almost impossible to distinguish.

Creative economic production, whilst long part of home productivity has become increasingly high profile in recent years, with the advent of new technologies that have woven home spaces into the spaces of the international creative economy. As well as reworking the spaces of the home, sometimes even to the degree of 'un-making' them, such practices have also reworked the ideas and imaginaries of work, and as discussion has demonstrated, participated in the formation of subjects. These themes continue to play out in the discussion that follows, which turns to the home as a site of vernacular creativity; shifting focus from economic production to explore creative practices as entwined practices of home-making and subject formation.

CREATIVE GEOGRAPHIES OF HOME-MAKING

Home within geography is increasingly understood less as a material location and more in humanistic terms as a set of exportable routines and practices, offering more permanence, more shelter than any lodging. Home is no longer a dwelling but the untold story of a life being lived (Berger, 1984, p. 64). The home is thus recognised as both a material and symbolic site of production and consumption, and a potential site for creatively making and remaking identity (Datta, 2008, p.4). In this section a series of creative practices are explored as practices of home-making. Exploring creativity as part of social reproduction in the home involves considering both the daily arts and crafts of keeping home – the domestic labours of home-making, as well as those more clearly 'creative' practices such as decoration and DIY and amateur creative practices (see also Box 3.1). Discussion unfolds through four sections, exploring variously the crafts of home-making, and creative practices of decoration and DIY. The home not only becomes made through these practices, but importantly subjects are formed too under a range of different socio-material conditions. Whether this be the early-mid-twentieth-century home as a site where the making of the modern home entwines with the making of the modern woman, or the mid-to-late-twentieth-century home in which DIY became a means for men to reshape their environments and retool their identities as fathers, husbands, and men. Across all these discussions we find once more complex slippages and negotiations of different forms of labour and creativity, with the same practices becoming in different times and places, domestic labour, economically productive and a hobby. Before considering these specific case studies, however, this discussion will explore the nature of vernacular and everyday creativities more generally.

Vernacular creativities: everyday creativities

As well as paying attention to sites other than the gentrified areas of the urban core, to attend to vernacular creativities is also to engage with the very ordinary sites and practices of culture; the 'city of people, of everyday life, common occurrences, small shops, bus stops, allotment and waste ground' and of course, the home (Miles, 1997, p. 257). To pay attention to vernacular creativity is to attend to how creativity occurs in the ordinary landscapes around us, landscapes that are 'parts of an ensemble which is under continuous creation and alteration as much or more from the unconscious process of daily living as from calculated landscape design' (Meinig, 1979, p. 6). In short, in attending to the mundane spaces and ordinary practices of vernacular creativities we need to appreciate how these creativities, as much as professional arts practices are both produced by, and productive of the sites, spaces and cultures in which they are produced and consumed.

Creativity in these terms, is not just the preserve of the artistic genius or a rare talent, rather we are all creative. In recognition of such a distributed creativity we find a growing body of work that attends to so-called vernacular or everyday creativities (Burgess, 2006, 2007, 2009; Edensor et al., 2009a). Whether we are thinking of practices of home-making and garden-tending, the place of creative practices in a range of hobbies, or more improvisational in the moment creative acts of use and reuse, it is clear that creativity is part of how it is we go on in the world (Hallam and Ingold, 2008). Often such everyday vernacular creativities have become understood as a site from which to develop critical perspectives on the dominant narratives of creativity that conjure up the urban-based creative economy. As Edensor et al. (2009b) note, vernacular creativities offer the chance to 'transform space and the everyday lives of ordinary people to reveal and illuminate the mundane as a site of resistance, affect and potentialities' (p. 10).

To understand vernacular creativities we need to appreciate how 'participation in cultural activities is initially not driven by career development motivations, but by a personal desire to engage with the affective, emotive, cathartic dimensions of creative pursuits' (Gibson and Kong, 2005, p. 544). Acts of creative home-making are very clear on such more-than-economic values. On the one hand, this might involve finding creativity amidst mundane routinised practices such as cooking, laying the table or cleaning, as well as ways of living such as 'make do and mend' or 'waste not want not' (Strasser, 2000). As discussion has made clear, there were worries that over the twentieth century such crafts and arts of home-making were gradually being eradicated by the efficiency drive of modernism's mechanisation (Hackney, 2006). The result being routinised practices that left little space for individual, creative practices. Recent times, however, have found space for such narratives once more, whether driven by fashion or need (see discussion in Box 4.1). On the other hand, creative home-making might intertwine with the fashioning of the self in what sociologists have called 'grounded aesthetics' (Willis, 1998). This is to re-engage with those strands of cultural studies that maintained that culture was ordinary and every day (Hebdige, 1979; Williams, 1958; Willis, 1998). There has been much work that has blurred previously distinct spheres of 'high' and 'low' culture, as Edensor et al. put it 'positively valuing local, popular, subaltern and everyday cultures and subcultures' (2009b, p.9). As such, the lines between the artistic myth and the possibilities of creative products and creative practices as practical situated activities within everyday culture, becomes blurred. As Willis observed there is a need to take seriously what he called 'grounded aesthetics', the 'vibrant symbolic life and symbolic creativity in everyday life' (1990, p.206), that is part of the 'necessariness of everyday symbolic and communicative work' (p. 208) that '(re)produces identities, places and vital capacities of humans' (cited in Edensor et al., 2009b, p.9). As Willis writes of the symbolic decoration of bedrooms;

> the extraordinary symbolic creativity of the multitude of ways in which young people use, humanize, decorate and invest with meanings their common and immediate life space and social practices – personal styles and choice of clothes; selective and active use of music, TV, magazines; decoration of bedrooms; the rituals of romance and sub-cultural styles; the style of banter and drama of friendship groups; music making and dance.
>
> (Willis, 1990, p. 206–7)

The research that has followed in this vein is rich and deep, and covers everything from the collected pursuit of hobbies (Hoggett and Bishop, 1986) to the value of a range of amateur practices and groups (Finnegan, 2007; Nicholson, 1997, 2015). What is often emphasised in these studies is the place of these practices as part of the progression of daily life they are embedded in and the routines they are enabled by. In comparison to other studies of creativity where the focus is resolutely on the thing produced – the model railway built, the amateur play created – here the concern is with what happens in the process of doing and being part of these activities. While contemporary theorists are interested in making as connecting (Gauntlett, 2011), such discussions sit within a much longer tradition of studies of creative hobbies and amateur practices that have identified the importance of these practices in fostering sociability, conviviality and affective pleasures.

Ideas of craft consumption have sat central to many of the discussions of domestic hobbies. Craft consumption has been described as the desire to 'engage in creative acts of self-expression' through the purchase, collection and assembly of particular goods (Campbell, 2005, p.24). As Campbell outlines, and Yarwood and Shaw (2010) develop, there are three key features of the craft consumer. Firstly, they develop the whole process of making from start to finish. This includes designing, sourcing materials and producing the final product and often involves the crafter 'invest[ing] his or her personality or self into the object produced' (Campbell, 2005, p. 27). Secondly, they will participate in 'choosing, buying and assembling different materials to produce a new item' that is primarily intended for self consumption, a jeweller making things for herself for example, or a model aircraft builder who builds for the enjoyment of building as well as the practice of flying. Thirdly, the process of collecting – including 'the accumulation, organization and display of goods – is as important as a creative and personal activity that is enjoyable, a significant driver of craft consumption and a way of narrating identity' (Campbell, 2005, p. 28). In identifying these three facets, Campbell (2005) acknowledges a complex relationship between identity, leisure and craft consumption and suggests that further study is needed.

Following Campbell's lead, Yarwood and Shaw (2010) examine the spaces and identities involved in the model railway building, involving the construction and running of the scale model trains on layouts. They explore the different ways of making the layouts. For some, models and materials are purchased from commercial mass producers or from cottage industries devoted to supplying these parts. For others, scenery is made from scratch, with pleasure taken in finding the pieces needed, in 'making do' with bits found around the home to construct buildings, bridges or vegetation. As they note each layout is unique, 'a combination of details and vignettes that bring the track to life and reveal much about the creator's enthusiasms, humour and earnestness' (p. 427). In exploring the creative and crafted nature of the hobby, they examine the shift from train-sets as ready-made kits, to those with more skill who work on layouts of their own design, minimizing the presence of ready-made elements, or adapting these to their own ends. Interviewing modellers and attending conventions, they discuss the process of making the layouts, explore the production and consumption dynamics of building the model railways, and engage the social networks that are associated with the hobby. From their ethnography, they draw conclusions about the hobby as a means of dealing with life-course changes, that is, children leaving home, but also a way of engaging with children as well as building relationships with others.

In the face of potential dismissal from economically determined value structures, scholars of vernacular creativities have rejected economic productivity and worth in favour of the possibilities that creative activities that appear frivolous, playful and thoroughly indulgent might in fact be productive of personal and collective identities. As a process that is carried out by someone, sometimes alone, sometimes as part of a collective, creativity plays a key role in shaping individual and collective lives and identities. This might be creative practices as part of everyday practices – from housework to lives lived with concern for style, for example – or it might be hobbies taken up at moments of transition such as migration, or at certain points in the life-course, such as pregnancy or old age. As well as having an important place in individual and family lives, creative practices have also become part of the shaping of community identity and cohesion, forming modes of conviviality across difference.

What such studies further emphasise is how, to the figure of the creative labourer who might be considered the creative professional, we should add the figures of the hobbyist and the amateur. As well as valuable in their own right, we might also want to reflect on the recent rise of what has been called the 'pro-am' (Leadbeater and Miller, 2004). This group are, as Leadbeater and Miller explain, 'innovative, committed and networked amateurs working to professional standards' (p.9). Citing the rap culture, Sims computer game and Linus (the open source computer operating system) as examples they note how influential pro-ams have been over the last few decades and how technology, such as file-sharing sites and crowd-funding sites will only continue to enable these practices. If the twentieth century was seen as the century of the professional, with amateurism being derided, then the twenty-first century, they argue is that of the pro-amateur. Whilst these discussions continue to evolve it seems the pro-am idea embraces a sense of co-production as well as 'mass creativity' (sourcing creativity more broadly). Further, it is seen as a way to evolve new and innovative designs to current environmental and societal problems as well as to create creative products. The remainder of this chapter will explore three examples of vernacular and pro-am creativity reflecting on how these practices make both homes and the subjectivities of makers.

Creating homes – creating subjects 1 – crafting the home and making the feminine modern

> There is [sic] surely few more valuable services to be rendered to the national character than an insistence on good craftsmanship, and especially on good craftsmanship in home-making.
>
> (cited in Hackney, 2006)

The mid-twentieth century home in the UK and the US was often a fraught space. The era was one of significant social tensions between the 'desire for change, progress and improvement and a need for stability, continuity and certainty' (Hackney, 2006). The space of the home and its increasing mechanisation became the location where many of these tensions played out, with discussions about the arts and crafts of keeping home and the place of craft in the home offering a lens for these debates. Home craft in the mid-twentieth century was about a feminine modernity that bound the modern to the domestic and, furthermore, it endorsed a wider ethos of handicrafts that integrated practices of 'making' thoroughly into everyday life (Buckley, 1990; Hackney, 2006). 'Worthy handicrafts' in the era, which included not just practices such as making clothes, sewing and knitting but also cooking, bread-making and even 'laying the table nicely' (Lethaby, quoted in Hackney, 2006, p. 16) have been understood as playing a key role in the production of the modern woman and the idea of the modern home.

For domestic scientists, house-craft had long been a science but also an art. Writing of the place of craft in women's magazines, Hackney (2006) notes how woman's housekeeping expert Susan Strong used scientific and managerial metaphors, such as 'working a house to schedule', but also emphasised the artistry of keeping a home. So alongside the mundane routinised cleaning there was creativity involved in activities such as making final touches to a bedroom, replacing flowers or rearranging articles 'where fancy wills'. Properly managed housework was described as fulfilling and expressive; turning out the bedroom was a 'joyous job' and newly ironed linen gave a 'thrill'. As the manufacturing industry produced more convenience goods and machines, it was argued that such tools often only added to the routinisation of housework. Thus whilst celebrating the efficiency such tools enabled, domestic scientists and home editors argued for the continuation of the home as a site of creativity and individualism in which women could exercise their taste and skills. We see this continuation in the discussion of the practice of make do and mend in relation to the contemporary austerity economy in Britain in Box 4.1.

Box 4.1 MAKE DO AND MEND: RETHINKING THE RELATIONSHIP BETWEEN AUSTERITY AND CREATIVITY

Bethan Bide

Austerity breeds creativity out of necessity. Since the British coalition government began its programme of austerity policies in 2010, this notion has formed the central premise of many discussions about home sewing, and is frequently linked to the phrase 'Make Do And Mend', a reference to a government-issued sewing manual from 1943. Today, 'Make Do And Mend' describes inspirational sewing classes, money-saving blogs, and has even been used as a television programme title. However, the original

---

Continued

wartime campaign actually had limited impact, because it over-looked the vast majority of the population who historically had little choice but to remake and reuse (Slater, 2011). In reality, London's home sewers had long been proficient in the process of adaptation and recreation, and were well prepared to meet the challenges of official clothing restrictions with a level of ingenuity and flair that outstripped the suggestions offered by this rather utilitarian instruction manual. As such, it is perhaps more useful to think of 'Make Do And Mend' in terms of an ongoing, grass-roots movement rather than official austerity policy.

'Make Do And Mend' is most commonly associated with wartime shortages, but in fact the period of austerity following the war presented Londoners with much greater challenges in terms of obtaining new clothes. In these immediate post-war years consumers contended with limited supplies as the government focused production on the export market, and clothes rationing persisted until 1949. However, after a period of relative stability in fashion design during the war, women's clothing underwent a series of major changes, and London's home sewers found them-selves at the forefront of a fashion revolution as radical as the social policies of Britain's new Labour Government.

London fashion during this time is typically discussed in relation to trends elsewhere. In particular, the city's post-war fashion changes are described as an import of Christian Dior's 1947 Corolle Line (commonly nicknamed the 'New Look') from Paris, at the time the self-declared world fashion capital. Characterised by its tightly structured and corseted torsos, and long, voluminous skirts, the New Look was widely criticised as impractical and restrictive to its female wearers (Beckett, 1947). In spite of this, fashion mythology tells us that British women were desperate to emulate this new trend, and attempted to alter their clothes accordingly, with even Princess Margaret adding a strip of fabric to lengthen the hem of an old coat (Behlen, 2012). The resulting outfits are gently mocked as weak imitations, but the clothes of Londoners held in the Museum of London's collec-tion tell a different story, one that demonstrates how creative amateur adaptations contributed to a new fashion line that belonged to London.

While home sewers in London followed certain aspects of the New Look trend, namely the lower hems, fuller skirts and softer shoulders, they rejected other, more restrictive features. In October 1947, *Woman's Weekly* ran instructions to show how a coat could be adapted by letting out the hem, removing the shoulder pads, and reshaping the waist (*Woman's Weekly*, 1947). Although adhering to fashionable lines, the resulting garment was still a practical length, with a lack of internal structuring that would restrict the body.

Examples of home dressmaking from the period demonstrate that this looser, shorter trend was not simply due to the restraints of altering an existing garment. Even when starting from scratch, only certain aspects of the Parisian New Look feature, for example, full skirts are created by clever pleating rather than a weighty excess of fabric or cumbersome petticoats, and hem lengths are short enough not to be tripped over when running for a bus. Oral history interviews demonstrate that these altered lines cannot be simply explained away as compromises resulting from austerity restrictions, but were positive choices made by the sewer, who saw their garments as an ideal, improved version of the new fashion.

Such shorter, freer styles were the London's fashions of choice, and this consumer demand can be seen reflected in the ready-to-wear creations of the period, which also demonstrate these features. Crucially, these developing trends in ready-to-wear and home-made garments were occurring simultaneously, defying the theory that fashion trends 'trickle down' from upper to lower classes (Veblen, 1994). The new London Look was not a compromise between Parisian couture and the practicality demanded by austerity conditions, but the result of a process of co-creation between professional dress designers and the home sewers of the city.

Since the 1940s, London's creative home sewers have largely disappeared – not from a lack of austerity, but due to the

Continued

changing nature of the clothing retail market. In a world of planned obsolescence and mass manufacture in fashion, where today's London consumer can pop to Primark and buy an entire new outfit for under £20, there is less financial need to remake and remodel old clothes, and there is certainly no shortage of products in shops. However, there is, more than ever, a need to promote sustainability in clothing consumption, and by looking to the creation of the 1948 London Look, a fashion evolved by home dressmakers to suit London lives, we can perhaps rediscover the creative possibilities offered by a 'Make Do And Mend' mentality for the promotion of sustainable fashions.

As well as the arts and crafts of keeping a good home, women across class boundaries have made and designed things throughout their lives, often through combinations of need and pleasure (Buckley, 1990, 1998; Hackney, 2006). Home craft was concerned with creating decorative and functional objects for the home and involved skills that were traditionally perceived as feminine. In this era magazines, kits and activity groups were promoting the practices of embroidery, rug-making, crochet, needle and appliqué work, flower arranging and some simple carpentry and woodwork (Buckley, 1990; Hackney, 2006). Handicrafts, for example, were intended for those who desired to 'use their hands intelligently', as 'handicrafts demanded significant levels of skill, commitment, time, energy, application and creativity from practitioners' (Hackney, 2006, p. 19). More than just creating nice environments, these domestic and 'feminine' crafts contributed to a culture of home-making that safeguarded moral standards in private and, by implication, public life (Buckley, 1998; Hackney, 2006). The feminine touch 'in each home, together with the personalized quality of the handmade item contributed to the 'saving value' of the home'; a humanizing and civilizing element that protected against the encroaching demands and pace of modern life (Hackney, 2006).

Despite playing such an important role, located in the home, for the most part, such activities remained 'on the margins' but are central, nonetheless, to the formation of feminine identities and female subjectivity as well as to the making of homes (Buckley, 1998; Hackney, 2006, p.14).

Where such practices sat within the wider scope of creative practices was debated, however. Oftentimes, they were seen as domestic skills; at best a leisure activity or a hobby performed by amateurs. In a famous damning of Women's Institute handcrafts as 'a rarefied form of household husbandry ... a vision of craft void of the original political commitment, a vernacular ruralism with pretensions to decorative art', Greenhalgh demonstrates that a hierarchy within the crafts continues to exclude and undermine amateur practice (cited in Hackney, 2006, p.13).

Scholars of such home-forms of crafting have sought to recover these forms of home-based creative practices through various means. One route has been to note not only their role in keeping home, but also as a valuable revenue source. As Hackney and others note during economically troubled times (i.e. late 1920s and early 1930s) the growing range of crafts magazines and groups were full of hints and tips for turning amateur skills into viable businesses (Hackney, 2006; Hughes, 2012). For example, makers were encouraged to fill the gap between the mass-produced and the expensive prices of hand-made products in locations such as London's West End (Hackney, 2006, p. 70). Tensions persisted, however, about the respectability of such income orientated craft, so often the focus was placed on ideals of leisure and amateur work to create/make the home, thus masking the economic significance of women's earnings in the home. Another route to valuing these home-based creative practices has been to recognise the intersections of the making of the modern home with the making of feminine subjectivities and the promotion of well-being (Grace et al., 2009; Hackney, 2006).

Modern women were often seen to be plagued by two neuroses; either the suburban neurosis, wherein the housewife was in danger of becoming obsessed with 'keeping house', while the business girl sacrificed all to become 'efficient with a capital E' (Hackney, 2006). Craft was attributed therapeutic qualities and often seen to be ideal for staving off the nervous strain of modern living. As Hackney (2006) explores, magazine editors underlined its therapeutic and relaxing qualities. A women's editorial enthused, 'this is the simple, effortless sort of needlework we all love doing . . . just the thing for the busy woman who likes to have a piece of work on hand which she can pick up at will', (cited in Hackney, 2006). Writing in the magazine ModernWoman in 1935, Professor Winifred Cullis observed, 'what is needed is an outflow for nervous energy into other paths, and it probably is a self-protective instinct that makes a woman pick up her sewing or knitting while sitting still' (cited in Hackney, 2006). Cullis celebrated the pleasure to be had when producing 'beautiful', 'original' or 'quality' work. In addition to the satisfaction of simply 'having accomplished something', such qualities were to be valued in a world where creative instincts were increasingly threatened by the 'mechanical' nature of work. Thus the editors of interior, home craft and domestic features aimed to foster a sense of agency and self-determination in their readers (Hackney, 2006). The diverse range of styles meant that women could not avoid developing their own judgement and taste. As such, as well as being part of the making/creating the home through making, adapting and transforming objects for the home, such modern home creativities are also bound up with the creation and maintenance of the modern woman.

Home-making 2: do-it-yourself – making home, making masculinities

Do-it-yourself (DIY) has come to be associated with two key categories of activity, 'the making of objects' and 'the maintenance of the home' (Atkinson, 2009, p.2). So DIY might include 'repairs or additions to the home or garden, including installing a new bathroom or kitchen, central heating, putting up shelves, fixing a fence, building a barbecue etc.', while decorating gets a separated category 'internal and external painting, staining and wallpapering' (ibid.). As Gelber (1997, p.82) notes, DIY is a practice full of contradictions. It is 'not-quite-a-chore, that is, something useful undertaken voluntarily', but it is also leisure that is work-like and a chore that is leisurely; it produces outcomes with real economic value that might actually cost more in time and money than the product is worth; it is performed by middle-class men acting like blue-collar workers and blue-collar workers acting like middle-class homeowners. Pinpointing the origin of these activities is debated. For some, DIY's roots lie in nineteenth-century handicrafts, for others the term can be dated to a 1912 US magazine encouraging homeowners to carry out their own decoration. By the mid-1920s DIY was established as a hobby in the US, with the market growing much more quickly in the US than the UK. The mid-twentieth century growth of suburbia and home ownership saw a rise in the practices of home-making amongst men and women, and even when tradesmen could be afforded many turned to DIY as a means of demonstrating a sense of ownership over their newly acquired home (Atkinson, 2009). For example, between 1971 and 2002 homeowner-ship in Britain increased from 49% to 69% with most of the increase occurring in the 1980s, and this gave people both the means and impetus to add value to their property and make it their own. It was in the 1960s that DIY shifted from being practiced out of need, to being part of leisure time and a form of material cultural practice fuelled by a rise in economic prosperity (Atkinson, 2009; Powell, 2009). The buoyant housing market of the late 1990s also saw a boom in property focused TV programmes that supported and fuelled the development of DIY (Rosenberg, 2011). In 2000 in the UK 13% of time spent on house-related activities was spent in DIY/decorating, and generated about £12 billion in sales of tools and consumable products (Powell, 2009).

As with woman's home crafts, there is a sense of the transformatory nature of these practices; transforming both those who do DIY and the physical objects and structures on which they work. In short, these are fields of practice in which consumers are actively and creativity engaged in integrating and transforming complex arrays of material goods and themselves (Watson and Shove, 2008). A number of studies have shown DIY to be good for mediating and maintaining relationships within the family, enhancing self-esteem, and constructing self-image and identity. Miller (1997), for example, writes about the creative and therapeutic enterprise of transforming a council–owned house into one's own through the process of physical engagement

> the transformation of kitchens was regarded as a positive move that changed the relationship from one of alienation from 'council things' to one of a sense of belonging within a home *created from one's own labour.*
>
> (Miller, 1997, p. 17, emphasis added)

For others, the gendered dimension of DIY is crucial. The centrality of tools and skills enabled DIY to be positioned as a legitimate arena in which men can respond to the expectation that they should play a more active role in the home in the early to mid- twentieth century. Indeed, as Gelber (1997) notes, by the end of the 1960s DIY was an important part of the definition of suburban house-husbanding. Through DIY, male householders were able to renegotiate the relationship they had with their wives and their residences. DIY was available to men in the mid-twentieth century, as whilst based in the space of the home, traditionally a female space, it was also seen to be a practice that reasserted traditional male dominance over the physical environment through use of heavy tools. This permitted men to be both part of the house and also apart from it, claiming part of their homes as workshops, engaging with their family whilst retaining some sense of spatial and functional autonomy. It offered, in short, a role that suburban men created so that they could participate in family activities while retaining a distinct masculinity. There was, as with women's craft work, a certain palliative aspect to DIY. By World War 1, it was a 'revivifying hobby for the affluent, the nervous businessman would return refreshed to the office after a weekend of puttering at his basement workbench' (Jackson Lears, 1981, p. 65). For Gelber it was DIY's categorical fuzziness that allowed it to become so central to domestic masculinity. As he concludes, the justifications and satisfactions were multiple, permitting men, depending on their circumstances, to rationalise it as money-saving, trouble-saving, useful, psychologically fulfilling, creative, or compensatory (1997, p. 83). Within these contradictions, however, the aspect of creativity was never far from the definition, as with women's crafts DIY was seen as a cure for boredom or repetitive jobs. The qualities and pace of traditional artisan labour was seen as an antidote to the white-collar work. Skill was seen to take the place of thought, and 'fixed values (such as 12 inches today is 12 inches tomorrow, and a good joint once learned is a good joint forever)', was seen as a 'tremendous consolation in a world where the most fundamental concepts are subject to change without notice' (Starr quoted in Gelber, 1997, p.102). Thus while DIY had work-like qualities, indeed for some it was work, for others it was an exercise in creativity and forms of productivity not found in their own work, but yet requiring of skills such as planning, organisation, knowledge and skill, the same values necessary for success on the job. DIY was thus both leisure and work, but it was also a creative practice, with fundamentally transformative possibilities.

A central question remains here for DIY but also for the material practices of home-crafting, namely how is it that we can understand this mutual transformation of self and home? For some this is explored through an engagement with the recursive relation between products, projects and practices; for others, a concern with the embodied practice-based characteristics of DIY has been central. As Watson and Shove (2008) describe, the sweat, dust and frustration that occurs through the active combination of bodies, tools, materials and existing structures are all implicated in repairing, maintaining or improving the home. As this and other ethnographic studies demonstrate, DIY-ers often emphasised the ongoing and dynamic interaction between themselves, the materials being worked with and the home that was being created. DIY practice also created social connections and bonds, with practical know-how and related forms of folk knowledge being passed through informal networks of friends, as well as being exchanged through groups of 'expert' amateurs. In some accounts the evolution of tools, techniques and competencies are crucial to the growing popularity of DIY. As Atkinson (2009) notes 'a combination of availability, advances in tools and materials (MDF) and fixing technologies (glues), on-going training in schools combined with television programs have ensured that properties can be customised and that renovation of homes is normal and legitimate, in short DIY is something that ordinary people might do'. In these discussions, the role of the cultural support around DIY, from the growth of manuals and the internet, but perhaps most often television programmes that present easy house make-overs were often noted as being important for the growth of this creative material practice (Powell, 2009).

Home-making 3: moving homes

> Moving home involves an active engagement with its material cultures – photographs, memorabilia, furniture and so on. Moving homes also involves an active building and inhabiting a built form that reflects one's journey – of movement and settlement across borders, territories and space.
>
> (Datta, 2008)

Amidst those studies that explore the making of the home, there is a clutch of studies that explore the role of everyday creative practices in making migrant homes. Such studies reflect on decoration and the making and placement of objects as an intersection of the making of home and the negotiation and renegotiation of identity in the new spaces. As Cloffir (cited in Datta, 2008) puts it, diaspora is not only about travel and movement but more powerfully implies 'dwelling, maintaining communities, having collective homes away from home'. In such contexts, discussions of creating home – in material, aesthetic and symbolic artic- ulations – take on an added significance. As such, the aesthetics of the everyday become a site for imagining and building collectivities and for constructing and expressing ambiguities, conflicts, multiplicities and contestations in our identities.

Objects and decorative practices are understood to support the past and provide a sense of cultural continuity between the pre-migratory and post-migratory life on a personal and generational level (Savaş, 2010; Tolia-Kelly, 2004). Tolia-Kelly (2004) in her study of British–Asian homes explores how South Asian women create a familiar visual landscape through composing collections of photographs, pictures and paintings as part of the way in which they create home and identity in their new domestic environments. Other studies focus on how objects become points of identity negotiation in diasporic home-making. Savaş (2010) discusses the Turkish home in Vienna, as 'collectively created through shared aesthetic practices and discourses'. As she explores, with respect to Turkish migration and resettlement, efforts to construct a collective belonging to the new place of dwelling intertwine with the aesthetic and material practices of making the homes that are lived in.

Oftentimes, these material cultures and creative practices of decoration are based in the cultivation of memories and ties with past lives and places. Savaş (2010) describes decoration that develops a mixture of cultural objects as a performance of complex states of multiple belonging and plural identities. Objects from Vienna and Turkey are combined, with some migrants making very Turkish homes, using curtains, carpets, decor- ative objects and bedding often brought from Turkey or bought in Turkish shops in Vienna. She describes how the community manage what object counts as a Turkish object, and how groups of Turkish women will shop together, swop objects and make things in order to 'create a truly Turkish home', and in their collective activities produce and reproduce this idea of home. Creating a typically Turkish home through assembling a series of objects, decorative ideas and creating a particular atmosphere evolves as an aesthetic and social medium 'through which a coherent collectivity is imagined and performed' (Savaş, 2010). Creating a home becomes there- fore part of the performance of creating a Turkish collectivity, and failing to create a properly Turkish home is often regarded as a disruption of

Turkish solidarity. Creative home decoration is, however, not stable and also reveals a highly contested practice. Including shifts over time, as people settle in Vienna, their homes shift to mix Turkish objects with more local objects, signalling the evolution of Vienna as the major and permanent ground of belonging. As such, home decoration serves as a powerful aesthetic and social medium by which Turkishness is performed, debated and positioned in a complex web of diasporic relations.

CREATIVE HOME-MAKING: A CONCLUSION

To consider the home as a creative site is to explore not just the geographies of home-making, but also to reflect on how the home is constantly made and remade as a complex contestation of socially and economically reproductive capabilities. It is to engage with wider questions about the intersections of private and public, of living and working that assert the relationship of creativity to everyday life as well as to ideas of livelihood, and identity formation, perhaps especially questions of gender. Studies of hand-made objects sold on Etsy have offered an interesting means to cast and recast complex geographies of domestic production and reproduction (Luckman, 2013; Pace et al., 2013). Not only that, but it is widely claimed that the role of networking and online dissemination has been important in the renaissance of craft practices, with the networked home becoming a normalised paid-work location in this as well as other industries (Luckman, 2015). In the midst of these negotiations emerges the figure of the pro-am, who negotiates the role of their creative practices as amateur and professional, often mediated through the routes to market provided by online websites.

To reflect on creativity in the home is not just to explore the home as a site of the creative economy, but also to consider how such creativities entwine with those of hobbies and of meeting practical everyday needs. If the penetration of the creative economy into the home might be empowering for some, for others it might threaten to un-make the home. Vernacular creativities, however, demonstrate how creative making practices can be part of the making of home, but are also often about shaping our subjectivities. In short, it is not just home that can be creatively made and remade, but individuals, and communities too.

5

CLUSTERS, COLONIES AND OTHER GEOGRAPHIES

Colonies, clusters, scenes, quarters, districts, valleys, alleys; agglomeration, or the clustering of people, businesses and services within distinct often urban areas, occupies a key place within scholarly and policy maker understandings of the geographies of the creative economy. This work is overwhelmingly urban and focused on internationally known sites. If the creativity economy is to be understood as one of our great economic hopes, then comprehending its geographies and, in particular, the spatialities that help support and cultivate it are rather important. This chapter will turn attention to the spatial patternings and imaginaries that have come to shape the geographies of creativity, focusing on the logics of agglomeration that have been understood as dominant. It will reflect on the 'other' geographies that so often get left out of these primarily urban-based stories of 'geographically proximate group[s] of associated institutions in a particular field, linked by commonalities and complementarities' (Porter, 2000, p. 254). It will also explore the new landscapes that are being created in the name of creativity as a 'forcing' ground for the creative economy. Such spaces for the cultivation of creativity vary in scale and scope, from the hot-housing of creative incubators often based in single buildings, to the larger-scale creative quarter and right up to the illustrious sounding Media City. What emerges alongside studying these geographies is the evolution of ideas of creativity. Reflecting on the logics behind clusters and networks requires consideration of whether creativity is an asset of people or spaces, the impulse of singular individuals or a more distributed social phenomenon, and how and why it has been linked to the development of particular spatial grammars, including those that challenge the proliferation of the dominant logics of agglomeration.

While scholars may be interested in other geographies of creativity, clustering remains key to those policy makers, planners and creative mediators who believe that creativity can be cultivated as a force of (primarily) economic gain. Indeed, clustering has become so popular and such is the force of argument behind its effectiveness at supporting and stimulating the creative economy that it has become central to much policy in this area. So UK policy, from national creative sector policy documents (such as *Creative Britain*, DCMS 2009a[1], or *Digital Britain*, DCMS 2009b[7]) to creative strategy within Regional Development Agency and Government Office contexts in the UK, Europe and the US have promoted clusters within both urban and rural areas in attempts to develop creative spaces for both economic and (less often) social development. Cluster ideas are not, of course, a uniquely UK phenomenon and can be seen around the world, from Mongolia to Dubai.

At the heart of many of these scholarly and popular geographies of concentration and agglomeration sit two intersecting understandings of creativity and the creative economy. The first concerns the concepts of creativity that sit at the heart of these logics of concentration. The second relates to the nature of products and services offered by the creative economy. The understandings of creativity that underpin explanations of and justifications for clustering quite often challenge the myth of the individual as a creative genius. Conceptualizing instead a creative products industry whose production processes are exchange-based, successful creative outputs are understood as the result of intersections and collaborations between individuals with different tools, knowledge and skills. So as Leadbeater and Oakley (1999, p.14) argue;

> Cultural industries are people intensive rather than capital intensive ... Cultural entrepreneurs within a city or region tend to be densely interconnected. Cultural entrepreneurs, who often work within networks of collaborators within cities, are a good example of the economics of proximity. They thrive on easy access

to local tacit know-how – a style, a look, a sound – which is not accessible globally.

It is not just the people-intensive nature of creative production, but also the nature of creative products themselves, as an expression of changing economic conditions, that has come to be seen as a driver for clustering;

> agglomeration and related increased returns effects not only enhance system efficiency but also creativity, and perhaps nowhere more so than in the case of cultural products complexes ... Creative production is also unstable, unpredictable and continually evolving-meaning that frequent access to a large variety of relevant skills is paramount.
>
> (Scott, 1999, p. 811–12)

Indeed, advanced forms of capitalism are characterised by a revalorisation of creative ideas as a significant source of value and of difference. For, as many theorists of the creative economy detail, the saturation of mass markets characterised by 'infinite choice and intense competition' and the fragmentation of demand has stimulated interest in differentiated commodities (Hracs et al., 2013, p. 1144). Products associated with the creative economy – film, fashion, art, advertising and so on – are often symbolic goods, their aesthetic dimension becoming one way in which they 'stand out from the crowd' in terms of functionality and price (Hracs et al., 2013; Postrel, 2003). As lifestyle capitalism and consumerism continue to value the differentiation between products, the intangible aspects of the things we buy come to the fore and 'singularisation' becomes a highly sought-after element of products as consumers demand more and more difference (Grabher et al., 2008; Zukin and Maguire 2004). For organisations that produce highly symbolic and aestheticised goods whose psychic gratification is high relative to utilitarian purpose, there are blurred lines between fashion, innovative performance and market dynamics (Grabher et al., 2008; Hracs and Leslie, 2013). Aesthetic qualities in combination with skilled forms of branding and marketing are coming to override material and labour costs as the key markers and makers of value for a product. Success in the creative economy thus often requires an appreciation of how the immaterial and experiential aspects of products have come to override their technical and utilitarian aspects. In the context of market-led demand, not only is innovation constantly required but so too is a fast and responsive turnaround, both in the development of new products but also in the supply of existing ones (Amin and Thrift, 1992). The result is an economic argument for the clustering of firms and individuals contributing, as we shall see, to single sectors (e.g. fashion or media) or a wider creative ecology.

The return to localisation is often associated with the growth of the knowledge economy, and with it a sense of the need to comprehend 'the embeddedness and embodiment of economic processes in networks of interpersonal relations' rather than conceiving of them as 'representing some kind of free-floating logic or rationality' (Coe, 2000, p. 394). Using case studies from around the world, this section will build on classic work (e.g. Christopherson and Storper's (1986) research on Hollywood), to examine ideas around localisation in the form of clusters. Beginning from the discussion of Hollywood, so influential for many creative policies and such a classic example in the literature, discussion will unfold ideas of traded and un-traded interdependencies. Respectively these concern visible and formal linkages, as well as the less tangible, informal and the unexpressed connections that exist between cluster components (Storper and Christopherson, 1987; Gibson and Kong, 2005; Thomas et al., 2010; Henry and Pinch, 2000; Scott, 2000). It will explore concepts such as 'buzz' in relation to both historical discussions of localisation as well as contemporary geographical approaches to affect and atmosphere (Currid and Williams, 2010a; Storper and Venables, 2004). Furthermore, it will emphasise throughout how, whilst creative clusters might be seen as expressions of localisation, they should also be understood as constituted through 'complex interdependencies' that work 'within, without, and across' the cluster (Pratt, 2006, p. 1884). Throughout, discussion will draw together scholarship from economic geography on clusters with that from cultural and historical geography and beyond on scenes and colonies in order to tell a rich story of agglomeration that moves beyond the confines of either of these bodies of scholarship read alone.

The chapter will end by considering the growing body of work on 'other' geographies of creativity (Gibson, 2010). These other geographies are rich and varied and encompass creativity beyond major urban centres, as well as understandings of clusters and agglomerations that challenge ideas of the spatial and temporal logics of fixity and permanence. While 'marginal' and periphery creativities (such as those in provincial and rural areas or suburban creativities) will be explored in Chapter 8, temporary 'clusters' formed by events and trans-local networks will be explored here (Bain, 2013; Gibson, 2010; Harvey et al., 2012; Power and Jansson, 2008). With such 'other' geographies come critical reflections on what creativity is and does, challenging the dominant economic lens through which explorations of creativity's geographies are often framed and ensuring we appreciate the multi-faceted nature of creative clusters beyond their economic imperative.

HOLLYWOOD: THE STORY OF A PLACE AND A CLUSTER

> the positive connection of product image to place yields a kind of monopoly rent that adheres to places, their insignia, and the brand names that may attach to them. Their industries grow as a result, and the local economic base takes shape. Favourable images create entry barriers for products from competing places.
>
> (Molotch, 1996, p. 229)

Hollywood, a district in Los Angeles California, USA has become synonymous with the movie industry. It has also gained the status of case study zero in both scholarly and policy maker explorations of the creative clusters (e.g. Mould, 2015; Scott, 2000). The area became the capital of the movie industry in the early decades of the twentieth century. Film-makers from the east coast moved west to escape Edison's moving picture patents and, drawn by the sun and scenery, settled in LA, with the first motion picture being made in Hollywood in 1910. Hollywood's so-called Golden Age (1920–1950) was founded on what has been called the Studio System, and indeed its whole story, together with its inter-related geographies, is one of the movement from a highly geographically concentrated and centrally controlled production system to a more distributed form of production that favoured regional production in the context of global networks.

During the age of the Studio System, a very hierarchical arrangement persisted, with film-making dominated by the 'Big Five' studios (MGM, Paramount, Warner Bros., RKO and Fox). Each studio was vertically integrated, having control over production, distribution and exhibition. Film-making during this time was akin to a mass-production process

(Christopherson and Storper, 1986; Storper and Christopherson, 1987) from an efficiency in shooting to a controlling mode of managing 'stars' and ensuring a studio's films were shown. Nowhere was this seen more clearly than in the way that the 'Big Five' would control the distribution and exhibition of their films. As well as practicing block booking (selling a year's films to the theatres as one unit, both the good ones and the less good ones) the studios would also own the venues. In an extreme example, Paramount once owned every single movie theatre in Detroit, thus having a monopoly on the films shown in the city.

Hollywood's story is one of the changing organisation of film production in the 1960s and with it the geographical organisation of the industry, a shift characterised as a move from the Studio System to the Project System. The latter was a production system characterised by the temporary coherence of producers and services around particular projects, in this case films, with one result being that the majors began to lose control. Two factors have been seen to be crucial in this shift in production system; the development and rapid growth of television in the 1950s and the 1948 Paramount antitrust decision (Storper and Christopherson, 1987). In the midst of such competitiveness, uncertainty and instability studio-based mass-production collapsed. A new system grew up in which the majors, instead of controlling the whole process, became nerve centres in the midst of production networks. These production networks consisted of skilled employees from across the production process – producers, directors, writers, actors, musicians, camera operators, and so on – who had become freelance agents of their own labour (Scott, 2002). Alongside these individuals were large numbers of small flexibly specialised firms covering all subsectors of the industry, from sound to costumes as well as services such as catering. Production thus became a function of numerous companies whose scale of output is small and whose activities were specialised. In this new vertically disintegrated system the majors had lost control but were still key players in deal-making and gate-keeping in the industry, as well as in certain parts of the process such as financing and distribution (Christopherson and Storper, 1986, p. 317). As many suggest, in cutting their overheads the majors were able to explore new and diversified forms of movie production, pushing forward innovation and evolving in response to a higher risk Hollywood.

This form of project based working – or a contractual model of business activity – is very common across the creative economy (Caves, 2000).

The reconstitution of production systems as transaction intensive groups of complementary firms reinforces the agglomerative logics of production. The result is that location and especially Hollywood and the Greater LA area become important again. The project system meant that for each project a team for production and pre-and post- production was assembled. For the many small, specialised firms being present became vital as a way of getting in the frame for each new project. As the industrial entertainment complex of Hollywood grew it spread across the Southern California region. The area eventually became known for its 'motley crew' assemblage of different suppliers, able to form ad-hoc production networks that can mobilise around a project at short notice (Townley et al., 2009). Agglomeration thus enabled film production to break down into a multitude of detailed tasks carried out by specialised firms and sub-contractors. In an industry known for short-term precarious labour, spatial agglomeration was seen to off-set the instability of short-term contractual work, with the continual process of job search and rehire made easier and more efficient for both labourer and employee (Caves, 2000). Hollywood, once the focus for the studios, was no longer their home; MGM was based in Culver City, others including Warner and Paramount, moved into the San Fernando Valley (Scott, 2005). What remained in Hollywood were three things. Firstly, the movie premiers, it remained a social centre of the movie industry (now displaced to Beverly Hills). Secondly, it was the residential centre for stars, agents and executives. Thirdly, it remained the centre for smaller firms, especially sub-contractors, 65% of which are in Hollywood. As a result, Southern California gradually became a geographic space criss-crossed with intricate networks of deals, projects and relations between film studios, TV shows, musical recordings and multi-media products.

Recently discussion of 'New Hollywood' has sought to explore how this industrial complex based in Southern California has shifted with the growth of other film industries around the world (Scott, 2002, 2004, 2005). In a further important paper on the region, Christopherson and Storper (1986) note a 'split location pattern' observing a pattern of 'city as studio: world as back-lot', In other words, LA and Southern California more widely are important as a location for production, but only in relation to an industry that increasingly goes further afield to film. LA has remained, what could be termed after sociologist of knowledge Bruno Latour, a centre of calculation. Being in LA enables firms to keep tabs on the evolution of key projects, to hear about projects are shooting where and what resources might be required. It enables them to get the 'face' time needed with key industry executives to get in on the deals. In short, 'Los Angeles is the headquarters and technological centre for an industrial complex that has the whole world as its back lot' (Christopherson and Storper, 1986).

The jury is still out on whether an intensified global distribution of movie theatres has strengthened the cluster or weakened it. Indeed, some have claimed 'Hollywood is now everywhere ... production moves almost at will to find its most ideal conditions, and with it go skills, technicians and support services' (Aksoy and Robins, 1992, p. 9). For others, however, Hollywood risks being hollowed-out. Such hollowing-out is principally the result of the LA-based majors being subsumed within huge media empires (Aksoy and Robins, 1992). Whilst such media conglomerations with global reach seem akin to the earlier studio system, what also persists is the sense of subcontracting that now covers not only movies, but also TV, news, multi-media and theme parks. As well as acknowledging the global nature of the cluster in terms of these global conglomerations, we might also look to the role of satellite locations that decentralise production from Hollywood along what have been called creative and economy runaways. As Coe (2001, p. 1754) notes, the 1990s saw an increase in runaway productions, as studios took advantage of 'foreign locations where exchange rate differentials and government incentives providing considerable cost savings'. Canada, sometimes nicknamed the Hollywood of the North, has long been the home for Hollywood's 'offshore' location shooting. Between 1990 and 1998 somewhere between 63% and 81% of all runaway productions were hosted in Canada. In Vancouver, a major location for these productions, the sector contributed C$808 million in direct spending to the British Columbia economy in 1998 alone, and provided 25,000 full and part time jobs (Coe, 2001). This made Vancouver one of the leading film production centres in the world, albeit one dependent on its southern relative. A trend that will perhaps advance this hollowing-out of Hollywood is the growth of globally competitive film industries from the Global South. Alongside Nollywood (coming out of Nigeria, see Box 5.1) Bollywood is, of course, the classic example, with India producing more films annually than Hollywood, although the production costs on the latter's films are usually higher.

Box 5.1 NOLLYWOOD THE GREAT FILM GIANT

Sallie Marston, JP Jones and Keith Woodward

No discussion of global creative industries would be complete without reference to the commercial film industry. While it is not possible to describe the whole industry in this short vignette, an industry that receives far less attention than the other more famous 'woods' provides insight into the variety that characterises global film production and consumption and aesthetics. This variety applies to the range of elements of film-making and viewing from the producers, actors and directors who create them, to the visualities that characterise them, to the ways audiences have access to and enjoy them. Nollywood is one such creative industry that defies the usual expectations of Westerners for whom, in many ways, it is a trivialised if not completely invisible entity. Most film audiences – whether Western or Eastern – are aware of Bollywood films, made most often in India on lavish production budgets. They are distributed and consumed widely and are almost always viewed, at least initially, through theatrical release. Their global popularity has much to do with the large global diaspora of Indian peoples as well as their kitschy appeal to Western audiences. Their aesthetic is distinct: colourful, energetic, syncopated and hyperbolic. The production values are visually rich and the stories, where good triumphs over evil and the hero gets the girl, are uplifting if often unrealistic. And yet, despite their formulaic nature, it seems fair to contend that no one would mistake a Bollywood film for one made in Hollywood. This difference has largely to do with the target audiences for the films, both of whom have different narrative and aesthetic expectations for filmic experiences. Less well known among Western audiences, but arguably more globally popular than either Hollywood or Bollywood, is Nollywood, the film industry centred in Lagos, Nigeria. Nollywood produces well over 1,200 video-films a year as VDCs (video compact discs) and DVDs (for home viewing) that are viewed avidly, not only in southern Nigeria but across the globe, by the large diasporic African population. (There is also a film industry in northern Nigeria, called Kandywood, which appeals to the largely Muslim population in that part of the country.) Unlike Hollywood films, with their high production inputs, Nollywood video-films are made on inexpensive Chinese cameras with short production schedules, often by actors who are in the films but have little directorial training or financing, and with a minimum of editing. Sound and lighting are of variable quality throughout the film, making the visual experience of them uneven. Actors occasionally forget their lines, but their misspeaks are retained in the final cut because of minimal editing. (Within about a three-week period the video-film is shot and edited and is available for retail consumption.) As with Bollywood and Hollywood, there are widely recognised stars that populate Nollywood films, but they may be outnumbered by the untrained actors in the film. Supernatural events occur routinely in even the most secular narratives. The films are only sometimes shown in theatres; mostly they are viewed by way of DVD players in homes in Nigeria – where they sometimes play in a loop for hours or days at a time – but also around the world. As well, they are frequently passed from one person to another after the person possessing the film has seen it multiple times and is ready to view a new one. In Nigeria the films are most readily available for purchase through large urban markets such as the electronics market in Lagos. They might also be hawked on street corners, especially when newly released. Beyond Nigeria, in Paris, New York, Montreal, or university cities and towns where Nigerian students live and go to school, those not acquiring the latest films through relatives or friends, can find them in small and occasionally large African 'video shops' where they may be rented or purchased.

The jumpy camera and uneven sound and lighting, and the occasion verbal and physical stumbling of the actors often cause Western audiences watching the films to complain about their

Continued

quality and their disjointed narrative style. But Africans, no matter in Nigeria or wherever the diaspora has assembled, enthusiastically applaud the video-films as mundane reflections on their own lives and experiences where mistakes, mis-steps and accidents are common features. Jonathan Haynes, an expert on Nollywood, has recently written about the 'new Nollywood', seeing a potentially significant shift, though still rather small and fragile, in Nollywood's production, distribution and consumption practices. The shift began to unfold in 2010 and is in part a response to a crisis of over-production (and widespread and free television viewing) of the 'old Nollywood' video-films. New Nollywood refers to an emerging strategy among a small group of Nigerian film-makers to work with higher budgets on films destined for theatrical release in cinemas both in Nigeria and abroad. New Nollywood films are also being entered into international film festivals and thus competing not just for popular but for critical attention. While the future of new Nollywood films is still uncertain, these celluloid products represent a return of Nigerian film-making to the production and screening standards that exists in other African countries at the same time that new movie theatres – usually multiplexes – open in Nigeria's major cities and internet streaming becomes the norm.

Despite these international challenges, Hollywood persists as a result of a series of unique characteristics concerned mainly with local conditions and history. Scott (2004, 2005) notes the overlapping production networks within which the majors continue to function as important nodes in global networks. The local labour pool also plays an important role, already the site of large numbers of skilled individuals where there is constant churn from New York, the wider US and the world, all seeking the Hollywood dream. These production networks are also supported by an institutional environment that monitors and controls movie production across the globe, setting terms and conditions for filming locations, workers' rights, and global distribution of products (Aksoy and Robins, 1992; Coe, 2000, 2001). Historically then a regional milieu has evolved that has led to the creation of Hollywood as a particularly powerful locale for movie making, but also situated this industry in synergy with Southern California's other cultural products industries.

What this extended discussion of Hollywood offers is a glimpse into the workings of a cluster that has formed the basis for much creative economic policy making, but it also flags some of the main themes of the discussion of clusters more generally, themes that the remainder of this chapter is going to draw out in more depth. These themes include the benefits of clustering for creative production and distribution processes, which can be talked about in terms of traded and un-traded interdependencies between firms and even individuals. Also clear in the discussion of Hollywood was the role of global connections, in other words it was not just about connections within the localised cluster but also about the links beyond the cluster too. These will form key themes of discussion going forward.

Traded and un-traded interdependencies

In the discussion about clusters and why such agglomerations offer competitive benefits, two key groups of reasons are commonly identified; traded and un-traded interdependencies. Such interdependencies are seen to offer productivity gains, innovation opportunities and to support the formation of new businesses (Porter, 1996). In exploring productivity gains, Porter notes the value of clusters for providing access to specialist inputs and labour, knowledge-sharing and key institutions (e.g. universities, specialist finance, industry governance). He observes how clustering presents innovation opportunities through both the proximity of buyers and suppliers and sustained interaction with others in the industry. The latter is both a positive force as well as an important pressure to innovate in an environment where most competitors face the same basic costs. Porter (1996) also notes how important clustering is for new business formation, providing, for example, access to information and enabling resources for start-ups including everything from a workforce to liquid capital. As well as reduced barriers for entry, clusters also enable reduced barriers for exit, that is, takeovers and mergers happen easily. While clusters are often situated as a product of the knowledge economy's trend for localisation rather than weightless, footloose production, for some theorists of clusters it is useful to turn to historical perspectives to think through these ideas.

Neo-Marshallian nodes in global networks

> If one man starts a new idea, it is taken up by others and combined
> with suggestions of their own, and thus becomes a source of further
> new ideas. And presently subsidiary trades grow up in the neighbor-
> hoods, supplying it with implements and materials, organizing its
> trades conducive to the material of the economy.
>
> (Marshall, 1920)

Nineteenth-century British economist Alfred Marshall developed a localisation thesis based on his experiences of small firm districts in Lancashire
and Yorkshire in Northern England, including perhaps most famously the
cutlery and crockery industries. From his observations these were based
on local product specialisation, together with the vertical disintegration
of firms and horizontal integration that saw lots of small firms coming
together across the range of production tasks to produce the finished
product. As Marshall famously noted, these often one-product towns had
'something in the air', for rather than a single monolithic factory, what
was observed was a supportive, cooperative network of smaller firms
often dominating the economic profile of the area.

Seeking a way to make sense of the return to industrial agglomeration –
clustering – in the context of twentieth-century capitalism, so often
thought to be footloose, Ash Amin and Nigel Thrift (1992) turned to
Marshall's nineteenth-century localisation thesis. They are concerned to
explore what they term 'Neo-Marshallian nodes in global networks'.
Drawing examples from the creative economy and from finance, they
examine how the conditions of late twentieth-century capitalism effected
a shift in production towards fashion-led goods, meaning quick turn-
arounds, and a demand (especially at the high end) for small orders that
often require specialist features rather than mass-produced goods. They
describe the breaking down of formal hierarchies – the vertical disintegration of large firms – and the evolution of horizontally integrated sets of
smaller firms. These horizontally integrated firms can be responsive to
evolving patterns and trends, responding quickly and efficiently to
producer needs and changing consumer demands. A further characteristic
of these forms of capitalism is their knowledge-intensive nature and the
often high levels of trust needed for these forms of working practices. The
result is a valorisation of face-to-face business rather than of transactions

done at a distance. The key difference they note between these twentieth-century Marshallian nodes and their nineteenth-century expression is that the more recent clusters are built of complex interactions and exchanges (of people, knowledge resources) within that cluster, but also without the cluster too. These often global connections beyond the cluster are vital not only for doing global business, but also for keeping ideas within the cluster fresh, well-developed and hence competitive.

A core example of a 'Neo-Marshallian' District Amin and Thrift give us is the one-product town of Santa Croce in Italy, which specialises in seasonal-based footwear and bags for the fashion industries. The $10km^2$ town area contains 35% of all Italian leather production and its contemporary success is based on the nineteenth-century legacy of the tanning industry in the area. The area has succeeded through flexible specialisation and horizontal integration that witnesses many small firms all contributing different parts of the 15 to 20 phases of the leather tanning process. In 2008 the town had close to 1000 companies, with an average firm size of 12 employees, employing over 10,000 people, generating a turnover of €2billion/year. This was a steady increase from a decade earlier when the figures stood at 5000 firms, 6500 jobs and 860 million euros.

This horizontal integration of small firms is a relatively recent occurrence in the economic history of the area. In the 1960s and 1970s production was concentrated in a few older tanneries, reflecting the wider region's artisanal history. Cultural and state policies came to increasingly favour small firms, rural Tuscany's strong tradition of self-employment and small-scale entrepreneurialism supported the development of many independent firms across the area. What is more, local administration was for a long time communist and thus restricted the development of large factory units, rural banks favoured risk-minimisation strategies in their loan portfolio, shying away from large loans in favour of making many smaller loans. Further, the concentration of firms all specialising in different skills was ideally suited to the leather process, and with the evolution in the cultural products industry, tanners came to call themselves artists and combined design skills with latest chemical and organic treatments.

There are a number of examples in the Santa Croce area of how the traded and un-traded interdependencies enabled by the horizontal integration and flexible specialisation of the small firm system in Santa Croce has led the leather tanners in this area to be globally competitive. Cooperation across local firms enables each to specialise in a small part of the process, so some firms remove hair and fat from skins, others slip the hides, and others will flatten and dry them and dye them, before they get taken to the workshops to get made into products. Cooperation happens in terms of shared resources too, so various strategies are developed to enable firms to stay small and for costs to be kept down without losing productive efficiency. Sharing the bulk purchase of raw materials across a number of local firms is common, as is pooling resources to employ export consultants from Milan, to help them better market their leather and finished products. What is more, this flexibly specialised system of leather tanning firms are now supported by a whole set of services and other firms. This includes warehouses of international leather traders and offices of major import and export agents and brokers and customs specialists, as well as the depots of multi-national chemical giants and firms supplying specialist dyes and paints. There is also a growth of firms who manufacture and service the machines needed during the process, providing skilled labour, parts and even second-hand equipment and a number of international haulage companies and agents have moved into the area. The firm owners have explored how to extract further value from the product chain by, for example, developing a glue factory to use left-over animal products and a water purification depot that converts waste products to clean water and fertilizers. For Amin and Thrift (1992) what Santa Croce offers is a very good example of a one-product town, a Neo-Marshallian node in a global fashion network. Clusters are not always about whole towns and, of course, are not always about single products. Indeed as discussion will show, ecologies of firms and individuals from different sectors can offer rich support to one another, whether strung throughout a city or focused on a single building.

Buzz: the economics of a good party or thinking about atmosphere?

If many of the benefits of clustering seen in Santa Croce are traded, it is important not to overlook the role of un-traded interdependencies. A key way of thinking about these is through the discussions of 'buzz'. Currently discussions of buzz integrate two sets of ideas. Firstly, buzz is seen as being about the ideas and know-how needed to keep not only up to date, but also ahead of the curve in these competitive fashion-led industries. Secondly, buzz is seen as being focused on the social scene, or what Currid (2007b) has described as the 'economics of a good party'.

The understanding of buzz as a reason to cluster is not exclusive to the creative economy. Rather, buzz has become associated with the geographies of the knowledge economy more broadly. Two key elements of the geographies of knowledge are understood to lead to a valuation of buzz and hence of clustering. Firstly, the tacit knowledge that is central to the creative industries. Tacit knowledge is a form of embodied knowledge based around skilled practice and learning (explored further in the Chapter 2 on the body). For Michael Polanyi, one of the foremost thinkers of tacit knowledge, 'we know more than we can tell' (1966, p. 4). As generations of crafts people and makers have suggested, and contemporary sociological studies have reinforced, we may not be able to tell, but we could show (Leach, 1940; O'Connor, 2005, 2007). As such, proximity is encouraged by the transfer of skills and knowledge as well as shared services. By contrast, codified knowledge is, in sociologist Bruno Latour's terms, an immutable mobile, and thus can travel in the form, for example, of text in books or emails. If codified knowledge can travel, tacit knowledge requires localisation, hence clustering. The second characteristic of the knowledge economy that is important is the recognition that innovation, knowledge-creation and learning in these tacit forms requires and builds dense social interaction, 'network ties link actors in multiple ways' (business partnerships, friends, agents, mentors, suppliers, buyers etc.). What accounts of clusters of advertising in Soho London or on Madison Avenue in New York, or TV production in Bristol, or fashion in New York or Toronto, share are descriptions of information and communication ecologies, often linked around the world as well as seen in intensive localisations (Bassett et al., 2002; Mould and Joel, 2010; Pratt, 2004, 2006; Rantisi et al., 2006). In this ecology we find the intersection of different modes of formal and informal communication, including chatting, gossiping, brainstorming, in-depth discussion, strategizing and local broadcasting. It is, of course, not just the formal spaces of business where this networking and interaction is done, but also the street, coffee shop, hallway, water cooler, visits to restaurants and bars after work, as well as formal spaces of industry parties, meetings and other business transactions. Across the cluster this buzz generates,

> a certain milieu ... vibrant in the sense that there are lots of piquant and useful things going on simultaneously and therefore

lots of inspiration and information to receive for perceptive social actors.

(Bathelt et al., 2004, p. 38)

Such networks are not always based on positive, affirmative connections, rather the buzz of clusters can develop rivalries that might also be productive. Grabher's (2001, 2002) discussion of the advertising business of Soho, London, and its cluster of activities – ranging from graphic design, photography, music to film production and post-production – emphasises these rivalries. This 'ad village' came to rival Madison Avenue in New York for its concentration of activity. This imaginary of the village is not, however, based only on commonalities in the community of practice, but also in productive differences, competition and rivalries. So Grabher (2001) describes the idea of the heterarchy, wherein within the ad village we find a diverse range of ownership and organisational forms and business models and philosophies. The village imaginary he develops is not one built on the co-existence of different, static organisational forms, but a village driven to evolve by the rivalry between these organisational forms. This rivalry fuels sustained engagement and ongoing innovation, leading to the creation of new ways to organise, interpret and evaluate business activities. The result is an increasingly diverse set of organisational ecologies providing a varied 'gene pool for the evolution of new organizational mutations' (Grabher, 2001).

If some scholars foreground the benefits of buzz to creative economy production processes, then for others the benefits of buzz lie in its shaping of consumption dynamics. Currid (2007b) might quip that this is the 'economies of a good party', but as her study 'The Warhol Economy' (2007a) demonstrates with respect to New York, this is to take parties very seriously. It is to understand the collective economic force of events in the production of a social scene or milieu. While the formal and informal social life of the creative economy is often thought to be a by-product of successful creative clusters, as Currid demonstrates with respect to New York, it is actually a 'central force, a raison d'etre, for art and culture' (2007a, p. 4). Shuttling between historical comparisons of Andy Warhol's era and the turn of the millennium scene in New York, Currid emphasises the importance of creative synergies, relating stories of how crossovers, synergies and mixes characterise the city. Mixes that not only intersect different forms of creative production – art, film, fashion music, and so on – but also create intersections across production and consumption and across formal and informal activities. Ethnographic work at these social gatherings, as well as across the cluster more generally, enables her to assert the importance of social spaces and interactions. They are vital she notes, not only for creativity and knowledge transfer, but also for constructing relations and negotiating between producers, for the development and careers, as well as for sales and promotion.

Studying scenes

Often overlooked in these studies of economic clusters are the possibilities of literature from within cultural and historical geography and even art history that seeks to understand place-based social scenes, or the histories of the formation, development and sustained role of art colonies (Jacobs, 1985; Lübbren, 2001). Such texts are often timely reminders that such concentrations of creative practices are a coming together of not just policy and economic forms, but also aesthetic traditions and evolutions, counter-cultural politics, and the wider diffusion of science, technologies, and ideas. Oftentimes, the result of these texts as a better understanding of how such clusters, whether it be New York's Soho or Greenwich village in the 1960s and 1970s, Milan's Fashion cluster, or Swinging London in the 1960s offer not just economic benefit but have potent imaginaries that circulate the globe.

Research on scenes tends to focus less on singular product clusters – such as fashion – and rather more on the cultural ecology of the place. Swinging London during the 'long' 1960s, for example, has been the subject of much study. Rycroft's book-length exploration emphasises the 'long front of this era' (2011, p. 49); in other words, an appreciation of culture and creativity that moves beyond merely sanctioned 'high' culture to be both expansive and inclusive. As a result, such discussions take in iconic individuals and artistic groups – for example, in Swinging London the Op Artists or writers group Angry Young Men – as well as underground and subversive practices, whether these be poetic or the artistic support infrastructure such as display spaces or presses. Indeed, these accounts often make space for discussing the evolution of the vital infrastructures that supported the scene. In Rycroft's account of Swinging London this includes temporary but iconic events such as the Festival of Britain; galleries such as the Institute of Contemporary Arts; and counter-cultural groups such as the Underground Press Syndicate (UPS). Rather than present atomised elements of a scene, Rycroft's account, like others, is rich in the discussions of the ecologies of these creative cities and clusters under study. Furthermore, whilst focusing on a specific location such accounts fully appreciate those imaginaries that attend to place as not solely understood through locational fixity. Rycroft's study, for example, is a *geographical* imagination of Swinging London that is as attentive to its relationship to the provinces as it is to the evolution of transatlantic sensibilities in the context of the geopolitical shifts of the post-war era. Further, his account of London and its rise to cultural iconic status during this era is also bound up with an intellectual history of what he calls 'cosmic nature' (2011, p. 8). As Rycroft shows, this mid-century cosmology is produced and reproduced across a range of creative practices. Featuring textual analysis of underground publications and examinations of light shows and synaesthetic cinema, this account explores the intersection of new ideas of science, technology, and communications. A rich terrain is traversed, from Marshall McLuhan to J.R.R. Tolkien, taking in anarchist ecologist Murray Bookchin along the way.

Such rich historical and ethnographic stories are complemented by the innovative quantitative studies that have been emerging. Using contemporary GIS technologies, Currid and Williams, for example, geocoded a large dataset of images drawn from the Getty images database in order to study buzz (Currid and Williams, 2010a, 2010b). Using over 300,000 photos of 6000 events between March 2006–2007 in New York, they mapped certain dimensions of the scene based on the frequency and popularity of cultural happenings: film and television screenings, concerts, fashion shows, gallery and theatre openings. The hot spots in New York City were located in the areas around Lincoln and Rockefeller Centers, and down Broadway from Times Square into SoHo, in Los Angeles, the hot spots are Beverly Hills, Hollywood, and along the Sunset Strip.

The geographies of cultural scenes offer an interesting point of intersection of creative economic cluster literatures with cultural studies literatures. What might provide a further useful bridging point for thinking these two bodies of literature together in the future is recent work on atmospheres (Adey, 2013). Atmospheres in social and cultural geography have come to be understood as material-affective ecologies. As McCormack notes, an atmosphere is 'something distributed yet palpable, a quality of environmental immersion that registers in and through sensing bodies whilst also remaining diffuse, in the air, ethereal' (2008b, p. 413). As such, to think of atmosphere let us explore that palpable yet immaterial 'something in the air' that Marshall talks about, that hype around a place or a product, both in terms of buzz, but also the form of cultural cool it can be hard to vocalise. This encompasses not only creative production but also consumption, it can be engaged within physical location or virtually, it is part of the imagination of the place as well as its practical experience. It is also temporally specific, maybe built on an historical imagination of a place or can be as temporary as the energy that surrounds a party or an idea. However these localisations are explored, it is important to also appreciate some of the challenges of too much localisation, to appreciate how such clusters often function best as nodes in global networks.

BEYOND LOCALISATION: PIPELINES

> some places are able to create, attract and keep economic activity . . . [particularly] because people in those places 'make connections' with other places (Malecki, 2000, p. 341)

Recent literature on clusters has sought to augment the discussions of internal drivers for creative economy clusters with the recognition of the need for pipelines as a means to circulate ideas and practices from other networks (distant or proximate) into the localised cluster (Bathelt et al., 2004, Maskell et al., 2006). We see this with respect to the Hollywood example, where New Hollywood becomes characterised by global linkages. Clusters need a balance; too much of an inward focus is not a good thing for cluster growth or sustainability; to prevent stagnation clusters need to be nourished by links beyond their immediate area (Balthelt et al., 2004). Such over-embeddedness is associated with under-competition, less innovation, not enough new blood or ideas and a limited pool of customers (Grabher et al., 2008). The knowledge-intensive nature of the creative industries requires the continual development of new ideas in order to retain a firm's edge. As a number of studies have noted, this can be hard to do internally within the firm or even the cluster, rather there is a need to develop networks that transcend firm boundaries. Such networks are viewed as critical for accessing the vital resources of innovation that prevent stagnation and the over-dependency on suppliers.

One of the most common means of accessing beyond firm knowledge is to engage with trans-local networks and global pipelines of ideas and personnel (Grabher et al., 2008). These are seen as an important source of novel ideas and also as providing exposure to practices in distant markets and industries. These pipelines are imagined as 'pumping' knowledge in from outside to generate a more dynamic local buzz. In contrast to buzz, however, that is understood as 'frequent, broad, relatively unstructured and largely automatic, pipelines function in a very different way' (Bathelt et al., 2004, p. 40). As they detail, controlling and obtaining the information is a negotiation and 'the establishment of global pipelines with new partners requires that new trust be built in a conscious and systematic way ... [this] takes time and involves costs' (p. 41). Global pipelines access new knowledge and then disseminate this knowledge through intra-cluster connections enabling the facilitation of cluster wide-learning and adaption. Non-local buzz, extra local linkages, are used therefore to create knowledge and produce local growth partly through an intensification of local buzz. As Bathelt et al. (2004) continue the 'more firms of a cluster engage in the build up of trans-local pipelines the more information and news about markets are pumped into internal networks and the more dynamic the buzz from which local actors can benefit' (p. 41). It would therefore be misleading to create a global-local opposition; such a dialectic belies the interlinked nature of the local and the global in developing and maintaining successful clusters. Furthermore, there is a sense that the pipelines need to come from outside the cluster, whilst for many that is a spatially distanciated outside that is often global in extent, for others, as the example of the fur industry in Montreal, Canada makes clear (Box 5.2), this can be a cognitive rather than geographical distance. It seems therefore that what is needed is a balance between the intra- and extra-cluster networks, as well as networks that operate with either geographical or sometimes, cognitive distance.

Box 5.2 OVERCOMING COGNITIVE DISTANCE: LOCAL PIPELINES IN MONTREAL, CANADA

Writing on the fur cluster in Montreal, Canada, Rantisi (2014) notes a rather different geography to extra-cluster networks, one that is less global and more local. In her studies she is concerned with extra-cluster local networks, which she terms local pipelines, connections with 'spatially proximate but cognitively distant industries' (2014, p. 956). Talking to the Fur Council, the institutional body for the fur industry, Rantisi notes how it has sought to link traditional fur producers with fashion designers in order to enhance the level of cross-industry knowledge exchange and help the traditional fur industry become relevant for the global fashion market. In this example, Rantisi argues that while distant knowledge has often been sought to nourish creativity, it can also be found in other geographical proximate industries. As Rantisi notes, the pipelines are especially important to this industry, given that market demand for fur has changed and the furriers have been slow to respond. Interestingly, she draws a distinction between a craft industry focused on tradition and an industry that might be more shaped by trends, so she notes the need for the furriers to work less with traditional designs and rather with more fashionable practices (2014, p. 960). Focused on producing good fur, aesthetic innovation and creative designs were long considered secondary to the industry, this alongside the small size of their forms and their tight-knit nature have limited possibilities for innovation and change in modes of design and production.

The kinds of local pipelines Rantisi describes encompass both public and private initiatives. She recounts initiatives such as an annual design competition for fashion students that sees the winner working with the furriers to have their design made and then shown at a design fair. This model evolved to include a workshop or a 'laboratory' for the winners, putting fashion designers into training with advanced furriers to learn skills and techniques of handling fur. As well as teaching students and established designers about fur, the Fur Council also sought to ensure that the furriers approached the fashion sector. They introduced a trend seminar so the furriers see how their craft could engage with trends in fashion and pioneered a programme 'Fur Works' and an eventual label, to help support the development of these links. Networks then are needed to prevent firm lock-in; if they are only very proximate, however, they can lead to stagnation as a result of insularity that limits exposure to new ideas.

'OTHER' GEOGRAPHIES OF CREATIVITY

Academic work on creativity has an often unacknowledged urban, western bias (Edensor et al., 2009a; Gibson, 2010; Sorensen, 2009). Recently, however, there has been a growing body of work on 'other geographies' that overturns the dominance of the urban-based creative cluster, and seeks to draw attention to other geographies of creativity, not as lesser imitations of clusters, but as alternative forms of arrangement that offer their own strengths (Cole, 2008). On the one hand, research has explored the spatially diffuse and often cross-border networks within which clusters operate (Coe, 2000; Kong 2000a; 2005). On the other hand, research has explored the potential of temporary clusters formed by fairs, festivals and trade shows (Maskell et al. 2006; Norcliffe and Rendace, 2003; Power and Jansson, 2008). A further theme of research has attempted to correct the urban bias of existing work by exploring how creativity and the creative industries emerge in small, rural, suburban and remote places (see summaries in Gibson, 2010; Harvey et al., 2012). Such an attention to what Gibson (2010) calls the 'tales from the margins' explores the range of social, cultural and economic transformations that occur in these localities, as well as reflecting on what the forms of the creative economy in such places suggest about understandings of creativity more broadly.

Temporary clusters

If the cluster discussions that opened this chapter focused on permanent clusters, involving by and large the permanent co-location of business, organisations and individuals, then a growing body of literature asserts the value of temporary event-based clusters to both the production and consumption elements of the creative economy (Norcliffe and Rendace, 2003; Power and Jansson, 2008). Within the context of such literature, trade fairs, shows, conventions and festivals and other temporary industrial gatherings, are both understood to offer the benefits of clustering over a shorter time-frame, but are also understood to be vital spaces in their own right. Trade fairs have been likened to temporary clusters 'because they are characterized by knowledge-exchanging mechanism similar to those found in permanent clusters, albeit in a short-lived and intensified form' (Maskell et al., 2006, p.2). This body of work flies in the face of the marginalisation of temporary and event-based economic phenomenon within economic and industrial geography. For those looking for differences between temporary and permanent clusters, these temporary spaces offer modes of knowledge exchange, marketing processes and value-making that have very different dynamics to those in permanent clusters.

Temporary clusters can be understood to offer another alternative to global pipelines. Futhermore, as Power and Jansson (2008) make clear, these temporary events are not isolated from one another but are instead arranged in an almost continual global circuit. The regularity and period-icity suggests to them that what should be explored less as temporary clusters but rather cyclical clusters, 'complexes of overlapping spaces that are scheduled and arranged in such a way that spaces can be reproduced, re-engaged and renewed over time' (p. 423). As they note, the events themselves are short-lived, but their presence in the annual business cycle has consequences for creative economies of production and consumption. This section will explore a cross-section of different cyclical clusters, from furniture fairs, to fashion weeks and film festivals. It will examine the different roles that these events come to play in the support and development of the creative industries and the creation of value for creative industry products.

Cyclical clusters: from trade fairs to film festivals and fashion weeks

It is important to understand that for many who study temporary clusters these events are not secondary to, or some lesser alternative of, permanent clusters. As Power and Jansson (2008) note of trade fairs in furniture industries, they are rarely sidelines to the ' "real" business of the milieu in which they are based, but rather are central to it'. Their centrality is understood to be a function of their continuously cyclical occurrence and the 'multiple overlapping spaces' that are created at such trade fairs. These spaces – including sites for negotiating sales and contracts, for the formation of network and symbolic capital as well as for knowledge exchange and recruitment – are of considerable value to the industry.

Whilst attention has often focused on urban clusters, Power and Jansson (2008) Harvey et al. (2012) and others note how the creative sector is often distributed throughout clusters of small and medium-sized businesses that are often located in rural areas. Indeed, Power and Jansson (2008), in their study of Milane Salone, the world's largest furniture fair, note its importance to largely rural-based Swedish furniture firms, their core study group, which were often located in small rural clusters. The Salone, held annually in Milan has over 2 million square metres of core show, with a series of fringe venues that cater for more informal shows and events. The fair started in 1961 to showcase Italian Design. Since then the fair has grown to be a global thing, attracting over 300,000 people a year to see the products of over 3000 firms, 700 young designers at a satellite event, numerous others in the pop-up events around Milan. It is a key part of the global furniture industry, which is worth 80.7 billion Euros in Europe alone, employing close to a million people.

Fairs can come to structure the business year, creating a series of deadlines for new designs and high-quality proposals. Oftentimes, the fairs are more than start-ups can afford, especially if they want to attend more than one, so a fair will be chosen based on business aspirations and visited annually, renewing contacts and building new ones. Many large firms do five or ten fairs a year. A huge set of costs are associated with these events, from securing the stand space, the design and then the event attendance, but it is considered worth it for the negotiation and knowledge exchange opportunities that they offer (Power and Jansson, 2008). Drawing on studies of trade fairs, film festivals and fashion weeks it is possible to identify a series of overlapping spaces and activities that ensure these events have a key place in the production and consumption of the creative economy.

Firstly, these spaces are often viewed as market places, where purchases and contracts are made or negotiated. Given creative products often need aesthetic appreciation and experience of the quality of materials and craftsmanship in order to be valued, being able to see the product at first hand is considered important. Negotiating for new business is often seen as a key part of the activity at these fairs and the face-to-face conduct of business at trade fairs was seen as especially beneficial for promoting a range of business connections.

Sundance Film Festival, held every January in the ski resort of Park City Utah, is one of the largest independent film festivals in the US, attended by nearly 50,000 people in 2012. Focusing on Indie films, the festival has evolved from a small event for low budget indie films, into a media and celebrity spectacle. Whilst the opportunities to accrue network and cultural capital are important (see below) one of the key functions of Sundance remains its role as a buying venue. Most films arrive at Sundance without a distributor organised, but rarely, it is said, do the good ones leave. In 2011 45 films acquired distributors during the ten days. In 2010, 9 out of the Sundance films nominated as ones to watch won 15 Oscars, and 4 out of 5 of the documentary winners were Sundance films.

A second role that temporary clusters play is within the creation of network capital. They present the chance to make new contacts, refresh existing ones and negotiate networks. Regular annual contact enables companies and individuals to maintain their relationships with customers, suppliers, competitors and the press. As Power and Jansson's (2008) study of the Swedish furniture firms visiting Milan noted, most would attend the fair with a series of pre-arranged meetings with buyers and sellers, in dedicated created spaces on the stands to enable sitting, talking and enter-taining. Such formal pre-arranged meetings mesh with the informal, chance introductions that are also an important part of the event. On returning home considerable time is spent following up on contacts made. As important as formal events are, the informal networking that happens at the off-site events and the after parties are also deemed to be a crucial part of these events. At the 2007 Salone there were over 200 sanc-tioned off-site events in the two weeks around the fair, including official exhibition openings, gala parties, dance nights, and cocktail evenings. Many of these were invite only or had select guest lists, or otherwise traded off a sense of underground and 'those in the know' to attract audi-ences. Important for the furniture fair, and especially as we shall see the fashion weeks, was the creation of symbolic capital that came with being involved in the fair. Catching people's eye in Milan was a good way to gain attention throughout the rest of the year.

Thirdly, network capital was not the only form of capital that would be gained from attending trade fairs. As Entwistle and Rocamora's (2006) study of the annual fashion week circuit demonstrates, symbolic capital is vital to these events. The twice-annual international circuit of fashion weeks begins in New York and ends in Paris, taking in London and Milan along the way, and often involve designers making a net loss. Their function is not principally sales but rather what they have called after Bourdieu 'field-configuring' events. As they note, these are key moments in the life of the fashion industry as well as key instruments in fashion branding a city, their 'focus is to produce, reproduce and legitimate the field of fashion and the positions of those players within it'. As showcases for the up-coming season's pret-a-porter (ready to wear) clothing, these are often spectacular events, sensory experiences which create atmospheres around particular collections with models, clothes, props, music, choreography and narrations.

Through ethnographic methods Entwistle and Rocamora (2006) mapped the key agents and institutions within the field of fashion, including designers, models, journals and buyers, fashion bloggers, styles and celebrities as well as students and others on the margins of the field. Based on their observations they argue, after Bourdieu, that the fashion weeks are the physical realisation of the wider field of fashion, bringing together participants into one spatially and temporally bounded event. In drawing these individuals together it renders visible wider field characteristics, especially as they note, 'field boundaries, position taking and habitus'. Such renderings of the field are seen as vital in the maintenance of the industry. The cat walks, locations and parties are arenas for the accumulation and consolidation of the value of social, symbolic and economic capitals.

A fourth characteristic of these temporary clusters is as spaces for knowledge diffusion, and as sites for the flows of ideas, knowledge and innovations. As Bathelt et al. (2004) note, trade fairs are 'a rich arena for inter-firm learning processes.' The following knowledge activities were found to be consistently important as part of the trade fair experience: seeing competitors' collections, new innovations in technology and materials and seeing and meeting new designers. As Power and Jansson (2008) note, the experience was not always positive, so at furniture fairs mass-market furniture producers were seen as spies.

Fifthly, trade fairs are often seen as important recruiting spaces; specialist sector trade fairs are crucial not only for flows of knowledge and skills, but also flows of employees and sub-contractors. For those working outside of urban or key production and education centres, it is not always easy to find the skills needed locally so international trade fairs offer a good chance to reach new talent, especially as so many of these fairs have special sections devoted to young graduates where young talent is showcased and people can be recruited straight from the fairs.

A further dimension of the discussion around fairs and festivals to take into account is the relationship between these events and place-making. Like clusters, it is worth engaging with the relationship between these events and the places in which they are held. Whilst in some cases these might appear to be free-floating events, often more culturally inflected studies demonstrate the co-constitutive nature of festivals and place. Events such as fashion weeks or music festivals add to the branding and cultural capital of a place, whether that be rural regions and music festivals such as Glastonbury, or the Sur Le Niger music festival in Mali (Box 5.3), or whether that be the mutual reinforcement provided by fashion weeks in the major cities on fashion circuits (Gilbert and Breward, 2006). As such, events can help support and cultivate a creative scene, boosting the profile of a place and helping attract other creative practitioners and visitors both during the festival as well as at other times of year.

Box 5.3 FESTIVAL SUR LE NIGER, MALI: CELEBRATING CULTURAL HERITAGE

Festival Sur Le the Niger – 'a folk festival for the people' focused around a rich program of dance, music and art as well as discussion forums and conferences – is held every year in early February. The five-day festival was started in 2005 by local civic leaders in the city of Ségou, in Mali; in 2012 it attracted around 26,000 visitors, around 4,500 of which were foreigners.[3] A core aim of the festival is to bring together traditional and contemporary forms of arts practice in a multidisciplinary cultural event. The festival has grown up around aims to 'create a unique event that highlights the attractions and potential of the city and the region: a festival that both supports the local economy and spotlights the unique arts and culture of Mali'.[4] There is a clear sense from the outset that creativity and place are tied together in a celebration of rich indigenous cultures. Traditions are combined with an evolving sense of the creative economy to provide the means to support local communities and economies. The localness of the festival is very important; as such it contrasts to those mega events, such as the art biennales, which have been critiqued around the world for their shipping in and shipping out of cultural elite, with little benefit to local practitioners or communities. In

Continued

2009 a foundation was started to support the festival and ensure that its founding intents – to recognise the social value of arts and culture and support the growing creative sector of Mali – continued to be realised.

The festival is now an internationally recognised event and it has put Ségou on the map nationally and internationally, enabling the development of a tourist infrastructure, including a series of hotels, restaurants and other attractions which are now popular during but also outside of the festival season. Over the past five years the festival has developed a housing system that encourages festival goers to stay in private homes (200 families were accommodated in 2008), boosting income for local residents and fostering international friendships.

As well as service jobs, the festival has also supported and helped develop existing creative businesses. A women's weaving cooperative of ten local women pooled money made from selling handicrafts during the festival to buy a loom. A few years on and their cooperative has grown considerably, not only being able to use the loom for other things, but also to work directly for the festival making costumes and backdrops. They emphasise, however, that they are careful not to monopolise the festival profits and let all textile workers share in the work.[5] The spirit of the festival that was developed along the lines of 'Maaya', a humanistic vision based in a combination of modern management styles with social, economic and artistic development.[6] Locally, and by the UN, the festival is seen as an archetype of sustainable development for the region, as economic benefits combine with the continuing support and development of traditional skills and benefits for the area, as well as creating a platform for political and social debate.

Importantly, too, local artists gained access to international and national peers and markets, and new markets were created for cultural products and local people were up-skilled as both artists but also as technicians able to organise and run large-scale events. A key part of the festival concerned international music exchange, helping Mali's traditional musicians become known in the international world music scene, and also to arrange exchanges with local groups from Mexico, Portugal, and so on.[7] Prior to the festival, locals say that Ségou was in the darkness; local musicians credit the festival with making them famous, enabling them to pursue their music full time. One musician later opened a record store and developed a training programme to teach young boys the traditional instruments and forms of music-making, now they want to learn they respect the traditions. The hallmarks of the festival are to see the fusion of traditions of these places with international trends and economic potential, that together will support both the creative sector in Ségou, and Mali, but also help promote wider economic and social development. There are clear challenges too, including worries over the hybridity of culture, the growing force of the very seasonal tourist market and the precarious nature of dependence on a festival. It is hoped, however, that eventually the festival will help broaden the economic base and develop a rich cultural ecology to support the region more widely.

BEYOND ECONOMIC ACCOUNTS OF AGGLOMERATION

Agglomeration, whether permanent or temporary, has become a key spatial logic of the creative sector. As has been discussed, it should not be considered the case that festivals, fairs and trade shows are considered secondary or a less desirable option to a permanent cluster, rather they enable a range of things related and unrelated to the functions of more permanent clusters. Not only is business done and contacts made but symbolic capital is accrued, configuring the key players and ones to watch in a certain field, with companies investing huge amounts of money to ensure that their presence at these events is a lucrative success. Given the policy focus and economic imperative of creative clusters within the creative economy, it should perhaps come as no surprise that economic and policy based scholarship has come to the fore in these discussions. Yet, as has been demonstrated here, there is a range of cultural and historical work on creative colonies and scenes that might add other dimensions to these cluster discussions; enabling consideration not only of the wider social and cultural context of these clusters, a sense of the relation-ship between the aesthetics of the products produced and the location of the cluster, as well as an appreciation for the wider cultural ecology that the specific cluster is a part of. Perhaps, more importantly though, such culturally inflected studies are more likely to attend to the mutually co-constitutive relationship of creative clusters and places than is other work. In other words, to consider both how such clusters are shaped by the places they are situated within, but also how such clusters shape the places and communities in which they are based. Chapter 6 will take up some of these ideas in relation to the intersections of creativity and community.

NOTES

1   DCMS (2009a) Creative Britain: New Talents for a New Economy. Available from: http://webarchive.nationalarchives.gov.uk/+/http:/www.culture.gov.uk/ images/publications/CEPFeb2008.pdf [last accessed 5/6/2016].
2   DCMS (2009b) Digital Britain. Available from: https://www.gov.uk/government/ uploads/system/uploads/attachment_data/file/228844/7650.pdf [last accessed 5/6/2016].
3   Information on the festival from entry in United Nations Creative Economy Report 2013: Special Edition - Widening Local Development Pathways. Available from: http://www.unesco.org/new/en/culture/themes/creativity/creative-economy-report-2013-special-edition/ [Accessed 25/7/2015].
4   www.festivalsegou.org [Accessed 28/7/2015].
5   Information on the festival from entry in United Nations Creative Economy Report 2013: Special Edition - Widening Local Development Pathways. Available from: http://www.unesco.org/new/en/culture/themes/creativity/creative-economy-report-2013-special-edition/ [Accessed 25/7/2015].
6   http://www.festivalsegou.org [Accessed 28/7/2015].
7   Information from http://www.doenculture.com/222/en/the-festival-sur-le-niger-brought-us-back-to-life [Accessed 25/7/2015].

6

COMMUNITY: MAKING PLACES, CREATING COMMUNITIES

Summer 2014 in Central London and it is pouring with rain. A summer fair by the Thames in the ward of South Westminster, a stone's throw from the Houses of Parliament is a wash-out, but still people come. The event is a riot of colour and activity including two tents that feature drawing materials, sound recorders and video cameras. 'Draw the change you want to see', visitors to these tents are urged, 'design a video game to improve your community', the attendant artists say to interested families and passers-by. Some stop and sit down; they draw a lack of open space, the closure of public services, the pollution of the roads and rivers, the growth of second homes and the decline in affordable property and the rental market in the area. As they draw they talk together about their places and communities, they worry in unison over the common problems and challenges they face and they strategise ways to help make their urban space a more liveable, happier and more practical place to live. After they have finished their drawings some go on their way, others join the group of artists in the Tate Gallery the following week to see the drawings on display and play the video games that have been made from their discussion of the problems of their communities and the alternative futures they imagine for them.

Creative practices have the potential to make places and build communities. Indeed, amidst the popularity of phrases such as 'making is connecting'

(Gauntlett, 2011), the drive of participatory art towards community building and the visions of planners to build creative places, are a range of ways that creative practices shape communities. Such debates are importantly not just about professional arts practices but have come to encompass other kinds of creativity too, including everyday and amateur creativities; such that everything, from collective gardening to amateur theatrics, is understood to offer the means to connect communities and forge senses of place. As this chapter will explore, there are interesting tensions between these different practices; from the planning of creative places and communities in recognition of creativity's value as an economic and social force; the different vernacular and hobby based creative practices that help build community relations, and a range of ways that different forms of professional arts practices might be understood to make communities and shape social relations.

Creativity, however, can also have a negative effect on place and communities. As discussions unfold in this chapter they will reflect on this darker side of creativity. Dark creativity is most often associated with terrorism; however, we might also think about how creativity policies might be less about building community than destroying it; how arts practices have become instrumentalised to address some social needs whilst turning away from and even covering over others. We might also think about the forces of gentrification with their homogenising effect and the complicacy of many public arts projects and artists within this. Or, we might think of class divisions that emerge around different forms of vernacular creativity, such that what is for one community a positive affective force, for others is a kitsch, tasteless display. Such tensions make, as this chapter will explore, for a more complex set of stories of creativity and the making of places and communities than we might expect. The chapter will begin by exploring the planning of creative places, before going onto offer critical reflection on the idea of making as connecting, exploring both collective doings and consumings of creativity. It will close with an exploration of art interventions, reflecting on changing modes of intervention and the benefits and challenges of such practices for making places and shaping community relations.

PLANNING CREATIVE PLACES – TAKING PLACE SERIOUSLY?

> In creative placemaking, partners from public, private, non-profit, and community sectors strategically shape the physical and social

character of a neighborhood, town, city, or region around arts and cultural activities. Creative placemaking animates public and private spaces, rejuvenates structures and streetscapes, improves local business viability and public safety, and brings diverse people together to celebrate, inspire, and be inspired.

(Markusen and Gadwa, 2010)

Discussions around creativity and place-making are complex. This is well illustrated in the difference between the potential of creative practices to make place discussed in the quote above and the results of studies of gentrification in which neighbourhoods once full of local colour are seen to be rendered bland and homogenous. The quote above opens a report co-written by geographer Anne Markusen on creative place-making commissioned by the National Endowment for the Arts, the United States Conference of Mayors and the American Architectural Foundation (2010). As well as being interesting to note that academics are increasingly working as creative industry consultants in such a way, the commissioning of the report signals the wholesale interest in the potential of creative place-making as a force for regeneration. As with other reports and articles the focus of Markusen and Gadwa's report is how strategies of creative place-making can develop the much sought-after combination of liveability, vibrancy and economic benefits, including helping cities become more competitive locations in the battle to attract the creative industries. While claims to heightened public engagement, community identity, environmental quality and increased collaboration between civic, not-for-profit and for-profit partners are all important, what tellingly gains most attention are the claims that creative place-making 'fosters American leadership in globally competitive industries' (2010, p. 3). Notably place-making has become a popular strategy around the world but especially in the US where it has formed the focus of much cultural policy since 2000. The top funders are National Endowment for the Arts, and Art Place (a collaboration between 13 foundations and 6 banks), who together made a total of 232 grants across 50 states, investing nearly $50 million (Nicodemus, 2013). So valued has the practice of creative place-making become that some US states have channelled all their arts funding into creative place-making projects and have even doubled the funding available (Nicodemus, 2013).

The benefits of creative place-making policies are often seen to lie in their uniting of the very traits of creativity. On the one hand, they broaden the scope of cultural policy through a focus on wider economic and social issues and, on the other hand, they expand the narrow focus of the creative economy through cross-sector partnerships. Such partnerships have grown hugely of late, challenging civic stakeholders, artists and arts organisations to forge collaborations and discussions that examine the potential of relations between the creative economy, cultural activity and community development.

Clearly related to the geographies of creative clusters and cities more generally, creative place-making at a range of scales tends to develop a 'decentralized portfolio of spaces that are designed to act as creative crucibles' (Nicodemus, 2013). These spaces encompass a range of functions and, ideally, as with the clusters discussed in the previous chapter, support both creative production and creative consumption. Thus flagship buildings such as art centres or art galleries are situated alongside the provision of live-work spaces and affordable studio spaces created from vacant lots or under-utilised buildings. Attention is often paid to the need to found and support facilities to train the next generation of local creative talent, and the support and development of social infrastructure designed not only to display and sell local productions, but also to support the fostering of social connections. Such planned projects range in size and scale, from the top down approaches found in Singapore (discussed below) and the Guggenheim effect (discussed in Chapter 7) to the more community orientated engagement based around cultural assets that meet local needs, for example the programme of social urbanism in Medellín, Columbia (Box 6.1).

Creative place-making is, however, rarely an unqualified success. Interestingly, often the failures of such strategies but also their successes can be traced to the engagements the plans have with the history and current condition of the places being 'made'. As Evans and Jones (2008) point out, urban redevelopment strategies can tend to treat place as a tabula rasa. What needs to be queried, they observe, is whether it is possible to celebrate existing dimensions of place within development strategies, rather than treat urban landscapes as blank spaces to be redeveloped. Effacing material landscapes and a community's embodied connections to place not only disrupts original communities, it also perpetuates visions of these new communities as sterile and placeless. By contrast we might want to emphasise the need for projects to demonstrate a commitment to place and sensitivity to its traditions and its distinctiveness. This distinctiveness might be constituted by a place's material fabric and its creative traditions, or its social and cultural characteristics, its histories and the imagined futures of its communities (Markusen and Gadwa, 2010).

The story of Singapore, as a city attempting to rebrand itself a Global City of the Arts is a good case in point here. Singapore, with its soft authoritarian government has, like so many other cities around the world, turned to creative place-making practices (Chang, 2000; Kong, 2009; Ooi, 2008). In the last decade it has invested over a billion dollars to support and develop cultural infrastructure and the creative economy, seeking to brand itself as a cultural hub in the region. But yet, as a number of scholars working in the region have made clear, there are some inherent tensions here. Creative cities are places known for their diversity, tolerance and creation of conditions for creativity and experimentation. While Singapore is known for its paternalistic authoritarianism, looking to persuade rather than coerce, it is still some distance from the forms of democratic governance more normally associated with cultural policy.

Singapore's creative place-making strategy was stimulated by a report prepared by the 2001 Economic Review Committee set up by the government. Seeking diversification away from an industrial economy towards an innovation-fuelled economy, what was valued were ways to 'fuse arts, business and technology ... the city-state must harness the multi-dimensional creativity of [its] people' for its 'new competitive advantage' (Economic Review Committee report cited in Ooi, 2011, p. 124). This document is understood to have catalyzed creative policy in Singapore, wherein the arts became viewed as the activity of responsible government, and were promoted as able and necessary to 'enrichen us as persons, enhance our quality of life, help us in nation building and enhance our tourism and entertainment sector' (cited in Ooi, 2011, p.125).

A diverse portfolio of activities has been developed to support Singaporean arts and culture including fostering arts infrastructure to produce a 'dynamic cultural eco-system and a culturally engaged community'. Unsurprisingly, investment in high-profile and iconic institutions and structures such as the building of Singapore Art Museum and Theatre on the Bay has been important. Elsewhere money has been plugged into supporting Singapore as the global media hub of South Asia. A total of 250 million Euros spent between 2006 and 2010 were

Figure 6.1 Singapore Art Museum

ear-marked to develop the digital media industry including attracting key international media players such as Lucas Film (who work on George Lucas's movies) and Electronic Arts (makers of computer game *The Sims*). The area has also attracted companies such as BBC, HBO and MTV. Such high-ranking names on the global scene, together with a 400% increase in arts events and a 700% increase in museum visits, have been seen as symbols of the success of Singapore's arts and culture drive. The creation of Singapore as a creative place was confirmed for many when the global press endorsed the city as a regional hub and an artistic gateway (Chang, 2000; Ooi, 2011).

While externally it seemed that Singapore had fashioned itself as a global city for the arts, for those on the local arts scene the story was rather different. For Singapore might have the semblance of a creative place and the financial figures and global success stories to back up such an image, but the reality was that for many artists the government situation was not conducive to the development of creativity; 'the soil is still not viable enough to encourage an artistic and creative sensibility' (Ong Keng Sen, quoted in Ooi Martin, 2008). As Ooi's (2011, 2008) discussion of fine arts in Singapore details there has been a cautious welcome of creative practices as long as 'disruptive' creativity is avoided. He contrasts this with furniture design where controversies in creative content are rare as the 'creative expression seldom incorporates any message that may threaten the social political order' (2008, p.293). In contrast to the volumes of money plugged into support the digital media industry, the creative arts received only 50 million Euros over a six-year period. Further, the arts are encouraged to look towards more financially lucrative sectors in order to make money. For many artists, this equation is not one akin to cultural development, where important art is not necessarily going to translate into economic success. If for some this is a function of the policies around supporting creativity in Singapore, for others it is also a function of the closing down of the arts out of a fear of the combative and controversial political and social statements artists might make. There has been an increased liberalisation of social spaces in Singapore but this is only part of the story. As Ooi notes, the 'Singapore government is hypersensitive to any threats against our racial and religious harmony' (Chua, 2008, cited in Ooi, 2011, 130). The arts community feel this hypersensitivity and in an environment where citizens are weary of forms of social political activism and where the government continues to control mainstream media, artists can be subject to censorship or are banned. Stories abound of theatre and drama groups whose funding was cut or whose plays were unable to be produced due to their critiques of the government on issues of race, religion and homosexuality (Ooi, 2008, 2011). In the light of such sanctions, artists continue to exercise self-censorship in order to maximise chances of their work getting funded and displayed and reduce the chances of the authorities deeming their work disruptive and unproductive. For Ooi (2010, 2011) the combination of focus on commercial success and political expediency has stunted the growth of arts and culture. Artists are discouraged from producing works that engage in forms of political activism, or that resonate with the public through social and political messages. There is a view that arts practice is for the economically desperate. This has led to the conclusion that creative Singapore is not a freewheeling experimental space that generates a vibrant creative scene (a potentially dangerous illusion anyway). Rather, it is a space managed in nuanced ways to ensure creativity fits within the existing social, political and economic conditions of Singapore (ibid.).

The relationship between place-making and the cultural and social circumstances of a place is a complex one. In Singapore, characteristics of place and government were seen to pose a challenge to the future development of a creative city, whilst elsewhere cultural traditions and heritage offer a resource and stimulus for place-making efforts. If Singapore was challenged by the importation of western creative policy practices on top of a local context with varying degrees of success, then elsewhere different scales and forms of policy have been adopted. The city of Medellin, Columbia (see Box 6.1) trialled a more bottom-up approach to creative planning and the possibilities it might offer for community development. As all these discussions show, there is a complex story here regarding how creative place-making works, to whom the benefits accrue and what the challenges might be for creative practitioners and other groups. As discussions of creative cities and gentrification make clear, there is much controversy over the universality of creative place-making. As many discussions of creative city plans show, the local creative scene often benefits little, whilst the existing local community often not at all. The relationship between creativity and urban practices of gentrification is a complex one. Gentrification, itself often stimulated by the presence of artists and creative practitioners rendering a neighbourhood cool, has become a force shaping creative neighbourhoods, and often not for the better (Ley, 2003, see Chapter 7 on cities). Instead, it has become recognised as a force associated with the destruction of the local cultures these areas once fostered, creating antiseptic spaces at the heart of cities that are often affordable only for the very wealthy. Thus creative practitioners are forced out of major urban areas and the very creative buzz that was often the attraction of the spaces is, if not destroyed, certainly diminished (Lees, 2003, 2008). On the one hand then, creative place-making can support, make and shape communities, but it is important to ask whose community and for what purpose when we consider these practices. The discussion that follows turns to explore how vernacular creativities might be understood as shaping communities and places. Again this is not a straightforward story, in which the force of creativity is inherently good, and the communities successfully shaped or represented, rather the creation of communities is more complex than this.

Box 6.1 REMAKING PLACES: MEDELLÍN COLUMBIA

'closing the door to crime and opening the door to opportunity'

Medellín, Columbia was for decades one of the world's most dangerous cities, an outpost in the global drug war and the site of ongoing fighting between the drug cartels and the state. At the height of the violence several car bombs a day were exploding and the murder rate in Medellín was 380/100,000. In autumn 2009 the city, under the leadership of Sergio Fajardo, developed an award-winning bottom-up strategy of 'social urbanism' that was focused on 're-conquering spaces that had been lost to violence'. As part of an 'urban renaissance' Medellín has situated the development of cultural assets within a wider programme of renovation and infra-structure development that has been dubbed 'the most remarkable urban redemption project in modern history' (Nolan 2014).

The project is based on the city spending a large proportion of the budget in the poorest areas; in 2014 this was $2.2 billion, 85% of the budget. The activity is based on community consulta-tion, holding a series of 'workshops for urban imagining' and at its heart sat a series of 20 'library parks' designed as 'environ-ments of conviviality' (Medellín Mayor's office, 2008'). These parks functioned as a state presence in those areas of the city often seen as more informal and dangerous, 'buildings which, for their scale, form, materials and colors contrast strongly with their surroundings, and clearly announce that the local administration is providing facilities, worthy of envy even in the wealthy areas of the city' (Brand and Dávila, 2011).

These parks included a range of cultural facilities and program-ming as well as training and education courses, spaces for business start-ups, social programmes and sports. The buildings and their complexes were designed through architectural compet-itions attracting ionic national and international architects. Unlike other projects however, their focus was not on bringing outsiders into the area; rather their aim was to remake place and community

Continued

with and for the local population. Pursuing the rather radical idea of 'our most beautiful buildings must be in our poorest areas' (Fajardo, cited in Brand and Dávila, 2011) good architecture has been seen as a force for creative place-making rather than a branding exercise, these high quality buildings linked literally to the building of 'a new social contract, through the provision of spaces of citizenship, places of democracy'.

These parks-libraries-cultural centres are part of a wider programme of social urbanism that represented a much-needed investment in the long-neglected poorer areas of the city. As well as the library parks and upgrades to housing and services, a major infrastructure upgrade was undertaken. This involved installing the world's first modern urban aerial cable car public transport system in 2004, developing a model for how to deal with high-density, informal settlement on high gradient land on the edge of cities. Nicknamed the 'Medellín effect' the city's ambitious plan of social urbanism earned it the title 'the world's most innovative city', beating out New York and Tel Aviv. The prize, sponsored by America Urban Land Institute and *Wall Street Journal* and Citigroup, major forces of global capitalism, was also voted for by the local citizens of Medellín, 70% of whom supported the vote.

As the development becomes part of the community, questions remain whether the library parks, which are popular with local communities, might serve to acerbate issues of inequality and associated unrest. Further, the cable car is expensive and can take longer than the buses at peak times due to the time needed to queue. What Medellín demonstrates, however, is how creative place-making around cultural assets might involve high-profile and large-scale examples like the Guggenheim museum, but it can also involve networks of smaller scale facilities. Such facilities take as their focus the requirements of the local community and are built around them and their needs, rather than catering for international tourism. As a result, we see the transformation of place, linked closely to the transformation of individuals and communities, in this case through symbolically collecting the community around the valuation of a new collective place.

MAKING IS CONNECTING . . . INVESTIGATING THE CLAIMS

In a rather different vein to concerns with creative planning, for others the community-making potential of creative practices lies in the actual material practices themselves, and in particular practices of making and consuming together. David Gauntlett (2011) for example claimed that making is connecting; while Richard Sennett found in the possibilities of crafts practices the building of ethical and community relations. Indeed, in some studies the practices of doing are more important for the forging of communities than the finished project. Such claims require verification and justification. In short, if making is indeed connecting then the questions of what is being made and what is being connected, how and to what effect must be addressed.

The discussion that follows will collect together a series of examples to explore how creative making practices have been seen to be practices of connecting communities. A number of different elements of the creative process emerge as important in this community formation, whether it be collective making practices or collective consumption practices. The examples discussed include both amateur and professional projects, ones that deliberately set out to create community and others where community practices are secondary. Further, some of these are vernacular practices, whilst others involve commercialised and professionalised creative practitioners. As with the previous section, these discussions – which range from class politics of Christmas lights, to Japanese weaving practices and the creative geographies of religious place-making – also foreground the challenges and contention around the community forming claims made for these practices.

Creative doing together

One of the key dimensions of discussions around making and connecting has been an attention to how the actual practices of creative doing, and doing together, can come to form communities. Ethnographic work conducted with the project *Play Your Place*, explores how in the practice of designing and playing video games individuals are encouraged to become citizens, to identify, debate and engage with issues common to their local community. As the discussion recounts, the processes of drawing and gaming together offers the means to encourage the group to imagine the changes they want to see for their place.

Play Your Place is a participatory on and off-line arts project led by Furtherfield Digital Media Arts organisation.[2] The project explores the potential of creative arts to engage communities with the future of their places. The slogan 'if you draw it will happen', captures the sensibility of future geographies and the political possibilities that sit at the heart of the project, fuelled by the belief that participatory arts practices can drive place-based social change. In practical terms the project involves communities – most recently South Westminster in London and Southend-on-Sea, on UK's east coast – identifying changes they want to see in their community – better bike paths, more recycling facilities, affordable housing, dog parks – and then taking part in the process of designing simple platform computer games based on drawings (Figure 6.2). These games introduce the local problems and then collectively suggest solutions. They are then uploaded onto the Play your Place website where they can be played by anybody with an internet connection.

Research suggests that Play Your Place does a range of things. Firstly, that the processes of imagining, drawing and gaming come to shape the subjectivities of those who participate, producing its users as 'citizen-subjects' (Figure 6.3). Once the local issue was chosen, the story boarding and drawing exercises began as a fictional local heroine or hero was identified,

Figure 6.2 A Play Your Place game

Figure 6.3 The collective drawing process, *Play Your Place*

along with their quest for change, what challenges they face, what the rewards and solutions might be, and what the outcome is. In Southend, for example, in this instance the 'someone' was a bike-rider, the 'goal' was a more bike-friendly environment, challenges included pot-holes, aggressive drivers and traffic, rewards were more friendly attitudes and more diverse animal and plant life. One solution was to increase the numbers of cyclists on the roads to develop a collective change in attitude. Participating in the process seemed to make the participants more politically aware of the place in which they lived. It empowered them with a sense of imagination regarding possible futures for their place and a sense of investment in place, whether or not that was linked to an ability to effect change.

As has been noted of the process of digital story-telling, 'being the author of your own life, of the way you move through the world, is a fundamental idea in democracy . . . there is power in storytelling to help everyone project their authority' (Lambert, 2012, p. 22). Furthermore, Lambert continues we should listen to other people's stories as a way of therapeutically and realistically moving beyond our predicaments.

Participating in *Play Your Place* involved individuals participating in group discussions and in collective drawing and game playing practices. In the process of these collective activities the participants became engaged citizens, enrolled in shared narrations of communities and places.

Secondly, *Play Your Place* was also a process of shaping communities. As the game play was developed there was a transitioning of experiences of place from an individual, sometimes quite personal experience, to a shared understanding and a group narration of change. In terms borrowed from digital story-telling, this is the movement from a 'me' story – a personal issue to an 'our story' (Lambert, 2012). One result of the development of the games was the collectivisation of a group of individuals, some of whom had known each other beforehand, some of whom had never met, around specific issues, but also a shared sense of investment in the future of their place. In the process of forming a community with something in common, the collective game design, making and playing, also offered a series of 'social goods' – happiness, confidence, self-esteem and the enhancement of social skills. As drawing and game making progressed, the rooms were full of laughter, chat and exchange as attention shifted to the processes of collective thinking and doing in the game production, over and above the discussion of the politics of the issue at hand. People became deeply involved in making things work – both the actual game levels and the group dynamic – with some attending both to the task at hand and to ways in which everyone in the room could contribute something.

The third aspect of the *Play Your Place* process is that of connecting different communities together. *Play Your Place* can be situated alongside other forms of public experiment, including using story-telling and workshops as part of local consultancy processes, wherein these public experiments become experiments in knowledge production. The project is looking, in the next phase, to try to find a way to connect with local government stakeholders and developers to share local thoughts on community futures. As such, *Play Your Place* has the potential to be a radical participatory method. Radical, because, following discussions of participation in science literatures, we can find in *Play Your Place* the potential to align lay knowledge alongside expert knowledge in a way that shifts the power relations in the political process. *Play Your Place* can be understood to be radical because the project is as much about creating a new public capable of making a political intervention as it is about producing the solution itself. In this process of knowledge generation there are real possibilities for the project not only to connect communities, but to reshape the nature of those connections in ways that make space for radically democratic practices.

> Politics more than any-thing else needs the magical touch of dream and desire, needs the shock of the poetic; left to professional career politicians, the political is always deemed to feel stifled and lifeless and apolitical; its always destined to induce a jet-lagged, deadening insomnia.
>
> (Merrifield, 2011, p. 386)

The politics of Play Your Place removes political ambition and action beyond the remit of the career politician, to resituate it in the practices of everyday individuals and their daily engagements with their place and community. This is to offer an alternative vision of politics, who can be political, and what it means to be politically engaged. The force of imagination as a sphere in which political battles can and are being fought is reinforced (Thrift, 2007) and the potential for creativity to do such work is heightened. Not only did Play Your Place connect communities, it also enabled a collective recognition of the politics of possibility: the possibility that something might be different. In terms of the impact it has on the moral and identity of the individual subject, this possibility of difference can be as important as actual material change. The result is the shaping of an alternative realm of politics that recognises political action and possibilities in the everyday practices of individuals and the textures of their life experiences. It recognises politics too in the poetic and enchanting practices of drawing, gaming and play. This is to enable the individual and the community to be active, engaged and empowered in the face of, and operate within, those more 'transcendental' 'official' and 'expert' political spheres that are often far distant from the lived experience of a particular place.

Collective consumption

It is not only in creative 'doing' together that we might think about communities being connected, but also in the consuming together. In a suite of papers that explore illuminations in the UK – both street-based Christmas decorations, as well as more permanent illuminations in the northern seaside town of Blackpool – Edensor and Millington (2009,

2012) address the issues of community and class raised by the production and consumption of these street displays. What becomes clear is how these forms of street-based creative practices might form social collectives and shared atmospheres of place.

Displays of Christmas lights are often not understood as 'creativity' but rather viewed through the lens of taste-based judgements and, as such, reveal ideas of creativity to be shot through with class-based concerns. Developing their case studies, Edensor and Millington note a clear aesthetic demarcation between the 'chic' blue and white displays of middle-class housing areas and the more colourful extravaganzas, deemed tacky or kitsch, that pervade working class districts (2009). The latter include various Christmas symbols, from religious icons to trees, Santa's sleighs, snowmen as well as toy trains, teddies, and flashing lights and rope lines. Transforming the British night-scape from late November until mid-January these light displays offer, from the point of view of the hangers at least, a point of collective coming together.

In interviews with the light-hangers regarding their motivations, what they emphasised over aesthetic concerns was the sense of community that the lights created. Design appeared in terms of a common sense of which colours were 'Christmassy', but the displays were talked less about as arrangements and more as organised chaos. Indeed, good taste and design were less important than the 'conviviality, neighbourliness and festive pleasure' that the illuminations developed for the displayers and their community (2009, p. 113). There was much talk of festive togetherness in family and community brought about by the lights, this was a time for sharing and the lights were a way of doing this, indeed as they show, 'the production of neighbourly conviviality is thus not merely generated by the displayers but becomes part of a collaborative understanding' (2009, p. 115).

The sense of a vernacular creativity bringing about collectivity and shared celebration is also found in the study Edensor and Millington (2009) did of the Blackpool illuminations. In exploring the annual light festival in this northern English seaside town, they found that amongst visitors what was important was less the look of the lights – aesthetic qualities or an exercise in discernment – rather, emphasis is placed on consolidating bonds with family and friends. These bonds evolved around nostalgia and shared conviviality and the experience of a good atmosphere beyond one's immediate group. The town had for a while been trying to 'improve' the light display, rebranding it a 'festival of light', and trying to appeal to a more middle-class audience. Fearing the lights were tacky and losing appeal, the local design team was augmented by invited artists and designers in order to evolve what was coming to be seen as a backward traditional style. The imported 'taste-makers' had to carefully negotiate the tension between the cosmopolitan design and 'local' or vernacular practices of illumination. As Edensor and Millington (2012) recount, the designer and lifestyle guru Laurence Llewelyn-Bolen was brought in to appeal to external audiences, but he was careful to celebrate Blackpool too, making a feature of the traditional elements of the place and its lighting displays. Thus rather than creative taste-makers riding rough-shod over local traditions, the latter was demonstrated as an evolving thing, carefully blending novel technologies and new designs with the traditional, in order to maintain the atmosphere of the lights so valued by locals and visitors alike.

The story of lights and community connection is not, however, quite that straightforward. Indeed for Edensor and Millington, the focus of their discussion is that of tensions and cultural politics. They observe how such 'local, vernacular creativities and specific cultural practice and meanings ... can contribute to rethinking and diversifying cultural politics' (2012, p. 146). Far from creating communities of togetherness, the residential light displays often divided communities along class lines. The differential class position of judgement that aligns these illumination practices not as vernacular creativity but as tacky and kitsch, often termed in the derogatory 'chav bling'. Reviewing popular discussions and internet chat rooms on these displays indicated a common discourse that understood them as aesthetically, environmentally and socially inappropriate. Far from celebrated forms of local creativity, they were determined as an immodest spectacle, excessive aesthetically, a waste of other people's money by poor, inept consumers and above all a waste of electricity. Such creative re-makings of street and home did not fit into the development of cultures of taste and lifestyle by the evolution of a set of cultural intermediaries who 'comment on lifestyle providing "tasteful" "makeovers" of homes, bodies, clothing and gardens, all the while reinforcing cultural values that masquerade as common sense' (2009, p. 109). As such, creativity, its production and consumption can be understood as creating communities of similarity and difference in cultural tastes, as well as creating community through creative practices. What becomes clear throughout many of these discussions is that more research is needed on the material and embodied practices of making together. For only then will we understand exactly what forms of connectivity – conviviality, generosity, animosity - collective creative making practices might form.

Box 6.2 FAITH COMMUNITIES AND CREATIVITY

Claire Dwyer

Geographies of faith, religion and spirituality have seen a recent resurgence, consistent with a recovery of the salience of the sacred, or a valorisation of the 'post-secular', in wider social science. While explorations of the spaces of religious identity and practice often prioritise social formations of identity and community, calls to reanimate geographies of religion urge taking seriously religion itself, recognising the agency of the sacred and allowing religion to 'talk back' (Holloway, 2012; Yorgason and della Dora, 2009). Researching the creative geographies of faith offers a parallel approach consistent with a materialist turn in religious studies, focusing attention on the embodied practices, spaces, objects and performances of religion and foregrounding the ordinary, mundane and 'everyday' lived religious practices of believers (Ammerman, 2007; McGuire, 2008). An exploration of the everyday material cultures and practices of faith communities might include artistic and architectural practice and forms of 'vernacular creativity' (Edensor and Millington, 2012) in the creation of religious spaces, whether congregational or domestic or more provisional and temporary, and attention to the creativity inherent in the performance and practice of religion.

Forms of religious architecture facilitate congregation and communion through shared religious practice and act as official sacred sites where the divine is presenced and encountered. Creative geographies of religious place-making involve networks of artistic practitioners alongside planners and faith communities in the realisation of material, sacred spaces. Take, for example, the recently constructed Thrangu Tibetan Monastery, opened in Vancouver, Canada, in 2010, funded by Hong Kong Chinese transnational migrants (Figure 6.4).

The monastery's architectural style is narrated as a faithful rendition of its parent monastery in Tibet. Artistic work was undertaken by visiting monks and community volunteers using traditional techniques. However, this is a hybrid creative form of

Figure 6.4 Thrangu Tibetan Monastery

architectural and artistic practice that had to accommodate local weather and topography within a Canadian context. While offering a semblance of solidity, the building incorporates fibre-glass ornamentation creating the *effect* of a 'traditional' monastery via creative invention. The Salaam Centre, under construction in London, provides a more explicit creative engagement with diasporic migration and place through a design that references the migratory histories of the centre's Muslim community (Dwyer, 2015). Alongside such purpose-built congregational spaces, are the creative and improvised spaces for religious worship, whereby homes, disused industrial spaces or public buildings and former places of worship are transformed and appropriated. Examples from current research such as the placing of an Islamic doorway in an office building at the West London Islamic Centre, or the siting of a baptism pool within the cavity formerly occupied by the organ in the old cinema now used as a Pentecostal church, suggest forms of vernacular creativity in the realisation of places of worship.

Material culture and practised creativity is often central to the performance of religious ritual and performance. Research on the

Continued

growth of festival celebrations for the Tamil Hindu diaspora in London, such as the Terotsava Chariot Procession, when the sacred deities or statues are taken out of the temple and paraded on specially built chariots (David, 2009) explores how creative practices of music and dance, alongside decorative traditions of flower garland making, shape new identifications in diaspora. Focus on embodied creative practices and performances offers a methodology for exploring religious identities both within congregational spaces and domestic spaces. Research on Irish-Catholic domestic devotional material cultures (Garnett and Harris, 2011) and Hindu domestic shrines (Tolia-Kelly, 2004) or the material cultures of 'everyday spiritualities' (MacKian, 2012) suggest the centrality of domestic material cultures in shaping faith identities. Forms of domestic creative practice associated with religious festivals or ceremonial meals provide insights into how faith identities and communal belonging is negotiated or transformed (Watson, 2009).

The making of sacred spaces and landscapes, often beyond the 'officially sacred' (Kong, 2001) particularly through practices of pilgrimage or acts of memorialisation (Maddrell et al., 2014) are also centred around forms of creative practice in the embodied and aesthetic performances of pilgrimage making or the material-isation of sacred landscapes and vernacular memorials. For researchers of religious identities, analysis of material religion and creativity prioritises 'everyday' or 'mundane' geographies but must also be attentive to the theological and contextual framings of creativity and the material.

Ongoing research on 'Design, material culture and popular creativity in suburban faith communities' explores the cultural produc-tion of space and the material cultures and practices of diverse faith communities in West London. It combines analysis of creative geographies of faith with a methodology which engages participants in shared creative practice. Shared creative practices, in partnership with arts practitioners, such as a participant photography project (Dwyer, 2014) provide opportunities both to access faith identities and to facilitate cross-cultural and inter-faith encounters.

ARTS, CONTENTION INTERVENTION AND THE SHAPING OF PLACES AND COMMUNITIES

A further way we can explore creative practices as practices of place-making and community formation is to explore how professional artists might develop such goals as a part of their work. The first part of the discussion will focus on the intersection of art and the history of places; this is key to geographers as such projects mark an early embrace of art by the discipline. While it will begin by thinking about sculptural practice, discussion will move on to think about socially engaged art (including participatory art and relational art) and the possibilities of art as a form of 'politics in action' (Hawkins, 2013a). It is worth noting a growing set of tensions across these interventional arts practices. Such tensions reflect the realisation that arts practices can both be a strategy for making and remaking communities differently, often adopted and adapted by grass-roots groups, but that these practices can also be commandeered by those in power. We see, for example, a growing place in the UK and US for arts practices as parts of wider social policy. If we are not watchful such projects risk art being made safe for investors, with goals watered down to the point where meaning is lost, where cultural expression is co-opted for existing interests, and chances are missed for dialogue with marginalised peoples. Furthermore, as art forms evolve to be constituted by and through relations built between people time must be spent reflecting on the form and nature of these relations. Reflecting not only on how art might bring about social relations, but also whether or not these should be about bringing people together around things in common, or about raising uncomfortable and some-times confrontational issues.

Places and their pasts

Since the nineteenth century public monuments have formed 'foci for the collective participation in the politics and public life of towns, cities and states' (Johnson, 1985, p. 51; Warner, 1985). Following a range of art historians and others, Johnson (1985) explores monuments as a material basis for the emergence and symbolic structuring of nationalist imaginaries. The imagined community of the nation needs to be sustained and, as many agree, nation building is an ongoing process, whose myths and imaginaries are created by elite and popular people alike. Within such debates architectural and memorial sites become flashpoints for the development and contestation of these imaginaries. David Harvey's (1979) study of the Basilica of the Sacre Coeur in Paris highlights the contested meanings of the urban landscape. Working to explore the politics underlying the development of the site makes plain how exploring the history and development of such buildings can enable us to let landscapes and urban spaces speak of the struggles over space. Opinions differ over the ability for monuments in general and figurative monuments in particular to 'engage the viewer reflexively with the past or future, and the anti-monument movement seeks to reclaim memory as part of everyday life' (Johnson 1985, p. 56). Indeed, a key category of studies has focused on who gets memorialised, how and where. In such studies social categories such as ethnicity and gender often become a focal point, noting for example that commentary on women in such monuments is largely allegorical, such that while 'men often appear in as themselves, as individuals, but women attest to the identity and value of someone or something else' (Warner, 1985, p. 331).

If exploring the memorial logics of some sculpture is one key theme of study, a second is focused on those works that are part of the aestheticised landscapes of urban regeneration. Of particular interest is the relationship between the art works and the populations that the schemes often displace or overlook, especially where cities are turning away industrial pasts in favour of creative or knowledge-economy based futures (Deutsche, 1996). Hall (1997) offers an extensive discussion of the sculpture – *Forward* – placed in a flagship location in Birmingham's (UK Midlands) city centre.

In Hall's study this piece of public art reveals the fractures between the city's now failed industrial past and the new urban identities that the city is trying to create. Raymond Mason, the Birmingham born but Paris-dwelling artist, famous for his sculptures of working class cultures under threat, was an interesting choice for the commission. A city's industrial pasts can be hard to negotiate in the context of contemporary narratives of place-promotion, forming a problematic and contested terrain from which to narrate new cultural geographies of the city (Hall, 1997, p. 58). *Forward* reaffirms industry as part of the identity of the city; it is a human history of industry, a representation of the ordinary working men and women of Birmingham. It is also therefore a sculpture focused on the very people who the regeneration project overlooks. Yet it was placed in the middle of a square in front of the International Convention Centre which was the symbol of the 'new' post-industrial, international, cosmopolitan, service-orientated Birmingham. The piece was destroyed in an arson attack in 2003.

If geographers have been interested in how artistic forms have framed and facilitated engagements with the past, there has been a move recently beyond questions of representation to query how art might have a more active role in the production of memory and community. To restrict discussion of public art to questions of representation is to risk fixing works as being reactive rather than understanding them as potentially generative of new identities, histories and communities. More recently, geographers interested in the intersection of art and place's pasts have explored how socially engaged art might offer means to intervene within and shape places and communities.

Across a whole series of these discussions artistic practice is seen as an effective way of exploring the 'fluid and fleeting connections between environment, place, identity and meaning' (Cant and Morris, 2006, p. 859). A series of geographers, art theorists and memory scholars have extended geographical work on memorialisation by examining how artistic practices can be part of a recognition of 'memory in motion', shaping people and places, rather than seeking to unearth repressed pasts, represent or preserve them (Till, 2008, 2012; DeSilvey, 2010). Creative mediations have been shown to open up alternative ways of knowing the past in place and may 'offer resources for the researcher seeing more subtle and oblique strategies for exploring the performative aspects of cultural memory work' (DeSilvey, 2010, p.492). DeSilvey unpacks this idea via the exploration of the process of making a piece of sound art in response to the 'unmaking' of the industrial landscape. The site she is focused on is a decommissioned hydroelectric plant in western Montana (USA) that was understood to have contaminated the soils and waters around the town of Milltown. In her example, art production and consumption has a place in exploring what happens if the link between memory and material persistence in place is severed. DeSilvey describes the development of the piece *Sheng High* by the Seattle sound artist Trimpin. The artist, working at the intersection of the ad hoc and the high-tech, assembled improvised musical instruments in order to reinstall a work based on an ancient Chinese reed instrument. The instrument plays a musical score created from a map of the dam remediation programme that the artist had converted into a wall mural. The mural contained a series of sensors that relayed water through pipes to force the instrument to sound. Reflecting on her observations at the exhibition, DeSilvey observes how these discordant soundings of place opened up conversations about Milltown's past between friends and strangers alike. Conflicts emerged, shared memories were discussed and as DeSilvey notes the 'encounter with *Sheng High* accommodated the ambivalence implicit in marking the loss of a familiar landscape and accepting its indeterminate future' (2010, p. 504). The piece on DeSilvey's approximation remained 'sensitive to multiple pasts and possible futures' and developed the multisensory as a way to do this.

If some art explores community-based discussions about places' pasts and futures, then other forms of arts practice are more interested in acts of direct intervention. Research and practice by geographer Karen Till (2012) on 'wounded cities' notes the multi-faceted possibilities that exploring arts and art works offers for place-based memory practices. There are numerous examples that confirm what Till (2008) notes when she argues 'artist and activist memory–work has much to offer the emerging field of memory studies'. She finds its principal value to lie in its challenging of the ontological assumptions that underpin much recent research on memory, including understandings of the site, social and body memory and the role of place in memory (2008, p. 102). Till is concerned not just to ensure that memory scholars are aware of this, but argues that this will enable us to develop more socially responsible research practices. She applies these ideas to her analytic and practice-based work on wounded cities, as she notes 'places described as wounded are understood to be present to the pain of others and to embody difficult social pasts' (2012, p. 3). Creative practices thus offer the means to address urban wounds, offering 'political forms of witness, to respect those who have gone before, attend to past injustices that continue to haunt contemporary cities, and create experimental communities to imagine different urban futures' (ibid.). In an extended case-study, Till (2012) explores the place of theatre as protest and intervention in the context of an urban renewal project in Bogota. Focused in El Cartucho, an area known for its informal economies and sex and drug trades, this urban renewal project aimed to reclaim this area for the public; in doing so, however, it ignored the 'public' it displaced when it razed to the ground 20 hectares of the area, displacing tens of thousands of residents without compensation. Project Prometeo was formed as a means through which actor–residents raised ethical questions around the politics of such public place-making in the name of progress; querying as they did so which public and what progress for whom (Till 2012). Creative practices in this example gave city residents, previously talked of as disposable, a means through which to make themselves audible and visible, and thus a part of the discussion. Describing the site-specific performances, Till notes how by performing in Project Prometeo the displaced residents challenged the city's understanding of them as invisible, calling attention to how the plans for a sustainable city did not include all its residents. Through their stories and performances, residents documented how they used and made their neighbourhood and city thus, as Till (2012) observes, asserting basic individual collective and temporal claims to having rights to that city. She notes how the residents communicated and enacted their experiences of place and the city as inhabited, performing a remembering of everyday life in an otherwise marginalised city community. Performing on top of their ruined former homes through their interpretations of the Prometheus myth the actors invited audiences to consider the violence done to their houses and lives. The stories, actions and creative practices of these local experts who live with violent histories of displacement, did not let the audience stay as audience, but rather invited and encouraged them to dance and remember in the ruins with them. Till (2012) suggests that projects like this one allow an exploration of 'how these open ended pathways of memory might offer possibilities of shared belongings'. Here we see artistic interventions advancing the challenging work of memory in wounded cities marked by violent and difficult pasts. As Till (2012) argues, such projects offer the possibilities for place-based mourning and care work across generations that build self-worth, collective security and social capacity. What emerges here, very clearly, is a sense of artistic practices as socially engaged and as offering us a politics in action. It is to an exploration of what such politics in action might mean and enable that discussion will now turn.

Participatory art and politics in action

For contemporary art theorists socially engaged art involves appreciating the value and potential of the social relations that can be instigated through arts practices. For French curator and theorist Nicholas Bourriaud (2002) this can be termed 'relational aesthetics', wherein aesthetics is found in the social relations formed between humans, and less often non-humans. These kinds of socially engaged art practices have formed the basis for wider understandings that have found art to be politics in action (Toscano, 2009) or as a 'technology of connection' (McNally, 2015, Box 6.3). Engagingly diverse, iconic examples of socially engaged practices include: the convivial culinary arts of gallery-based cook-offs; the orchestration of parades and festivals; and the mobilisation of aesthetics as part of direct action and political protest (Jackson, 2011). The claims made for these forms of work have triggered a return to questions of the possibilities of arts practices as figures of opposition. Moving beyond state forms of public art and the corralling of creative expressions to state or capitalist-sponsored ends, arts practices are understood to 'actually be ways of living and modes of action within the existing real' (Bourriaud, 2002, p. 13). In the discussion that follows, a series of examples will explore some of the different ways arts and creative practices have been understood to intervene in places, transform subjects and connect communities, whether positively or negatively.

Box 6.3 ART AS A TECHNOLOGY OF CONNECTION: NOMADIC SCULPTURE

Danny McNally

With a defined 'social turn' emerging in contemporary art practice from the 1990s, where the creation of social relations is placed as a central element in the cultural-aesthetic form of artworks (see Bourriaud, 2002), how might this be of interest to us as geographers? By thinking of this sort of artwork as a 'technology' of connection, interesting links to contemporary geographical debates are made possible. The following participatory art project offers up an example of how geographers can think about art as a technology of connection. Rather than championing this artwork, as such, it instead offers ways 'into' the sort of encounters contemporary participatory art creates, and for what cause.

Amalia Pica's *I am Tower of Hamlets, as I am in Tower of Hamlets, just like a lot of other people are* 2011–2012, hereafter *I am Tower of Hamlets* (Figure 6.5) was an off-site, year-long, exhibition through Chisenhale Gallery in East London. Part of a wider project called *A*

Sense of Place, Pica's project saw a hand-carved pink granite sculpture tour the borough of Tower Hamlets in London. Any resident of the borough could request to host the sculpture for one week. Following their hosting period, the resident then had to deliver the sculpture to the next host. The travels of the sculpture were recorded on 'lending cards', detailing the name and address of the host(s) and the dates they hosted it (Chisenhale.org.uk). The project was pitched as an exploration of public sculpture and interventions into public space (see Zebracki, 2013; Miles, 2004 for related discussions), as well as experimenting with more intimate encounters with an artwork (Chisenhale.org.uk). Pica believes in the capacity for objects to act, to 'make things happen', and to become increasingly meaningful through their circulation in place-based networks. Accordingly, the aesthetic of this exhibition was not just the sculpture; it was the connectivity and encounters that this object created through its nomadism through Tower Hamlets.

Pica's project prompts an interesting connection to Human Geography's recent interest in spaces of 'meaningful contact' in times and spaces of 'super-diversity'. *I am Tower of Hamlets* facilitated encounters between people who would perhaps otherwise not have met, in an area of London renowned for its diversity. Eschewing appeals both to placed communities and neo-tribal collective socialities, Pica's work had a nomadic geography. Moving from studio to gallery, then from home to home it not only engaged critically with the siting of art in galleries or public space, but also with the potential for things and their biographies to (dis) connect those touched by them. Human Geographers have recognised for some time the connective aesthetics that can be performed through material forms: whether in engagements with the art of following things from production to consumption (Cook et al., 2000), in domestic art objects as repositories for diasporic memory (Tolia-Kelly, 2004), in identifications of the forceful role of objects in socially engaged art (Hawkins et al., 2015), or in the account of the 'ethical aesthetic form' of sociality produced partly through materials (McNally, 2014). Further, and more broadly,

Continued

the 'relationality' of objects, has been engaged within and beyond geography. The example of *I am Tower of Hamlets* builds on such concerns by looking at how an art object stimulates particular encounters within a contemporary urban context such as London.

Figure 6.5 I am Tower of Hamlets (Photo: Amalia Pica)

Importantly, these encounters should not be limited to the human. Research based on interviews with the participating hosts identified that the sharing of the sculpture, the fact they could touch it, and its temporary domestic residence created an 'enchanted' connection with the sculpture itself, something that could in fact be understood as 'meaningful'. Thus when we think about art as a technology of connection the importance of encounters with 'things', not just people, must also be taken into account.

However, and finally, when thinking about art as a technology of connection it is important to look critically at the *type* of encounters that are created rather than rest on the assumption that 'to connect' is inherently good or 'ethical' (see Bishop, 2004 for more on this). Although *I am Tower of Hamlets* did stimulate encounters in Tower Hamlets through the circulation of the sculpture, it did so predominantly within a pre-existing network, that of Chisenhale Gallery's constituency. A more interventional and politically astute approach could have extended the reach of the project beyond the gallery's network and created encounters that connected the vibrant diversity of Tower Hamlets.

Art as intervention into places

One common gathering point for geographers thinking through the liberatory and oppositional political potential of arts practices (Bonnett, 1989; Loftus, 2009; Pinder, 2008, 2011) has been in the politics and poetics of urban practices that often owe a considerable debt to the Situationists (see Box 7.3). In these performative walking practices, the subversive force of art is often located in its promotion of an aesthetic engagement with place that is other than the normative apprehensions and uses of those places (Hawkins, 2015a). Such interventions hold within them the potential to remake our experiences of places differently. In a special issue of the journal *Cultural Geographies*, David Pinder (2005b) draws together a series of art works under the title 'the arts of urban exploration'. Here we find everything from sound art walks, to treasure hunts and Situationist-style drifts through the city dictated by chance and contingency, opening out experiences of urban spaces to enable us to explore them otherwise. Whilst some of these artful urban occupations are explicitly political, others, like the work of Richard Wentworth (Battista et al., 2005; Hawkins, 2010a) are subtler in their forms of urban intervention. What these different forms have in common, however, is their intention to enable audiences to experience the city differently, reshaping their considerations of its pasts, presents and futures.

A recent preoccupation for many urban artists interested in intervening in urban space has been the guarded, surveilled and regulated nature of many of our global cities. The Belgium-born artist Francis Alÿs has made such explorations an ongoing theme of his work, making spatial interventions either through his own practice, or the carefully choreographed actions of others, whether this be Guards marching around the city of London (*The Guards*, 2005), or the movement of a sand dune by thousands of volunteers outside Lima, Peru (*When Faith Moves Mountains*, 2002). Alÿs often frames his work through provocative phrases such as 'sometimes doing something poetic can become political and sometimes doing something political can become poetic'. In a recent paper on Alÿs, geographer David Pinder (2011) pushes at these questions to ask how exactly these forms of poetics are political. This is an important question to ask of all arts practices for which transformatory claims are made. Indeed, these sorts of claims can be seen in the discussion of Handshake 302, an arts space in the urban village of Baishizhou in Shenzhen South China (Box 6.4), where arts practices are claimed to connect communities within and beyond this local area.

Box 6.4 HANDSHAKE 302: ART IN SHENZHEN'S URBAN VILLAGES

Handshake 302 is an urban arts group based in Baishizhou, one of Shenzhen's (China) largest urban villages. The organisation founded in 2013 aims, as its co-curator Mary Ann O'Donnell describes, to offer an experimental and ethnographic space to engage the living history of the urban village of Baishizhou and Shenzhen more widely. In doing so, it aims to 'reimagine urban possibility via Shenzhen's urbanised villages' (O'Donnell, 2014, p. 2).

In January 2013, these urban villages housed almost half of Shenzhen's 15 million registered inhabitants and most of its unregistered ones. One of the most densely populated cities in the world, Shenzhen didn't exist as a city prior to 1980, when the area was chosen as a special economic zone, legalising industrial manufacturing and foreign investment, much of which came from nearby Hong Kong. As a result of the boom in population new forms of housing were needed. Residents from villages who owned the original land built high-occupancy buildings that they leased to the new migrants and other low-status citizens. For a large proportion of the 140,000 residents of Baishizhou this means low-cost high-density housing (600 to 3000 yuan/ £80 to £320/ month in rent) with many buildings nick-named handshake buildings (Figure 6.6) because they are so close you can reach out and shake your neighbour's hand across the alleyway. In the 7.4 km² area the population density is 18,000/km², over twice that of the city's average of 7,500 people/km², a figure that earned Shenzhen the designation of second most densely populated city on the planet. As O'Donnell (2014) describes, these handshake buildings and the area more generally should be seen as a site where 'economic practices encompass both the formal and informal and home loyalties create a vibrant society seen to operate in parallel to mainstream Shenzhen' (p. 4).

These urban villages tread the line between being formal and informal settlements: informal because the vast numbers of residents have no rights to the city and because as informal forms of housing the city government is able to choose whether to supply services; yet formal, because they are often close to the city centre, have their own metro and bus stops and are able to access electricity and internet. Furthermore, the area is mixed, with recent migrants living alongside young professional families and recent graduates, including those who work in the nearby OCT Loft area, an area of renovated factories dubbed the Shenzhen Soho, which has become a haven for creative industries in the city (see Box 7.1).

Handshake 302 is named after its physical space, a 12m² efficiency apartment, room 302 of a handshake building, in building 49 Shangbaishi Street in Baishizhou. On the ground floor of number 49 there is a noodle kiosk and a flip flop shop. Opening on

Continued

Figure 6.6 Handshake houses, Shenzhen (Photo: Harriet Hawkins)

October 20 2013, less than 18 months later the organisation won the creative synthesis category of the first annual Shenzhen Design Award (O'Donnell, 2014). Growing out of CZC special forces (urban village special forces), Handshake 302 was part of the organisation's plan to discuss and support creative engagements with Shenzhen's urban villages. Led by an American anthropologist Mary O'Donnell, they wanted to explore what can be gained from returning to urban villages, how to repurpose the area's handshake efficiency apartments, densely crowded streets and small plazas as cultural spaces (O'Donnell, 2014). They wanted to give young Shenzhen artists a venue for showing new work as well as create locally relevant work, creating interventions that 'motivate Shenzhen residents to cross cultural and economic differences and discuss our common urban condition' (O'Donnell, 2014, p. 2).

Putting on five projects in their first six months, these included installations, performance projects, children's workshops and discussion groups. Each project engages local communities and other residents of Shenzhen in the hopes, dreams, challenges and daily life practices of those living in Baishizhou. For example, 'Playground' asked local children to explore their lived experiences of the urban village and then develop and perform plays about these experiences. The results were performed for students who came from schools from other parts of Shenzhen, helping introduce them to life in the urban villages they may never visit. Another project was entitled 'Superhero'. Taking the form of an installation in Handshake 302, the project celebrated the everyday superpowers of the residents of Baishizhou. Like one of those seaside cut-outs you put your head in, audience members could take turns to be transformed into a series of local superheroes that made life possible not only in the urban village, but also in the wider city. The local superheroes included wonder granny, who looks after the children and does the housework so both the husband and wife can earn money to support the family; Stir-fry fly, a hero food hawker who feeds the urban villagers, or village guardian, a local fire-fighter.

These projects like many other Handshake 302 endeavours have developed through ethnographic research based on interviews with local residents and spending time living and working in these urban villages. In developing these projects, the group are connecting with local residents, connecting residents with each other in a communal celebration of the area, but also helping

Continued

connect them with others living elsewhere in the city. As such, art here functions as what McNally (2015, and in Box 6.3) describes as a 'technology of connection'. In this case, the work aims to engage people with local challenges and issues on their backdoor step, helping them see these urban villages not as sites of urban blight, as problems to be hidden and eventually eradicated by gentrification, but rather as locations to be celebrated and sites of possibility and transformation.

For many artists and practitioners, the political questions around the relationship between their work, place and community are what drives their development. Alÿs develops his work as an explicit query as to how something poetic becomes political. Where these ideas become most clearly developed is in the piece 'The Green Line'.[3] In 2004, Alÿs punctured two holes in the base of a can of green paint, which he then carried 24 km through the city of Jerusalem. During his walk the paint dribbled over the streets and markets, through boarders and no-man's lands and across the everyday life of the city. Alÿs' green line is made in dialogue with the demarcation lines set out in the 1949 Armistice Agreements that divided Jewish and Palestinian territories. Known as the Green Line, after the green ink used, this is a line Israel has tried so hard to erase and here Alÿs is making it visible in a critical and subversive action that is both a performance of space and also a performance in space (Weizmann, 2010).

Displayed as a video piece, viewers could choose 11 soundtracks from Palestinian, Israeli and European intellectuals and activists that Alÿs invited to comment on the footage of his walk. Across these commentaries the 15 mm wide green line becomes a poetic microcosm of Jerusalem's politics. For some of these commentators and other critics, the ease of the artist's passage through the city risks masking the complexities and conflicts of this urban space. For others, the materiality of the paint line, its complexity, splotches, loops and breaks, its engagement with the materiality of ground and its gradual erosion signals the arbitrariness of the original line and its grounded complexity. To draw this paint line is to bring into dialogue the legitimacy and illegitimacy of all past, present and possible future lines drawn on this city. As Weizmann points out, however potent the line is it is also to render the city as a surface, whereas the complexities of the territorial boarders negotiated across bridges and tunnels and in the vertical scale means that Jerusalem must always be rendered in three dimensions.

In querying the relationship between the poetic and the political in his work, Alÿs is encouraging open discussion as to the political possibilities of art interventions. Pinder (2008) takes up these questions when he asks us to consider whether these interventions truly 'bring about an unforeseen way of thinking', whether they can, 'translate social tensions into narratives that in turn intervene in the imaginary landscape of a place' and 'even bring about possibility of change' (Pinder, 2008, p. 733). For Pinder it becomes a question of a need to remain attentive to the 'varied abilities of these practices to challenge – or not – the prevailing norms and power relations, rather than to succumb to the[ir] romance' (2011, p. 688). To ask what is actually being done through these creative practices is the concern that should preoccupy future work in this field.

Connecting communities, shaping places

The force of creativity for making places and shaping communities has been explored throughout this chapter. As we have seen this is far from the avowedly positive story that is often presented. It is not just, of course, in relation to human communities that we might ask about the effects of the connections formed through making but also (as the landscape and environment chapter will explore) how we might think about human-non-human connections. Indeed, whilst there are many positive ways in which creative practices and their consumption can be understood to create communities and shape places, so too can creative practices and their planning do damage to existing communities. They can exclude and side-line those who lived in the area, can neglect to attend to the social and cultural characteristics of the location and can be carried out at the behest of the powers that be. What is clear from studies across the spectrum of work, whether it be in creative city and place-making discussions or arts and crafts practices, is that further empirical work is needed to explore claims made for making as connecting and to develop evidence on what kinds of relations and places are made, for who and importantly how.

NOTES

1 https://www.medellin.gov.co/irj/portal/medellinIngles [Last Accessed: 5/6/2016].

2 I researched *Play your Place* with Furtherfield since 2013, the text upon which this entry is based was originally written as part of a report for Furtherfield and is available online at http://localplay.org.uk/wp-content/uploads/2014/03/Cultural-Geographies-of-Play-your-Place.pdf [Accessed 25/7/2015].

3 http://francisalys.com/greenline/rima.html [Accessed 28/7/2015].

7

CITY: FOUR CREATIVE STORIES

while no one kind of city, or any one size of city, has a monopoly on creativity or the good life ... the biggest and most cosmopolitan cities, for all their evident disadvantages and obvious problems, have throughout history been the places that ignited the sacred flame of human intelligence and the human imagination.

(Hall, 1997, p.7)

if cities have been essential to artists, artists have been essential to cities.

(Solnit, 2000, p. 19)

Imagine a city, a concentration of creative practitioners – artists, fashion designers, musicians, graphic designers – at work in studios, a profusion of old and new galleries and museums to offer inspiration, to display work and to attract visitors. This culture supporting and supported by a network of educational institutions, low-rent live–work spaces, start-ups and informal pop-up venues that display emerging talent. Imagine another city, where knit-graffiti, tagging and street art create a vibrant street-scape, where skateboarders and parkour practitioners find a home and where tactical urbanisms turn street corners and vacant lots into small gardens with benches and social spaces. Imagine a third city, where the old warehouses are clad in steel and glass and occupied by those able to afford price-tags in the millions, where every corner shop has been replaced by a tapas bar or a boutique coffee-shop, and where the only traces of industry remaining in the deadly quiet streets are those repurposed to make attractive flower beds, or the shipping containers craned in to serve as pop-up cocktail bars, barbers or cinemas. Imagine a fourth city, a city of possibilities of sensory suggestion, where the unexplored and overlooked are ripe for discovery, where surveillance is an invitation to subversion and where the city is a space to be hacked, repurposed and reengineered. Far from antithetical, these contrasting creative city imaginaries coexist in many of today's urban spaces.

Cities have long been well-springs for creative practices, with large cities seen as the richest melting pot for creative practices. In recent years, however, the imaginary of the creative city has come to dominate urban regeneration and become a key way in which cities distinguish themselves and attract people and capital (Florida, 2005; Landry, 2006). The creative city offers an intoxicating vision, promising both social and economic regeneration (Mould, 2015). Such a winning combination has ensured the potency of the creative city imaginary and enabled its justification the world over.

This chapter is going to examine the rich and evolving discussions of the creative city through the telling of four creative city stories. Focusing on a range of cities around the world, big and small, discussion will negotiate the reductive nature of the 'globe talk' often associated with popular economic global traits like the creative city. Instead, it will 'open up globe talk by animating certain agents of global capitalism and cosmopolitan ideology, by highlighting their distinctive and bounded territories, and by filling out more fully their distinctive subjectivities' and by exploring the local contexts of the particular locations being engaged with (Ley, 2003). The first narrative addresses the most pervasive idea of the creative city, the planned creative city. Exploring cities from Detroit and Bilbao to Shenzhen and Dubai, it examines a range of planning models such as the creative quarter and the media city alongside the ideas and practices of urban studies professor and creative city guru Richard Florida. While creative city policies operate across a range of scales and forms of city, with various effects, the potency of Florida's ideas and the currency they have come to have, have resulted in particularly strong critiques of his ideas. Such 'fast' urban policy (Peck, 2005) has been critiqued as enabling economic gain for private and public stake-holders, rather than making it possible to tackle root causes or systemic local issues. At its worst, the creative city paradigm has been understood as a linguistic tool that valorises existing infrastructural projects and justifies politics of beautifying public space, real estate development and the financialisation of previously non-business cultural activity (Peck, 2012). Others have offered lengthy critiques based on a lack of evidence, and concerns with the class-based analysis of much of the work and its lack of awareness of inequalities (Markusen, 2006; Peck, 2005). Indeed, Wilson and Keil powerfully argue that the creative class 'flagrantly configures an elitist theme for change that feudal lords and bourgeois captains of industry in the past would have hesitated to do' (2008, p. 844). Indeed, the story of the planned creative city is for some an antithesis of urban creativity, offering little support for creative individuals and causing further decay in the social fabric and cultural diversity of cities around the world.

The three other stories told here narrate three alternative creative cities to that of the planned creative city. The second focuses on the gentrified city and thus offers a less positive spin on the effects of creative city politics. Taking the classic case of SoHo in New York, the story unfolds some of the debates around creativity and gentrification, exploring the tensions around creative practitioners as both drivers and victims of gentrification. The third story takes inspiration from Marxist Geographer David Harvey to explore how artistic practices have offered the means to both investigate our urban spaces but also to rewrite our city scripts, to intervene within our cities, critiquing them and creating them anew. The fourth and final story takes to the city streets to explore the subversive creative city, querying how vernacular creative practices such as skateboarding and yarn-bombing might offer a different understanding of the creative city. In their telling, however, the third and fourth stories seek to negotiate the potential for romanticism that is often found in the discussion of critical and subversive creative practices. As such, they unfold in such a way as to explore how such practices both stand against but can also become folded within the capitalist creative city.

CREATIVE CITY 1: BE CREATIVE OR DIE . . . PLANNING CREATIVE CITIES

> 'Be creative – or die . . . cities must attract the new 'creative class' with hip neighbourhoods an arts scene and a gay–friendly atmosphere – or they'll go the way of Detroit'

'if you think of a place that was close to death and is now entering into a new life, that's Detroit. Why does that happen? Well there's great space available, there's affordability. But cities attract different people . . . Detroit is a place where anything goes. It's a place that's open to people'

In 2009 in the midst of Global Recession, Richard Florida proposed what he called the 'Great Reset'. Floated first in a piece for the magazine *The Atlantic* and later developed into a book (Florida, 2010), the Great Reset proposed the development of a very different economic order shaped by the recession. In Florida's telling, Detroit, the one-industry town of auto-manufacturing, was one of those places that was going to get left behind. As he wrote 'perhaps no major city in the US today looks more beleaguered than Detroit, where in October [2009] the average home price was $18, 513 . . . the city's public school system, facing a budget deficit of $408 million was taken over by the state . . . in December (2009) the city's jobless rate was 21%' (Florida, 2009). In his discussion, Florida noted the idiocy of attempting to prop up urban areas based on failing industries 'different eras favor different places, along with the industries and lifestyles those places embody . . . we need to let the demand for the key products and lifestyles of the old order fall, and begin building a new economy, based on a new geography' (Florida, 2009). For some such an argument was odd coming from Florida, who had for so long argued that even ailing cities could turn their fortunes around by way of attracting the creative class. Florida argues, however, that his intent was to speak out against those economic strategies that focus on rescuing failing industries. Four years after the piece and just after Detroit had filed for urban bankruptcy, Florida noted that all was not lost and indeed 'beneath its fiscal problems . . . lie the seeds of rebirth for the city' (2013). Exploring the ongoing repositioning of the area for the knowledge economy he cites economic growth that produces $200 billion in economic output, more than New Zealand and not much less than Hong Kong or Singapore.

In exploring the seeds for Detroit's regrowth, Florida (2013) noted his classic three Ts; talent, tolerance and technology. Talent wise – equated here to college degrees – he noted a concentration of college educated young adults in Detroit's downtown core (42% of the population, higher than the national average of 9%). He also observed 'substantial concentrations of talent; about 34.5% of the area's workers are members of the creative class, slightly above the national average'. He further went on to note how smaller creative and tech firms are coming back to the city, how start-ups are taking advantage of the provision of loans and the excellent technology and infrastructure including high-tech hubs like the investor developed M@dison building, home to Twitter amongst other tech firms. Indeed, as well as creative industries, hi-tech employment – which comes under Florida's creative class – is also seen in the area. Jobs in computer systems design and research are growing at a rapid rate in Detroit, up 7.5% per year (2010–2013) compared to a US national average of 2.7%. There is also a growing creative-industries sector within the city, not only interestingly linked to tech, but also web designers and advertising agencies and a growing arts scene with New York galleries such as Galapagos and others buying up space in Detroit for both reloca-tion and expansion. The city also has a burgeoning maker movement (see also Box 7.1). Dougherty, founder of the *Make Magazine*, set up the maker fair in Detroit in 2010 four years after piloting the idea in San Mateo California to prove that it had currency beyond Silicon Valley. He cited the legacy of manufacturing and making and the DNA of a place like Detroit as a reason for starting in the area. In Detroit some of the maker move-ment's key participants are ex-automotive workers who have transferred their skills in metal work, bodywork and so on, into crafts such as making furniture from repurposed industrial material and signage. As well as such possibilities in repurposed skills, for Florida the rich creative past of the area also sows vital seeds for its continued economic growth.

Richard Florida sits alongside Charles Landry as perhaps one of the most influential thinkers on the creative city, shaping both academic debate and also policy and governance strategies around the world. In 1995 Landry, together with Franco Bianchini, published one of the first books on the creative city, the currency of this concept clearly indicated by Landry's (2011) follow-up publication five years later by *The Creative City: A Tool Kit*, aimed specifically at policy makers. The text proposed the value of cultural resources as the intellectual and practical solution to the struggles facing cities as they negotiate the transition to a post-industrial landscape. As well as the economic benefits accrued through investments in physical infrastructure, such cultural resources brought added benefits in terms of their building of urban brands and images. If Landry focused on a range of infrastructural and urban planning solutions for Florida, the focus was on creativity as a people-based asset. So for Florida, creativity is not just an attribute of the urban landscape, rather key to his sense of the economic potential of creativity is a particular group, the 'creative class'. In the context of the huge volumes of discussion and engagement with Florida's ideas that are well summarised elsewhere (Markusen, 2006; Peck, 2005; Pratt, 2008; Wilson and Keil, 2008) this discussion is most interested in the ideas of creativity perpetuated within Florida's text. Clearly creativity is understood here for its creative possibilities, but creativity is principally associated with being an asset of people. For Florida the wider economic and social benefits of creativity evolve from attracting the right people. These creative people, who are attracted in turn by the right services, living spaces and aesthetic fuel the 'new' urban economy. In his book *The Rise of the Creative Class: Revisited*, Florida sets out his thesis:

> The distinguishing feature of the Creative Class is that its members engage in work whose function is to create 'meaningful new forms.' I define the Creative Class by the occupations that people have, and I divide it into two components. What I call the *Super-Creative Core* of the Creative Class includes scientists and engineers, university professors, poets and novelists, artists, entertainers, actors, designers and architects, as well as the thought leadership of modern society: nonfiction writers, editors, cultural figures, think-tank researchers, analysts and other opinion makers. . . . Beyond this core group, the Creative Class also includes 'creative professionals' who work in a wide range of knowledge-intensive industries, such as high-tech, financial services, the legal and health care professionals, and business management.
>
> (2012a, p. 38–39)

In an interesting inversion of the economic principles of people following jobs, Florida suggests that people and their creativity are desired by cities the world over. Such that cities will follow Florida's prescription for upgrading their cultural facilities and aestheticising their built environments in order to create the spaces and atmospheres desired by this class of individuals. Florida (2012a) notes three urban characteristics that are important in attracting the often highly mobile creative class; the three T's – technology, tolerance and talent. For Florida, technology denotes both the sorts of supportive infrastructure needed for many creative businesses, including high-speed internet connectivity, as well as the presence of high-tech businesses. Talent is measured through higher education achievement, drawing an equation between highly educated people and jobs in the knowledge-intensive economy. Perhaps the most controversial 'T' of Florida's trio is tolerance, measured through a combination of the 'boho' and 'melting' pot indices. The former draws together a group Florida thinks of as bohemians, including artists, designers, makers, writers, musicians, dancers, performers, and so on, with the gay community, whilst his melting pot index focuses principally on only one form of diversity – ethnicity. To score highly on these two indices is to find an area tolerant of diversity, an area in short that is desirable for his creative class, as Florida notes areas 'seething with the interplay of cultures and ideas; a place where outsiders can quickly become insiders' (Florida, 2002, p. 227).

If Florida's idea of creativity is one that is focused on people, his creative class is perhaps not the sort of creative people that we might expect them to be. Far from being focused on creative occupations, Florida's 'creative professionals' include an amorphous grouping of consultants, technicians, firm managers, financiers, realtors, as well as members of a political elite including public administrators, politicians and think-tank members. For those critiquing Florida's ideas of the creative class, his grouping is too wide and baggy, and he has gone too far when his calculations begin to equate creativity with higher education achievement. For his critics, this is not demonstrating creativity, but rather is measuring high levels of human capital as linked to educational achievement (Markusen, 2006, p.1922). In some people's eyes, his is not really a theory of creativity and a creative city so much as one based on those who have high educational capital and are focused around a lifestyle and experience economy.

Florida has quickly risen to guru status, with his creative class becoming the latest and perhaps most successful of all the positionings of creativity at the heart of urban development and regeneration plans (Peck, 2012). Around the world we now see a wave of projects, policies and processes put into practice by urban governments determined to attract creative people to their city. Core to such projects is a meshing of 'soft' marketing strategies with 'hard' infrastructure-based development plans – the creation of public spaces and aestheticised landscapes – that reinforce one another (Mould,

2015). Much sought-after the world over, cities will pay thousands of pounds to have Florida come and assess what is limiting their creative potential and suggest ways to climb up his much publicised city rankings.

Of course, Florida's notion of the creative city is just the tip of the creative urban planning iceberg. The planning of creative urban spaces around the world has taken a range of different forms and scales. From Florida's focus on the cultivation of 'people climates', to planning that focuses on supporting specific forms of creativity – for example, media cities – or specific zones of creative practice – creative zones or cultural quarters. Indeed, as Oakley noted of the UK,

> no region of the country, whatever its industrial base, human capital stock, scale or history is safe from the need for a . . . 'cultural quarter'.
> (Oakley, 2004, p. 68)

The cultural quarter has become a ubiquitous feature of urban forms around the world, being found in major metropolitan areas, as well as small market towns. Such spatial sites of creative cultivation whilst historically present, as the example of Birmingham's jewellery quarter demonstrates, have become increasingly seen as an effective way to redevelop ailing urban areas. Indeed, they rose to popularity in the UK and other cities around the world (from Tallinn Estonia to Shenzhen in China) as a means to redevelop post-industrial landscapes. As such, cultural quarters display a varied scale and constituency. They range in size from a city district to a street, a small cluster of buildings or even a single building, and vary in focus from those predominantly production-orientated and those that are based on mixed-use to those whose focus is creative consumption often highlighting the night-time economy.

With the nick-name 'City of a Thousand Trades', Birmingham in the UK Midlands has a history of small-scale craft production, and from the mid-eighteenth century onwards was known for its industrial village of jewellery makers, who also made boxes, trinkets, buttons and buckles. In its heyday the area employed over 50,000 people, but by 2005 this had fallen to 3,100. Despite this the area remains a significant force in the trade in Europe (Hughes, 2013, p.4; Pollard, 2007). Recently, the Quarter has shifted from production to consumption and continued to fight against the negative imaginary associated with the provincial industrial city in which it is based. It has become increasingly acknowledged as a valuable asset to the city, 'a vibrant atmosphere which not only attracts people to work, live, play and visit, but acts as a honey pot for creative businesses ranging from the current jewellery business base to the arts and media' (Hughes, 2013, p. 10). It is also worth noting that in this case the area had a strong history of being invested in this form of cultural production before it became supported by policy intended to strengthen the profile and ensure the survival of this cultural quarter.

As creative production complexes, creative quarters bring the artist's studio complex, incubator spaces and live–work spaces for creative workers together with support institutions, and commercial outlets. Sometimes these might find their form around a single industry, such as Birmingham's jewellery quarter or focus on mixed ecologies of products, such as in Currid's (2007a) 'Warhol Economy' in New York. Those creative quarters driven by cultural consumption often feature a high-profile locational anchor, principally a key museum or gallery or less often a site of cultural heritage (Mould and Comunian, 2014). Such flagship institutions are the drivers for the growth of other services and entertainment facilities, including often strong night-time economies (Evans, 2009; Mommaas, 2004; Pratt, 2008). Perhaps the most famous example of such a flagship institution and its potential effect on creativity-driven regeneration, not only within a city district but across the whole city, is Bilbao's (Spain) Guggenheim Museum. Many cities around the world have sought to emulate the 'Guggenheim effect' (Plaza et al., 2009), from Oslo, to Amsterdam, Hong Kong and Abu Dhabi. Often at great expense 'starchitects' are employed to design iconic structures that will ensure the city stands out from the urban crowd, redeveloping industrial areas economically and socially and rebranding cities. As well as trickle-down direct economic effects, it is hoped that such projects will also kick-start wider projects of gentrification within the city.

The Guggenheim Museum opened in Bilbao, in Spain's Basque country, in 1997. It is credited with transforming a deindustrialised port into a thriving cultural centre, increasing visitor numbers by 43% and adding 168 million Euros to the local economy in 5 years, the equivalent of creating 4500 jobs. The museum has become the centre-piece of the city's arts district, which also comprises maritime and fine arts museums and an arts centre. Designed by Californian 'starchitect' Frank Gehry, it joined the metro system designed by Sir Norman Foster and Zahah Hadid's proposed master plans for the city's ailing industrial area.

It is important though to not fall under the sway of this 'resurrection by culture' myth. It might, for example, be more accurate to term the effect the 'Bilbao effect', in recognition of situation of the museum as the flagship project of an integrated programme of urban redevelopment that linked a revitalised cultural sector with new transport policy and general urban redevelopment. Furthermore, it was clear the Guggenheim effect was also severely limited in spatial and temporal scale and scope. While it brought an immediate influx of tourists, the effects have been localised, failing to engage the wider region's economy or solve regional poverty and unemployment issues. Another key question regarding the Guggenheim effect was whether it would move beyond 'American imperialism' to support local artists. In other words, was the whole project to be focused on creative consumption, or was it also going to bring benefits for creative producers too (Plaza, 2000; Gómez and González, 2001)? There has been, for example, a growth in commercial arts ventures around the gallery area, developing a new cluster of galleries around the Guggenheim that have transformed the spatial layout of arts in the city. This cluster of galleries is, however, not really linked to the museum, but rather engages a more exclusive clientele. Their existence, however, speaks to the wider positive impact of Guggenheim as creating an image of an arts city. Elsewhere in the city there is a dense cluster of producers, as well as festivals and fairs related to local crafts and handicrafts. While this is a different form of creative reproduction to that being displayed in the gallery, it does receive a boost from arts engaged tourists who now visit the area.

More though than cultural consumption, a series of positive effects on the local arts of the region have been identified, especially on the website for Bilbao as a City of Culture. Initially the focus of the gallery collection and exhibition remit was on the key American masters, so the Rothkos, Warhols and others associated with Biebao's sister gallery in New York. Further, the gallery's original attraction was the offer of the site and the need for propping up its own finances rather than any pool of local art talent, as we see in the UK Tate's Cornish regional gallery in coastal St Ives. Until 2007 no local artists were shown in the Bilbao gallery, either in the permanent collection or in the temporary exhibitions. This gradually changed and, as well as putting local art on the exhibition agenda, by 2017 the gallery will have invested up to 150 million Euros in local and regional art. The ongoing support of the creative industries in the region, in particular the arts scene has also been recognised in the founding of a municipal exhibition network which is committed to finding and supporting other venues across metropolitan Bilbao to enable artists to exhibit work. Equally important is the establishment of Bilbao Art, in 1998, a year after the gallery opened. The foundation is focused on supporting artistic production and has done much to rebalance the dynamic in a city where increasingly art was being consumed but not created. As such, it provides an incubator space for artists, grants for material, space for exhibitions, library access, leaflet and book publishing contacts and so on.[1]

The Guggenheim effect was based on a set of circumstances and changes that brought a world-leading institution to a de-industrialised port city and debates continue as to whether or not it is a model that can be relocated elsewhere. These debates don't seem to deter those developers and local governments for whom flagship architectural projects by the likes of 'starchitects' such as Frank Gehry, Zahah Hadid, Norman Foster and Daniel Liberskind remain popular. In 2013 *The Economist* reported over $250 billion being spent on two dozen major new cultural centres around the world, many designed by key figures, including locations such as Inner Mongolia, Saudi Arabia and Kiev. It will be interesting to see how these projects evolve and what lessons they learn from Bilbao.

General lessons learnt from a range of creative planning projects seem to focus on the desirability of mixed-use sites that bring together creative production and consumption:

> An essential pre-requisite for a cultural quarter is the presence of cultural activity, and, where possible, this should include cultural production (making objects, goods, products, and providing services) as well as cultural consumption (people going to shows, visiting venues and galleries).
>
> (Montgomery, 2003, p. 296)

Further, the branding of the area and the creation of a distinct spatial identity seems important to many projects. In some places this is achieved through the development of a flagship institution, whilst in others it is sought through the creation of aestheticised spaces. This is especially the case with more consumption-focused cultural quarters and those that have been 'manufactured', rather than those that have evolved from an embedded sense of creative production that has spanned generations (see, for example, OCT Lofts, Shenzhen in Box 7.1). Indeed, studies of creative quarters identify the importance of 'on-brand' public space that adds to the 'atmosphere' of an area:

> In the more successful quarters this design ethos is carried through into architecture (modern, but contextual in that it sits within a street pattern), interior design (zinc, blonde wood, brushed steel, white wall) and even the lighting of important streets and spaces (ambient, architectural and signature lighting, as well as functional). All of these reinforce a place's identity as modern and innovative.
>
> (Montgomery, 2008, p. 307–308)

Box 7.1 OCT LOFTS, 'CREATIVE CULTURAL PARK' : ARTISTIC GENTRIFICATION IN CHINA

Li LeiLei and Harriet Hawkins

Shenzhen Overseas Chinese Town Lofts, known as OCT Lofts, in Shenzhen, South China, has gained the nick-name 'Shenzhen's SoHo', after the concentration of creative activities that have come to characterise this high-end area (Figure 7.1). Shenzhen rose to fame as a manufacturing megalopolis after the designation of the former rural area as a special economic zone in 1980. The city, known for fast building practices that created an entire urban area in less than 30 years, was built on the back of the 'made in China' labels that appear in consumer goods around the world (see Box 9.1). Yet as the city adapts to changing forms of capitalism and competition from manufacturing zones even cheaper than it is, it has begun diversifying into the creative economy. Not only is creativity attractive for potential economic value (the creative industries) but also due to the increasing recognition in China of the potential of soft power (see Chapter 10). As such Shenzhen, long an urban lab for the nation, has sought to become a 'lead site for creativity', and in 2008 successfully bid to be a UNESCO city of design, one of six around the world and the only one in China.

Figure 7.1 OCT Lofts, Shenzhen, China

As the city mushroomed in the 1990s the electronics industry moved from what had become a central city location to the suburbs, leaving large empty factory spaces. The area of these to the east of the city centre in Nanshan were bought by the private real estate company OCT. With the help of local architects Urbanus, OCT set about to transform over 2000 square feet of communist austere factories into a creative cultural park. This park combines characteristics of a creative industries hub with new-build luxury apartment complexes and takes advantage of the creative 'cool' factor to leverage the creative sector to create the vibe for the area.

OCT Lofts combines a lot of different facets of clusters; the production values of the creative cluster; the credentials offered by the presence of an anchor institution in the form of OCAT art gallery, as well as a vibrant commercial and entertainment space. The main occupants of the studio lofts are a series of both mature and start-up creative companies, predominantly architects, graphic designers, ad agencies and photographers. The repurposed industrial buildings

Continued

offer cheap spaces to these companies, offering tax incentives for them to relocate in the large bright spaces (Figure 7.2). For many, they take advantage of the space to develop open plan studio spaces from which they can both produce but also sell their products. As well as spaces in which to make work, these spaces are often fashioned as attractive spaces for clients to visit. When not taking meetings or in their studios, the local coffee bars and restaurants also offer sites for creative work or for doing business with colleagues and clients.

Figure 7.2 OCT Lofts studio space

OCT Lofts are committed to supporting young and emerging designers and makers. It is home to Chaihuo Makerspace, one of the first hacker spaces in Shenzhen. Here as Box 9.1 details further, innovators and creative makers can access (for a small fee) work-space but also technical equipment such as 3D printers to develop their ideas. As well as supporting the production of younger makers with facilities and space, OCT Loft also hold a creative makers faire, 'T Street Creative Faire' for several week-ends a month. During the fair local makers and up-and-coming designers sell their wares from street stalls. Jewellery, hand-made leather goods, fans and hand-made paper and bound books are all popular.

As well as a space for production, OCT Lofts creative cultural park displays the quintessential creative 'atmosphere' during the day but also at night. It is known as the place to see and be seen, with a cluster of bars and restaurants serving local delicacies, as well as global cuisine at some of the highest prices in the city. There is also a live music scene, taking advantage of the warm nights to let people spill out into the spaces between the bars, restaurants and galleries. The area is also home to a growing number of boutique hotels and commercial outlets. Clothes, books and pottery are the most common products sold, interest-ingly few sell products made in the lofts, even 'The Loft Shop' specialises in international lifestyle brands, such as Hay (a German furniture designer) and Freitag, a Swiss company who manufactures bags from truck tarpaulins. Elsewhere in the complex shops sell clothes, hand-made tea sets, as well as expensive and iconic lifestyle brands from around the world, including Tom Dixon (London) and Vitra (Switzerland).

OCT Lofts is also a site for display. As well as numerous commercial galleries the space is home to the OCAT, the contem-porary art terminal. OCAT Shenzhen is the head-quarters for the OCAT group of five museums, with branches across China, including in Shanghai and Beijing. OCAT is generally considered to be a national-level institution and one of the prime venues for display of contemporary art in China. The group specialises in the

Continued

practice and research of contemporary art and theory within China but also in the international arena. As well as exhibitions of Chinese and other artists' work, often curated by international curators, OCAT Shenzhen hosts the sculpture biennale as well as events in the Shenzhen/Hong Kong bi-city architecture and urbanism biennale. They also curate the performance art series 'OCAT performs' and 'OCAT screens' that focuses on screening lectures, documentary films and media work. OCAT Shenzhen also has a library and publishing division, developing exhibition-related publications but also other work, and 'OCAT youth' a programme aimed at engaging and stimulating the interest and supporting young practitioners. Many of these activities take place in two huge exhibition halls that sit in old renovated ware-houses able to hold large works and major travelling shows. In addition to these locations on the main OCT Lofts site, they also have a further location OCT Art and Design Gallery, which repur-poses a further factory into a gallery.

The Creative cultural park that OCT has created has become very popular, setting the model locally for a successful mixed-use urban setting. It brings together a range of different elements that have come to characterise creative clusters around the world, all concentrated here into one space.

Whether it involves renovating existing ex-industrial spaces or building anew, what is clear is that the cultivation of creative quarters can come in a huge range of forms and scales. While in the UK this sort of development seems to have been kick-started by the 1980s decline in industrial manufacturing, in other global contexts the development of these agglom-erations has a very different history and feel. We might think for example of Dubai Media city (Box 7.2) and its grand vision of a creative city that is thoroughly in-line with the privatisation and securitisation of urban space more generally.

Media Cities are gaining ground as an addition to the landscapes of creative development, from Dubai Media City, DR Byen in Copenhagen, Digital Media City in Seoul and MediaCity UK in Salford. Media cities are highly planned, developed urban areas, designated specifically to concentrate media and creative industry production in its broadest sense (Mould, 2014). Indeed as Goldsmith and O'Regan (2003, p. 33) note, the media city has become 'recast as a form of commercial property/ industrial park development'. In contrast to other creative landscapes that might seek to integrate within the wider city fabric, media cities tend to be more isolated and self-serving, being described as 'islands of investment'. Another feature of media cities, indeed a defining feature, is the vast level of financial investment that such cities need. As Mould (2015) notes, the 'technological capacity, luxury condominiums and hotels and high-end office provision needed is way beyond the price range of many governments'; however, many in the Middle East have sought to invest in these cities in order to grow their economies.

Created as one stop shops for the creative class, media cities purport to offer high-specification working facilities, luxury living accommodation often in gated locations, state-of-the-art cultural facilities and high-end retail outlets. They are also usually served by super-fast broadband connectivity in order to transfer the large (high-definition quality) content, and support state-of-the-art production and exhibition facilities and global connectivity (Evans, 2009). Dubai Media city (Box 7.2) is a classic example, with huge amounts of money invested to create infrastructure under a 'build it and they will come' mentality. As well as serious business investments these media cities are also seen as significant global branding exercises, similar to 'starchitect' projects such as the Bilbao Guggenheim.

Box 7.2 DUBAI MEDIA CITY: FREEDOM TO CREATE

The architecture of Dubai Media City (DMC), like Dubai itself, is a manifestation of dreams and aspirations. Rising from the sand dunes 20 km south of the CBD of Dubai, the first phase transformed 200 acres of desert into lush landscaped estates, including a lake the size of two football fields and crowned by three five-storey towers of glass decorated inside with marble and fake wood panelling. Housed within the city are 1400 media and marketing, publishing, broadcast, production, graphic arts

Continued

companies, occupying 33,000m² of the world's most highly networked and digitally equipped space. Taking only a year to build, the city opened in 2001 at a cost of between $700 million– $1 billion (Seib, 2007), its towers and offices can be understood as the latest incarnation of the high-tech fantasies that were the go-to strategy of the late 1980s and 1990s (Mould, 2015).

The built form of the DMC represents Dubai's fears of a limited future oil income and its investment in creating a world class media centre that they want to act as a meeting point between east and west. DMC is described as having 'field of dreams feel', and is one of a triumvirate of media cities in the Middle East (Jordan Media City and the Egyptian Media Production City) that have been tasked to drive the knowledge economy in the area; their key role being to act as magnets for human and financial capital and to assemble a critical mass of talent, money and technology, (Quinn et al., 2003).

DMC is well supported by the government. Occupying land donated by the crown prince, the city was built to provide 'an infrastructure, environment and attitude', that will enable new economy and media enterprises to operate locally and globally out of Dubai (Seib, 2007). The DMC vision was to form a global base for broadcasting; to bring this about strategies drawn from retail were used to encourage anchor firms to locate in the hope of attracting other key names. Reuters, CNN and Middle East Broadcasting Center all have named buildings but have taken very little actual space. They have also attracted Saudi Research and Publishing, the largest Pan Arab Publishers. DMC benefits from its location in the wider TECOME area, Dubai's new economy hub. Standing for 'The Technology, Electronic Commerce and Media' free zone authority, TECOME is seen as a 'symbol of potential for the knowledge economy in the region'. In practice this means sophisticated infrastructure and incentives for relocation offered to desirable global enterprises. It sells itself as 'providing a tax free environment, transparent relationships with government and legislative authorities and a one stop shop for all services' (Quinn et al., 2003). Other benefits include the

> permittance of 100 % foreign ownership within the DMC; outside
> the zone you need an Emiratee co-owner. The free zone also
> exempts all businesses from all taxes for 50 years and reduces
> sponsorships and visa requirements (Quinn et al., 2003). The
> sense is that these cities will capitalise on the area's most valuable
> resource, human capital through a synergy of business, govern-
> ment, technology and finance. They are also seen as shining
> symbols of modernity in the Arabic diaspora and an expression of
> hope (Seib, 2007).

Perhaps unsurprisingly, the raft of planned creative cities has not been without their stringent critiques. Indeed, the critical literature almost matches the positive literature in terms of its volume and vociferousness. The remainder of this chapter will explore three further stories of the creative city. Each of which, in its own way offers a critique and / or an alternative creative city to that of the planners.

CREATIVE CITY 2: THE GENTRIFIED CITY

the notion that cities must become trendy, happening places in order to compete ... is sweeping urban America ... a generation of leftist policy makers and urban planners is rushing out to implement Florida's vision [just as] an admiring host of uncritical journalists touts it ...

(Peck, 2005, p. 471)

While the belief in creativity as a panacea of economic and social ills, as a solution to the challenges of contemporary urban development is strong, vigorous academic debate has also explored how creativity as a narrative of development and regeneration does not always live up to its promise. Indeed, often it acerbates ongoing issues of social inclusion and economic inequality rather than easing them. For some, it is just another example of the sort of one-place-after another policies that move from city to city, an economic and political model that overrides the characteristics and contexts of localities – their social customs, histories, cultures and economic backgrounds.

It is not possible to talk about links between creativity and the city without considering the relationship between creativity and processes of gentrification, indeed the story of the planned creative city is often one of gentrification. Ruth Glass, on observing the struggles of London's working class around housing in the mid-1960s coined the term gentrification – literally denoting the overtaking of city spaces by the elite – an urban version of landed gentry. As Ley outlines, 'gentrification involves the transition of inner-city neighbourhoods from a status of poverty and limited property investment to a state of commodification and reinvestment' (2003, p. 2527). Neil Smith, one of the foremost geographers writing on gentrification observed,

> The crucial point about gentrification is that it involves not only a social change but also, at the neighbourhood scale, a physical change in the housing stock and an economic change in the land and housing markets. It is this combination of social, physical, and economic change that distinguishes gentrification as an identifiable process or set of processes.
>
> (1987, p. 463)

Gentrification has now become a central part of urban planning, with fierce debate over its effects. For some gentrification leads to segregation and social polarisation, for others 'it will lead to socially mixed, less segregated, more liveable and sustainable communities' (Lees, 2008, p.2449).

Key to Smith's analysis of gentrification was his development of the Rent Gap theory (Smith, 1987, 1996). In short this theory captures the difference between the value of a piece of land in its current state and what it could be worth. As a production-orientated approach focusing on cycles of property disinvestment and reinvestment as driving the process, Smith's theory stands in contrast to consumption-orientated approaches that understand consumer preference to drive gentrification (Mathews, 2010, p. 661). Whereas once these two perspectives were in competition, more recently they have been brought together. Not least in discussions of the arts and gentrification, which mesh real estate and the cultural industry (Zukin, 1989) and which explore what Smith (2006) describes as a 'class remake of the central urban landscape', in which artists as well as broader creative processes of gentrification play a key role.

Of course gentrification is a very diverse process, not all projects involve the eviction of former residents; some take over ex-industrial areas. Further 'gentrifiers' are not a homogenous group but also need to be differentiated, whether such differentiation be between creative workers or non-creative sector workers, or between 'ordinary' middle-class gentrifiers and the 'super' or 'hyper' gentrifiers (the super-wealthy identified in parts of London and New York) (Lees, 2003; Butler and Lees, 2006). The role of artists and creative workers in urban change is perhaps one of the most controversial elements of the 'creative city' (Mathews, 2010; Zukin, 2010). Creative workers are often seen as key agents in gentrification and drivers of the aestheticisation of the urban space that often accompanies it. Yet creative workers are also one of the sets of 'victims' of the process (Mathews, 2010; Ley, 2003). In her famous study 'Loft Living' (1989), Sharon Zukin explores the transformation of SoHo in Lower Manhattan from a blighted industrial area to a luxury neighbourhood, citing artists as the 'first gentrifiers' of the area. David Ley, writing of the process in Vancouver Canada described these artists as 'colonizing art' for the middle-class, 'opening up new spaces for the inner city through the image and identity attached to their lifestyle and production (Mathews, 2010, p. 665). In 1961 the New York authorities rezoned Manhattan, relegating heavy industry to the outer boroughs. The industrial buildings, now empty, formed the perfect site of the live–work lofts of thousands of artists, musicians and fashion designers. Attracted by what Bain (2003) terms these 'improvisational spaces', these run-down spaces were ideal for experimental practices and open to multiple functions. For others, especially more recently, it was as much the built form and its cheapness and flexibility, as well as the social conditions of an area. As Cameron and Coaffee (2005, p. 40) explain 'what the artist values and valorises . . . more than the aesthetics of the old urban quarter . . . is the society and culture of a working-class neighborhood, especially where this includes ethnic diversity, [it] attracts the artist as it repels the conventional middle class'. This concentration of artists working in the area shifted its focus as services grew up to support them. Whereas once they squatted the spaces illegally, in 1964 the city set up a programme – AIR – Artists in Residence, enabling the artists to legally access the spaces as work spaces. By 1971 the city had recognised the positive force of creative workers in reshaping the face of the industrial area, so created SoHo as an arts district. Eventually, the artist residency requirements were relaxed and the area became a general residential space, with incomers less likely to practice art and artists' gradually being priced out, as the value of the real estate was realised. As Patti Smith, who once frequented these lofts said to young artists 'don't move to New York'. David Byrne, a former Talking Head recently wrote of Manhattan, 'there is no room for fresh creative types. Middle-class people can barely afford to live there anymore, so forget about emerging artists, musicians, actors, dancers, writers, journalists and small business people'. As Ley (2003) documents 'typically, social and cultural professional and pre-professionals are early successors to artists, including such cultural producers as intellectuals and students, journalists and other media workers, and educators, to be followed by professionals with greater economic capital such as lawyers and medical practitioners, and finally by business people and capitalists'.

SoHo has become an archetype for the power of artist-led gentrification, and cities the world over were following the mantra 'bohemian today, high rent tomorrow', long before Florida described his creative class thesis (Podmore, 1998, Shaw, 2006). Indeed a SoHo Syndrome has been identified, 'a spatial and cultural process that involves more than simply copying the aesthetic of SoHo (New York) as a redevelopment strategy ... cities are 'locales' ... [and Soho Syndrome is] more than a universal valorisation strategy; it is a socio-cultural process that involves a complex web of relationships between place, identity and the media, that is diffused to, and (re)produced in, divergent inner city locations' (Shaw, 2006, p. 182). The question of whether artists and creative workers are victims in this process or active participants has generated and continues to generate a significant amount of debate within the literature (Mathews, 2010, p. 665). For Deutsche (1996) artists at best were complicit in wider processes and at worst rendered the working class and homeless invisible in their occupation of New York's Lower East Side. For others, it is more complex than simple complicity, especially as artists themselves often later become displaced. Drawing on Bourdieu's theories of cultural capital, Ley (2003) suggests it is inappropriate to blame artists as they alone do not account for the social valorisation of the[ir] cultural competencies and the cultural capital of art which attracts those with higher economic capital. As discussion goes back and forth between artists as 'anti-capitalists', as victims, as complicit, as agents who can make profit, Bain (2003, p. 305) sounds an imminently sensible note when she proposes that 'if artists are to be understood as anything other than urban pioneers and initiators of urban revitalisation efforts, they need to be appreciated more fully in their own right, as a social group with a distinctive occupational identity and a heightened awareness of the availability, regulation and character of urban space'. Such characteristics can enable artists to intervene actively in processes such as gentrification, highlighting the ongoing in equalities that such processes perpetuate. While Deutsche offered a swinging critique of how public art had come to be situated hand-in-hand with the development of cities, this is not the whole story and, as the remainder of this chapter will go onto develop, artists and other creative practitioners offer a range of ways of not only intervening within, but also remaking urban spaces, practices and politics.

CREATIVE CITY 3: ART AND THE REWRITING OF CITY SCRIPTS

Urban geographer David Harvey (2009) calls for a rewriting of city scripts in order to take back our 'rights' to the city in the context of the privatised urban space, the increase in surveillance cultures and the corporatisation of everyday life. For Marxist spatial theorist Henri Lefebvre, creativity and aesthetic practices might provide a path towards achieving these 'rights to the city'. While for some, Lefebvre's arguments around creative practice might have earned him the title of romantic revolutionary (Grindon, 2013, p. 219), he does offer the basis for a sustained engagement with the place of aesthetics in social change. Perhaps most famously he explores a 'revolution of space' (subsuming the 'urban revolution'), in which he foresees not only a place for aesthetics, but also for 'great inventiveness and creativity' (Lefebvre 1991/1974, p. 419).

The street has long been an overtly political site, as Lefebvre notes on the Parisian riots of 1968,

> it was in the streets that the demonstrations took place. It was in the streets that spontaneity expressed itself . . . the streets have become politicized . . . political practice transferred to the streets sidesteps the (economic and social) practice which emanates from identifiable places

As a political space, a site in which to act, rather than a static backdrop to action, the street becomes a site through and in which to practice. Central to this premise is an understanding of the street and urban space more generally as sites of possibility, as David Pinder proposes,

> Critical urban interventions and spatial practices are based on the refusal to accept current conditions as inevitable and natural. Through imaginative means, they explore possibilities and enter the register of as if: 'as if I were another, as if things could be otherwise'.
>
> (Pinder, 2008, p. 734)

This story and the next will outline a range of ways that creative practices have been understood to enable and bring about such 'as if' moments, offering the chance to rewrite city scripts through a range of creative practices. Further, as this story will explore, if David Harvey calls us towards creativity as a means to rewrite urban scripts then it is not just the creativity of others that might offer the means to rewrite these scripts. Indeed, in recent years geographers have explored a range of creative practices and methods as a means to both research and to live differently in cities.

Creative practices have long offered a rich means to enable us to explore and engage differently with our urban environments. Examining art 'taking place on the street . . . exploring and intervening in the city' through guided walks, maps or ludic activities is to take urban spaces 'seriously as a sensuous realm that is imagined, lived, performed and contested' (Pinder, 2005b, p. 385). For some, these are urban interventions, creating ways to intervene in city spaces and processes by rewriting these urban scripts to 'challeng[e]ing norms about how urban space is framed and represented and where they might help to open up other possibilities' (Pinder, 2005b, p. 385). Arts practices thus become part of critical cultural geographies of the cities, often intersecting 'rights to the city' with the 'writing' but also the visioning and sounding of cities'. This 'includes practices of studying, representing and telling stories about cities; it also involves ways of sensing, feeling and experiencing their spaces differently, and with contesting "proper" orderings of space to allow something "other" to emerge' (p. 386–7). A key lens for understanding these creative urban critiques has been the sensuous, psychic, and subversive urban engagements of Situationist International, and work made in the shadow of this group that are discussed in Box 7.3.

Box 7.3 SITUATIONIST INTERNATIONAL: 'WE DEMAND GAMES WITH GREAT SERIOUSNESS'

Situationist International (SI), a European avant-garde group of artists, philosophers and activists formed in 1957, enable us to appreciate the spaces of the city, in particular the streets as sites of creative practice (Sadler, 1998). Indeed, this group are often credited with some of the most creative reclaimings of the streets. Their self-appointed leader Guy Deborde was the author of *The Society of the Spectacle* which became, after its publication in 1967, the unofficial text of the movement. From Deborde's perspective, capitalism had created life as a spectacle that hid the reality of its degradation of life and social relations. One of their solutions was to creatively reclaim the streets through an aesthetic project of radical urbanism that aimed to establish 'a new form of geographical investigation that can enable the revolutionary re-appropriation of the landscape' (Bonnett, 1989, p. 136). They did so via a range of practices, including developing alternative forms of urban planning, experimenting with new forms of mapping, as well as creating new ways to explore and experience the street. Key within these urban explorations was the 'derive', a psycho-geographic seeking out of hidden and off-the-beaten-track urban spaces and experiences through the movement through often random urban ambiences. Importantly, this was work created outside of the studios and, while it has often been the subject of exhibitions, it was originally conceived to be experienced within the city itself.

Naked City (1957) is one of SI's best-known map works. Made from 19 cut-out sections of a map of Paris connected by red arrows, it aims to 'map' urban space as it is performed and experienced rather than by using conventional cartographic methods. The result is the replacement of a fixed and static view from 'above' with an emancipatory understanding of the city as subjective, fragmentary and mutable. Through experimental methods like the collaged map of *Naked City*, 'detoured' advertisements, and the practicing of *dérive* (a mode of experiencing

Continued

space that reflects that of the everyday user) the Situationists understood and performed space, after Henri Lefebvre, as a 'socially produced category where social relations are reproduced' (McDonough, 1994, p. 66). This contrasted with understandings of space as a fixed and static backdrop that conditioned social relations. Thinking and occupying space in this way gave the SI resources to challenge the fixity of the 'spectacle' of capitalist society and its uneven spatial relations. SI's practices and legacies have formed a point of intersection for geographers and artists around theories of space, politics and resistance. Their practices have been enrolled in broader explorations of the politics of urbanism and the production of social space, developing under-standings of space that deploy aesthetics to resist closure and elude determination and representation (Bonnett, 1989; Bonnett, 1992; Massey, 2005; Miles, 2004; Pinder, 2005a).

SI's 'interventions' into urban space have formed an important legacy for artists who 'seek to better understand the city, whilst also changing it' (Loftus 2009, p. 329; Pinder, 2008). A growing body of geographical research has explored the work of contem-porary artists who engage the legacy of SI practices (see Loftus, 2009, and a special issue of *Cultural Geographies* on the 'Arts of Urban Exploration', 2005, 12, 4). In these explorations creative practices offer a tactical resource through which to create altern-ative urban futures, especially in the context of 'everyday' life (Loftus, 2009; Hawkins 2010). As a result, the city becomes a laboratory of experiment with creative practices offering the means to occupy urban spaces as sites of political responsibility and arenas for political engagement (Deutsche, 1996; Miles, 1997; Miles, 2004; Pinder 2005b; Pinder 2008).

It is against the backdrop of these creative urban interventions that we see creative and arts-based research methods coming to play an important role in how geographers research our contemporary urban environments. Visual methods – especially photography and video-making – have become a crucial part of rethinking how it is we might do fieldwork in urban environments. These images are positioned less as factual records of places been, and rather offer a means to get to grips with multiple dimensions of our urban environments (Simpson, 2011b). As Latham and McCormack (2009, cited in Garrett and Hawkins, 2013) note, images enable an attention to 'the everyday ecologies of materials and things; for thinking through the rhythms of urban environments; and for producing affective archives' (p. 252). Images are in short understood as part of the 'elaboration of ecologies of non-representational ethico-aesthetic practices' (ibid.).

Such valuations of the visual, together with the knowledge gained through the attentive processes of making, analysing and reproducing images is crucial to Garrett's research on the practices of urban exploration (2013, 2014). Image-making is central to this practice, both in terms of the explorers for whom photography and video-making are part of their culture. Sometimes these serve as records, trophies of heights achieved and depths reached or documents of fantastic places seen, taken back 'home' and circulated, often via the internet, for more general consumption. In this latter form they serve to expand our contemporary imaginaries of urban space (Garrett, 2015). These images served to literally make visible those elements of the urban environment (roof tops and

Figure 7.3 Imaging urban exploration (Photo: Brad Garrett)

urban subterranean) that are invisible to many of us, in doing so they begin to build a valuable aesthetics of these charismatic urban spaces and query the aesthetics of geography's verticalities.

For Garrett as a researcher, image-making was a central part of his investigations. The images he and others made enabled him to inquire into the embodied practices of urban exploration, to query visually what it was that this practice did and also render in visual terms some of the accounts of why explorers do what they do. Image-making became key to investigations of the explorers' embodied engagements with urban space, the entwining of the materialities of cities and bodies, in rushes of adrenalin, in spikes of hormonal machismo, and in the movements of bodies across often dangerous urban territories (Garrett and Hawkins, 2013). The power of these images lies in the attentiveness to situations cultivated in their making, their editing, their continual production and reproduction, and the distribution and circulation of these images. These processes enabling an attunement to the relations between bodies and the urban fabric, to the rhythms and routines of exploration, as well as to its more obviously representational visual cultures.

Figure 7.4 Imaging urban exploration (Photo: Brad Garrett)

Taking the use of creative methods to engage urban space in another direction, Phil Jones (2014) develops an artistic performance of a sustainable ride across Birmingham, experimenting with GPS as an artistic device. Jones's trek is literally traced out in the word RIDE written on a map by the GPS track. In describing the act of performing the writing across the city, Jones emphasises the embodied experience, fearful because of forgotten lights and impending darkness, sweaty hot, hard work. This act of mapping enabled a rather different way of experiencing the everyday mobilities of a banal daily practice for the author. The tension between the scientific disembodied technologies of the GPS and the sweaty, emotionally intense and embodied experience of cycling made stark as the author narrates and maps out his journey.

This tension between mapping and emotional experiences was explored in a rather different way in artist/philosopher Christian Nold's experiments with emotional cartographies and the technologies of self. Nold has developed a series of projects exploring emotional cartographies based around the visualisation and geocoding of people's intimate biometric data and emotions using pervasive technologies. Biomapping emerged, as Nold explores, as 'a critical reaction to the dominant concept of pervasive technology', which aims for computer 'intelligence to be integrated everywhere, including our everyday lives and even our bodies' (2009, p. 3). The Biomapping project investigates the implications of creating technologies that can record, visualise and enable us to interact with each other in our intimate body states. He describes the creation of a Biomapping device, a portable wearable tool that combined a GPS with a biometric sensor measuring galvanic skin response. He records talking to people who tried out the device and the detailed way they would talk through their emotional routes.

Nold has made a series of emotional maps with communities around the world. Rather than showing static architecture and borders and boundaries, visible or invisible on the ground, these maps engage with the sites and their communities to explore the emotions, opinions and desires of local people. In the case of Stockport (a large town – around ¼ million people – in northern England near Manchester) Nold worked with over 200 people over two months in summer 2007. As recounted on his website, he organised six public mapping events that involved drawing provocations and emotional mapping. The drawing provocation involved asking people to sketch their responses to a series of queries about their daily lives. Examples Nold offers include what annoys them about their town, where they meet their friends, sites they fear, and so on; people created huge volumes of drawings that were then used to create a local map. The second activity saw participants walking through Stockport wearing the biomapping device Nold created. This device measured emotional arousal in relation to their geographic location; the result was a series of lines with peaks and troughs over certain locations, such as frustrating crossings or dangerous places. This participatory process not only generated the emotional map but also identified five issues for Stockport; the marginalised history of Stockport, the hidden river Mersey, the issues of shopping spaces, semi-public space and isolation of young people. As the artist notes, they hoped the map and text would 'stimulate personal reflections for people and then lead to a larger communal discussion that refines the issues of concern'.

Geographers have followed a number of artists in deploying creative methods as a means to sense the city. For example, David Pinder (2001) examines how the work of sound artist Janet Cardiff enables us to explore the value of listening to the city. Literally taking place on the streets, Cardiff's work requires her audience to listen to a sound track that directs them through the city. View and voice do not easily mesh, rather what is presented is an appreciation of how urban experience involves engaging with a multiplicity of perspectives, as information and memories jostle amid a shifting present (Pinder, 2001, p.7). As Pinder discusses, Cardiff's work does not just help us understand cities in multiple ways, but helps in the realisation of 'the difficulties of reading and knowing the city' where 'representationality has been thrown into doubt and become a focus of anxiety' (Pinder, 2001, p. 6). As well as exploring how artists' work can help understand the multi-sensory nature of cities and environments, geographers have also experimented with these creative practices themselves.

Exploring their experiences of being co-creator and audience for the street theatre production 'Do you see what I mean', Johnston and Lorimer (2014) recount how this site-specific work enables audiences to relate differently to their bodies in space, opening up 'possibilities for visceral encounter in an urban terrain reconfigured as an assemblage of bodies, surfaces, smells and acoustics'. This contemporary collaborative street theatre developed an urban choreography that saw audiences engaged in an immersive 2.5 hour tour of urban Vancouver, Canada. Blind-folded, audience members were guided through a series of public and private spaces and encouraged to encounter a range of people by way of telling stories, sharing personal objects, touching, smelling and eating foods. The tour ended with participants being led in a dance 'designed to explore expanded movement and to extend a heightened sense of embodiment gleaned through the guided tour'. They were also invited to listen to a live feedback loop developed by two artists. There is much in common here with the discussion of gallery based work and its encouragement of the exploration of the senses, only here these experiments become as much about knowing spaces and places as knowing the body.

The claims made for the sounding of cities have much in common with those made for writing and drawing places and cities. Tim Cresswell (2014), for example, issues a firm call for 'a renewed practice of place-writing'. He is not alone in these sentiments, indeed a number of geographers have challenged academic writing practices and conventions in search of ways to do things with words (in written and spoken form) that equip us in the face of the challenges posed by contemporary ways of 'thinking place'. Cresswell (2014) conducts written experiments with how to think about the becoming and dissolving of city spaces (a market in Chicago) in terms of both the long durée and on a daily basis, experimenting with 'gatherings', montages of created and found texts as a means to explore these ideas. Here, and in a debt he acknowledges, we find resonances of the work of geographer Allen Pred, who brought language and writing to geography as a conceptual form. Acknowledging, in turn, his own debt to key writer of urban space Walter Benjamin, Pred (1995) experiments with montage as a literary method, overlaying information and ideas to bring about a relayering of multiple pasts and ongoing presents. For others, experimenting with narrative form has become central to geographical experimentation with writing. Both Caitlin DeSilvey (2012) and Frasier MacDonald (2013) have explored juxtapositional and fragmented narrative forms as a means to deal with conflicting histories and uncertain futures in beyond urban spaces. Whilst for Patricia Price (2015) telling multiple, sometimes conflicting, stories of place offers a telling of tales that disrupts ideas of stable, unitary, univocal places. Cohering all these different writing projects is a detailing of specific places that is bound with a conceptual thinking of place that finds form in experiments with writing, whether this be with individual words, their fonts, the layout of the spaces of the page or performative soundings and the orderings of narrative form. In short, attention to dimensions of the composition of writing has gained conceptual force with respect to thinking about urban places and landscapes and how it is we know and understand them and might come ultimately, to intervene within them.

CREATIVE CITY 4: ON THE STREET – SUBVERSIVE CREATIVITIES?

Artistic practices are not the only creative practices we can understand as being critical spatial practices. This section will explore how we might find critical urban practices and the chances to rewrite urban scripts in a wider range of creative activities including skateboarding, parkour and urban exploration. It will examine how these vernacular and subversive forms of creativity 'question, re-function, and contest prevailing norms and ideologies to create new meanings, experiences, relationships, understandings and situations' (Pinder, 2008, p. 730). It will explore how in their doing, if not always in their conceptualisation by their practitioners, these practices come to offer critical purchase on our thinking about streets and urban spaces.

Recent years have seen analyses of subversive creative practices – skateboarding, parkour, or urban exploration – becoming understood as creative solutions to make cities more liveable (Borden, 2001; Garrett, 2013; Mould, 2015). Some have taken this to the extreme to argue that such forms of creativity are somehow more 'authentic' than other forms. Others take the opposite view, in the belief that such forms often sell out to capitalism anyway, thus reducing any sense of their subcultural or subversive possibilities (see discussion in Daskalaki and Mould, 2013). As this section will explore, to think about these street-based practices as somehow pure, as the locus for true critical forms of creativity is rather problematic. As too is any suggestion that such forms of practice are able to escape the narratives of the creative city, or subvert those of capitalism. This does not mean, however, that we cannot think through the politics of these kinds of urban practices and the basis from which they develop, just that it is important, as Mould discusses in Box 7.4, to appreciate the complexities of these creative subversions.

Box 7.4 THE COMPLEXITIES OF SUBVERSIVE URBAN CREATIVITY

Oli Mould

As many urban studies scholars will tell you, not only are the world's cities becoming increasingly populated, they are also becoming more privatised, commercialised and securitised. As urban development becomes the remit of corporate entities, urban space is more heavily protected, given the vast capital outlays that are spent on producing it. Therefore, how people use urban space is scrutinised, and any usage that does not conform to the most efficient and productive use of that space (from the perspective of those who produced it) is called into question and singled out for scrutiny. Usage of urban space that is considered 'alternative' (or not in-keeping with what those who built the space see as conducive) is often swiftly and strictly marginalised, or perhaps even criminalised. This alternative use of space is conducted most readily by subcultural activities, such as skate-boarding, parkour, graffiti and other collective forms of artistic interventionism; but also more direct political activism such as protests, sit-ins and rallies. Such activities are sometimes resisting their own increased marginalisation, while, at other times, they are undertaken for more artistic and creative purposes. Moreover, the prevailing forces of urban development may look to these subcultural activities to further enliven urban space, but in so doing create new forms of activity that are culturally and politically distinct from the original forms of the subculture.

Therefore, urban subcultural creativity is a complex and contested process that has political and social variance that cannot be contextualised by a dualistic binary of resistance against a dominant urban development discourse; us versus them. We only have to look at the recent events around skateboarding and the South Bank in London. In March 2013, the South Bank Centre announced as part of its development plans, that it was to turn the undercroft area, an area that has been used by skateboarders since the 1960s, into retail

Continued

units. The history of the undercroft is one very much about subversive creativity in that the skaters reused the space as a skate spot, despite the strategies of the South Bank Centre to marginalise and criminalise them. Skateboarding therefore is a creative subculture because it takes urban space that is designed for one (commercialised) purpose, and imbues it with new meanings, new ways of behaving, new (sub)cultures. Because of their strong attachment to the undercroft, the skateboarding community rallied together to form the Long Live South Bank campaign, which undertook a range of activities to actively resist these plans. Some of these activities were formal, such as delivering the largest planning petition in UK history, writing to MPs, collecting petition signatures. Other activities were more creative; art was produced in the form of photography, videos, blogs, exhibitions, books and sculpture; and even some clandestine activity such as protesting at closed planning meetings, undercover filming and other subversive activist procedures. So the subversive creativity of the skateboarders includes the reuse of the urban space, the more artistic creativity of producing critical art forms, but also the more political creativity of activism.

Figure 7.5 The undercroft, South Bank, London

But all these practices of the campaign were very much about saving the undercroft from being lost. In the course of the campaign, the skateboarders created institutions (i.e. the Long Live South Bank campaign) that began to produce codified procedures, branding and, at times, even hierarchy. Such 'institutionalisation' itself was resisted, but realised to be an important and necessary step to challenge their displacement. However, the South Bank Centre also began to engage in more activist politics. It campaigned against the perceived heteronormativity of the skateboarding culture (i.e. criticising them for being largely white, middle-class and able-bodied), and produced art works that championed their own reasons as to why they wanted to recapture the undercroft space. So far from being a clear-cut case of resisting the dominant forces of hegemonic urban development, the 'battle for the undercroft' represents how such binaries begin to break down, and new 'cross-border' assemblages are created that create new forms of creativity that are neither resistive nor dominant, but part of the ongoing dialectic processes.

The 'battle for the undercroft' came to a conclusion in September 2014 when a joint statement by the South Bank Centre and the Long Live South Bank campaign was released saying that the undercroft would be spared from any future development. Long Live South Bank has now changed their name to Long Live Skateboarding and now campaign for the rights of skateboarders across the UK, involved as they are in campaigning against the recent skating ban in Norwich (as well as against the destruction of many skate parks around the UK). But also, the South Bank Centre has utilised their skateboarders in their planning processes, and begun to articulate the community, cultural and creative benefits of having a thriving skateboarding community on their site.

So, while activism against commercialised urban development can be resisted, the very act of resistance by the skateboarding community required the use of formalised, institutionalised and in some cases commercialised techniques. In addition, urban development forces can loosen, and take on more activist qualities

Continued

themselves in attempts to continue the development process, albeit with different physical characteristics. This process is redrawing the contours of subversive creativity less towards an 'us-versus-them' ideal, and more towards a complex, conflicting and variegated process of urban creativity.

'If graffiti changed anything it would be illegal', reads a piece of graffiti apparently by the artist Banksy, prompting a whole series of reflections on what it is that is meant by the idea of 'subversion' being developed in these creative subversive practices. Interestingly, at the heart of many of these critical spatial practices sits a close engagement with their embodied nature. This is not a form of practice whose subversion comes predominantly from ideological slogans, or direct action, but rather from the often daily practice of these ways of knowing and engaging urban space.

For many, the modern city especially in the wake of post-war planning is full of unused, overlooked SLOP, space left over after planning. This sequence of public spaces – underpasses, spaces under bridges, around the bottom of structures, and so on – have offered a fertile backdrop for a set of creative practices, an 'open canvas' with buildings conceived 'as building blocks for the open minded' (Borden, 2001). We see this reuse of urban terrain very clearly in the work of urban theorist Iain Borden (2001) for whom skateboarding is a critical spatial practice; 'skateboarding is a challenge to our popular everyday concepts of the functions of buildings and the closed world we create for ourselves out of the massively unlimited city' (Borden, 2001, p.5). For some, the politics of skateboarding as critical spatial practice lies in the reclamation of space and the subversion of the laws and rules put in place to control the use of that space (Karsten and Pel, 2000). For Borden, however, what is key is the 'micro-experience' of that space, and its intersection of three elements, the urban fabric, the body and the technology. The politics comes therefore from a city re-engaged by the intersection of bodies, technologies and architecture. If previously such urban concrete arenas were once understood as empty and meaningless, in the practice of skateboarding, meaning is returned to them often in a subverted manner. Street furniture is no longer an object of control or safety, rather benches provide grinding surfaces, a handrail's function is inverted to offer obstacles and balance beams, spaces to show off tricks. As such the urban space becomes a site of display, but also a playground for these kinds of practices.

Skateboarding is a thoroughly embodied practice based in an improvisational knowledge of the urban space that comes from the skateboarder's experiences of engaging with the embodied and material practices of the urban environment, the adoption of the body and balance to the board as it slides along different surfaces, the sensing of textures and adjusting of posture to compensate for the camber of a surface. Skateboarding is thus understood as 'a total focus of the individual, body and environment to a level way beyond that of dead consumers interested at best in money, beer and the lads'. In his account, Borden (2001) emphasises how the skateboarders' skills will evolve in relation to their experience of the city, changing the construction of the body and the experience of the urban realm. Thus the process of becoming skilled at skateboarding is one that involves the evolution of the sensory and cognitive mapping of city spaces and a reorientation of the physiology and psychology of its inhabitants. In the wake of the growing popularity of such practices, including the deployment of parkour and skateboarding in popular films, there is a growth of skateboarding parks, prompting questions about how subversive or subcultural these practices are and what indeed it actually is that they do in terms of rewriting these city scripts (Dakalaski and Mould, 2013).

If embodied engagement with urban space has become a core means to understand the politicality of these practices, as Mott and Roberts (2013, cited in Garrett and Hawkins, 2013) rightly ask, what of those bodies that are unable to occupy those places? What does it mean for these forms of politicality when disabled bodies are excluded, or when bodies of certain colours might find getting caught crawling through sewers has more serious consequences? These are important questions that need addressing if the subversive nature of these practices is to be realised.

Geographers have long been interested in narratives of exploration, and so unsurprisingly one of the forms of critical creative spatial practice that has captured geographers' imaginations is urban exploration. Defined by Brad Garrett (2013, 2014) as a practice of entering, and often photographing, off-limits spaces, the practice of urban exploration is one that has been engaged with for its political and also apolitical ends. These practices of 'recreational trespass' see participants scaling the unfinished and largely publically inaccessible buildings that have come to populate our neo-liberal cities, alongside exploring the ruins that populate others, and going deep into the bowels of the city to develop intimate engagements with urban infrastructures. We see protagonists running tube lines, wading through sewers and walking miles and miles of tunnels filled with cabling and ducts that enable us to live our daily lives (see Figure 7.3). For some, the thrill lies in an attraction to derelict spaces and the beauty of ruins, for others the practice is exciting, treading close to physical danger and skirting the lines of legality. As Garrett makes clear the reasons for engaging in recreational trespass are many and varied and, as such, his accounts offer myriad ways to query where the political and subversive possibilities of these practices lie (2013).

A reoccurring idea across Garrett's work is his concept of the meld (Garrett, 2013; Garrett and Hawkins, 2013). Owing much to phenomenology and feminist theory and akin to Borden's account of skateboarding, the meld describes an embodied experience of urban spaces where explorer bodies and architecture are folded together to create an in-the-moment experience of urban space, which can then be explored through images and video as discussed above. From such experiences, new orientations towards the city evolve. Layered into this embodied account are other more conscious forms of politics and resistance, in particular the neo-liberal privatisation of urban space, a way to take back, literally and imaginatively, spaces within the city that are being sold off. The effect of what they do was dramatically demonstrated in the images the group took during their scaling of the Shard in London, one of the European Union's highest skyscrapers, just before the building work topped it out. For author and cultural commentator Will Self, who came out in defence of the explorer's actions, their practices of 'place hacking' draw attention to 'how physically and commercially circumscribed our urban existence really is' (Self, 2014). They perform, he argues, a valuable public service in 'reminding us that the city should, in principle, belong to its citizens' (ibid.). We might think then of urban exploration as making visible a surveilled city, and a rendering material the vertical geographies of city imaginaries, creating potent ways in which we might imagine and occupy our cities differently.

If many of these forms of urban subversion are thought of in terms of a hyper-masculinism, then yarn-bombing is often figured as an interesting alternative. Yarn-bombing, sometimes described as knit-graffiti and often situated as part of a wider understanding of 'craftivism' 'generally involves the act of attaching a handmade item to a street fixture or leaving it in the landscape' (Moore and Prain, 2009, p. 17) (Figure 7.6). As

Figure 7.6 Yarn-bombing in Long Beach, California, USA

Price notes, 'yarn-bombers complicate the archetypal performance of graffiti by intersecting multiple femininities and creative practices that are intergenerational, with complex histories of empowerment, disempowerment and relationships to public space' (2015, p. 89).

Like other creative urban practices, not all practitioners subscribe to their practices as political, indeed, it can be 'difficult to locate yarn-bombing as an intentionally feminist subversive or craftivist practice' (Price, 2015, p. 89), especially as it is often termed the 'fluffy face of graffiti' (cited in Price, 2015, p. 89). Yet such material qualities are often the very source of the forms of politics that this practice might be thought to engage (Price, 2015). Further, yarn-bombing can be seen as 'an intervention that enacts a playful and enchanting politics in urban space and re-negotiates, or reweaves, assumptions about city spaces' (p. 82). Investigating the intersection of wool, body and urban fabric, Price notes knitting offers 'a vibrant, creative act that speaks to the lively material making of urban forms, spaces and relations' (2015, p. 84). Whilst these politics might be understood as 'softer' or even a form of whimsy (Mann, 2015), they are, through these means, an important way of making and remaking urban spaces and communities (see also Elden and Hawkins, 2016). Such politics might lie in a re-enchantment of urban living offered by encounter with the yarn-bomb in the street, in the convivial experience of making together, or in the playful ways in which the yarn-bomb encourages urban engagements (Price, 2015). Drawing on the work on ludic geographies, Price notes how yarn-bombs might not only work through more representational political registers, but can also 'change the feel, touch and texture of the urban vernacular . . . offering the potential to enact ethnical generosity to things and experiences that we would otherwise engage without thinking' (2015, p.87). For some they are 'gifts to the city', for others just part of the ongoing maintenance and repair of urban space (2015, p. 89).

For many thinking about the politics of such vernacular creative practices, the writings of Michel de Certeau have formed an important start point, and in particular his idea of tactics (1984). Tactics, often unconsciously carried out activities within everyday life, hold within them the power to reclaim, resist and react. The preserve of the everyday person on the street, tactics are local and situated, and work against strategies, those visions from above that organise city spaces and activities. Writing on tactics and place de Certeau notes, it must . . . make use of the cracks that particular conjunctions open in the surveillance of the proprietary powers. . . . it creates surprise in them, it can be where it is least expected. It is a guileful ruse.

(de Certeau, 1984, p. 37)

Not unrelated to de Certeau's tactics is the growth of experimental practices that have come to be called tactical urbanism. Otherwise termed DIY, guerrilla, everyday, participatory or grass roots urbanisms, these practices usually denote low-cost small-scale often unsanctioned community-led improvements to urban environments that are often relatively temporary (Mould, 2014; Iveson, 2013, p. 941; see, for example, pop-up places Box 7.5). They are often understood to mobilise creative, skilful and playful highly localised practices in the name of the functional improvement of experiences of urban living (Douglas, 2014, p. 531).

Box 7.5 POP-UPS: CREATIVE PLACES

Ella Harris

Pop-up places are a type of temporary urbanism that has become increasingly prevalent in the past decade. Driven, in part, by rising vacancy rates following the recession, they involve the occupation of empty spaces for temporary sites of consumption, commerce, artistic practice and for social and charitable projects. Although found worldwide, pop-ups cluster in 'creative' cities such as London, New York and Berlin and have become central to the contemporary creative and cultural industries. Within my research, pop-ups are identified as demonstrating three different modes of creativity that converge towards a fourth kind of *creation:* the production of pop-up's specific spatiotemporal imaginary. My approach holds that unpacking this spatiotemporal imaginary is crucial for understanding the contemporary moment, because pop-up culture is both emergent of and instrumental within the shifting assemblages of the post-recession city.

Continued

One: places *for* creativity

Pop-ups are often pitched as low-cost spaces for micro-enterprises and start-up businesses within the creative industries. At a time of recession a 'DIY' attitude has been explicitly promoted through government incentives like the meanwhile lease and through private sector promotions such as the development of pop-up insurance. Premised on the idea that it's possible to 'craft your way out of a recession' (Hracs et al., 2013), pop-ups provide young, educated, people with a low-risk test bed for hands-on, creative businesses such as micro-bakeries, urban farms and handcrafted clothes shops, which are promoted as the centrepiece of economic recovery.

Two: places made creatively

As well as being spaces *for* creative practices, the term creative, when applied to pop-up is often used to denote creativity in the practice of place making itself, where this creativity is located in the thinking up of creative concepts for pop-up events, in the material design and decoration of spaces and in the finding of 'unusual' venues. With regards to material design, pop-up place making is hailed as creative because of its use of make-shift and repurposed materials, including crates, slats and shipping containers, to build temporary places. In terms of event concepts, pop-up culture involves running 'normal' businesses like cafes, restaurants and cinemas in *creative* ways, often theming events as, for example, a secret garden party or a speakeasy style bar. And finally, the creative selection of 'unusual' locations such as public toilets, decanted tower blocks and motorway flyovers heightens the appeal of the pop-up, giving the sense that these events are counter-cultural; skirting authority by moving unpredictably through forgotten, interstitial and precarious nooks of the city.

Three: places (re)made creatively

As well as making places, pop-ups are also instrumental in processes of *re*making. Many of the 'unusual' locations pop-ups inhabit belong to an urban imaginary typified by the grungy, the run-down and the working class. Pop-ups take up derelict or deteriorating places and, through the creative processes listed above, 're-imagine' them; turning flyovers into cinemas, public toilets into restaurants and tower blocks into spaces for perform-ance and exhibition. In this sense, the pop-up is a form of urban catalyst, brought in to create transformations.

Four: creation of a spatiotemporal logic

Lastly, pop-up places create a particular rendition of spatiotem-porally through their promoted emphasis on the fleeting, the DIY and the interstitial. Much like the practices of Archigram in the 1960s, the pop-up city is concerned with 'a shift away from the static, rooted and monumental towards movement, flexibility, transitoriness and indeterminacy' (Pinder, 2011, p. 168). There is an implication that, through their flexible and nomadic forms, pop-ups leave room for ongoing re-imaginations of space, making the city a platform for creative reinvention and that, likewise, their DIY production model enables participation in the creation of the city on behalf of individuals. Indeed the shipping container, as the emblem of pop-up architecture, embodies this sense of ongoing and easy alteration. Formerly constitutive and representative of globalised homogeneity, shipping containers within pop-up culture become the basic unit of a modular design in which they are customised, transformed and repurposed for 'creative' makings of place, producing a temporal logic of unpredictability, versatility, and transformation.

As David Pinder suggests in his discussion of Archigram, such 'visions of mobile cities' also 'function pedagogically' (2011, p. 182) and, in the case of pop-up places, their spatiotemporal imaginary is arguably normatively orientated. The various creativ-ities of pop-up places project spatiotemporal ideals which func-tion prescriptively to create and to recreate a political and economic status quo in the aftermath of a recession that was a shock to the city assemblage. At a time of extreme urban uncertainties, the performative short-termism of pop-up works to glamorise

Continued

unpredictability, and their DIY organisation and make-shift aesthetic normalise 'making do' with limited resources and shift the onus for economic recovery onto individuals. Equally, pop-up has become one of the key contemporary mediums through which gentrification is enacted: pop-ups nomadically traverse the city, 're-imagining' places in their wake, and create a landscape within which spaces of relative social deprivation are seen as 'unusual' venues for middle-class cultural and commercial events to occupy, not as the homes of demographics at risk of being forced out.

The modes of creativity evident within pop-up places therefore offer an insight into the varying roles of creative practice. While creativity is archetypically conflated with an inventive open-ended-ness, pop-ups are an insight into the directive role of creative practices within the creation and recreation of order.

Critiquing subversion?

The street clearly has much potential as a site of creative possibility where empty spaces – whether car parks, gaps between buildings or small green sites – are commandeered through practices of guerrilla gardening or pop-up retail. The unsanctioned nature of tactical urbanism is often part of its raison d'etre, yet it has increasingly become subsumed through its 'cool factor' into contemporary creative city narratives, 'effectively negating . . . the reactionary and tactical nature of urban creative acts and urban subversions' (Mould, 2014, p. 533). Indeed, in his reflections on the tactics of tactical urbanism, Mould concludes that so capitalist friendly is tactical urbanism, so 'on message' with the creative city more generally is its hip deployment of the everyday, the off-beat and the precarious – its hipster form of 'rough-lux' – that such spaces and practices in fact cease to be tactical and become instead part of normative city strategies. Indeed, for Mould (2014, 2015), such tactical urbanisms might be considered the latest in a long line of creative and so-called alternative activities to be co-opted by the new creative city.

Thinking through the formation of an alternative creative city narrative through street-based creative practices offers the chance to trouble simplistic understandings of urban subversions and any too-easy sense of the street and its practices as sites of alternative urban narratives. Whilst clearly full of possibility, there is much that needs careful engagement if we are to understand how it is that creative street-based practices build a politics of urban spaces and practices. The risk is that, as with Lefebvre, creativity becomes romanticised as a site for potential revolution and we lose grip on some of the more complex facets of how exactly it is that creative practices can be political, for those involved as well as their audiences, and how they can be enrolled within the wider development of creative city narratives.

CONCLUSION: CONFLICTING CREATIVE CITIES

As these four city stories have hopefully demonstrated, the creative city is a complex and multi-faceted idea, to consider any one of these, or the many other creative cities alone (such as the marginal creative cities considered in Chapter 8) is to miss out on the complex and often tense relationships between different ideas of creativity in the city. As these stories unfolded, what hopefully became clear was that this is not a straightforward narrative of the perils of the planned creative city, as somehow destructive of creative practices, whilst artistic, tactical and subversive creativities offer effective means to rewrite creative cities. Rather, as the stories hopefully made clear, planned creative cities can support diverse creative practices as well as create creative atmospheres that facilitate urban branding and support regeneration. We should, however, caution against either the wholesale support, or derision of creative planned cities that often come with a series of issues around who and what is creative.

Just as complex are the creative cities constituted through critical spatial practices, whether artistic or more vernacular creative practices. For some, these practices are cast as 'grass roots', authentic forms of creativity, practices that are both overlooked by creative planners but which also provide the antidote to their hollow, economically driven politics and their production of urban inequalities. As the examples developed here, those explored in Chapter 6 on community and those in Chapter 8 on margins, indicate that such an understanding is fraught. These critical spatial practices are often cast within, funded by and even mobilised within creative city planning, as such it is hard to see them as separable from these understandings of creativity. Further, their subversive potentials are no guarantee of social equality and critical engagement. As such, whatever stories of creative cities we are considering; whether planned creativities, critical spatial or subversive creative practices, what seems key is to attend to the local expression and manifestation of creativity as a means to understand both their problems and possibilities.

NOTE

1   http://www.bilbaointernational.com/en/bilbao-arte/ [Accessed 26/7/2015].

8

THE MARGINS

'Every culture proliferates along its margins. Irruptions take place that are called 'creations' in relation to stagnancies. Bubbling out of swamps and bogs, a thousand flashes at once scintillate and are extinguished all over the surface of society . . . creation is disseminated proliferation' (Michel de Certeau, 1997, p. 139–142).

Margins seem to suffer from a sort of schizophrenia of the imagination. On the one hand, the space of the margins summons up a kind of dank industrial waste-land, a forgotten space of flyovers, of out-of-town retail parks, or a monotonous suburbia that offers a place to sleep but little more. On the other hand, the margins are an edgy exciting place to be, an inspirational in-between zone, a site for illicit thrills enabled by the chance to operate under the radar. One thing that is clear is that marginal locations and positions have long been considered to be creative ones. Whether it be the edges of cities or the periphery of society, for many there is creativity to be found in sites seen as an escape from the dominant norms of life and work. For some the margins offer cheap places to live and work, and make space for alternative ways of living that enable and support creative lives. Of course, the margins are not always a romanticised place, embraced as a positive location for those seeking freedom and the space to develop subversive creative practices. The margins are also sites of precarity, where cheap spaces and a host of informal living and working arrangements are sought out as a necessity rather than a choice. Often to study and occupy such sites as scholars is problematic. Marginal activities, atmospheres and practices can be resistant to the perceived structures of studies and any potential exposure and the transient and informal nature of marginal practices and communities makes prolonged studies harder. Indeed, as the discussion of subversive creative practices in previous chapters has made clear, often the very 'edginess' or otherness of these spaces to wider narratives of the creative city or normative creative practices is what constitutes their very attraction for the creative sector; but is also what eventually undermines their very marginality. This discussion will reflect on a range of different intersections of creativity and the margins, considering geographically marginal locations as creative sites, as well as reflecting on the creative practices of those who exist on society's margins.

Any discussion of margins and their negotiation requires a centre. In many discussions of marginal creativities the imagined centre is the creative city, with the margins being those areas at a distance from major metropolitan centres and those practices that are not seen as central to the creative economy discourse. Studies of vernacular creativities layer together multiple forms of marginality as creativity is relocated from busy city centres, sleek cultural quarters and trendy studios, to garages, sheds, gardens, council estates, suburbs, trading estates, as well as a series of unofficial spaces around the city. Such spaces might include wastelands and industrial ruins, SLOP – spaces left over after planning, below overpasses or between buildings, for example – semi-abandoned infrastructure, or those spaces awaiting completion (Borden, 2001; Edensor, 2005b). As studies have shown, such sites form fertile loci for creative practices and play, whether this be graffiti or other forms of urban subversion or alternative forms of living. Further, these are practices conducted not by a hypermobile creative class, but by groups often overlooked in studies of the creative economy; unemployed youth, blue-collar workers, social tenants, street children. Oftentimes, such studies require a revisiting of dominant ideas of creativity, querying whose creativities are studied, where these creative practices take place as well as how.

In other discussions, the marginal might be less a geographic designation of space and rather more a social distinction. There has perhaps always been a mythologised figure of the artist and creative genius as the outsider. Often driven by a combination of a choice to reject dominant ways of living as well as by economic necessity, taking a place on society's margins has been a common site for artists. Of course, creativity too has long been associated with subcultural and counter-cultural practices, where forms of inventive stylings of self and body, as well as creative practices of living off the grid, become characteristic of certain cultural groups. Not only might we consider subcultures and counter-cultures to offer marginal forms of creativity, but for those who live precarious lives, marginality blends survival strategies with a set of collectivising cultural-identity practices.

In short, the relationship between creativity and the spaces and practices of the margins is complex and must be carefully negotiated. Not least as once marginal and edgy practices and spaces all too easily become subsumed into dominant capitalist forms, requiring their original practitioners to once more negotiate their places and identities. Importantly then, the constitution of creative practices in the margins often does not simply replicate that of the centre in less successful, less lucrative ways, but rather reconfigures these practices requiring that we reflect critically on those practices that have become the 'norm'. As with the city stories told in Chapter 7, this discussion offers four explorations of creative margins. It considers first, 'other' geographies of creativity in relation to the creative economy. The second story explores vernacular creativities as marginal creativities; whilst the third narrative of creativity in the margins turns to the subcultural creativities of precarious groups of individuals. The final story of marginal creativities is one that explores the creative practices that take the margins as their subject, both as part of the wider turn towards the margin as a fertile cultural territory, as well as the development of creative geographical research practices in and for marginal spaces.

MARGINAL CREATIVITIES 1: 'OTHER' GEOGRAPHIES

Academic work on creativity has an often unacknowledged urban, western bias (Edensor et al., 2009; Gibson, 2010; Sorensen, 2009). Recently, however, there has been a growing body of work on 'other geographies' that overturns the dominance of the urban-based creative cluster and seeks to draw attention to other geographies of creativity, not as lesser imitations of clusters, but as alternative forms of arrangement that offer their own strengths (Cole, 2008). Such an attention to what

Gibson (2010) calls the 'tales from the margins' takes a range of forms. On the one hand, research has explored the spatially diffuse and often cross-border networks within which clusters operate (Coe, 2000; Kong, 2000a, 2005). On the other, research has explored the potential of temporary clusters formed by fairs, festivals and trade shows (Maskell et al., 2006; Norcliffe and Rendace, 2003; Power and Jansson, 2008). A further theme of research has attempted to correct the urban bias of existing work by exploring how creativity and the creative industries emerge in small, rural, suburban and remote places (see summaries in Gibson, 2010; Harvey et al., 2012). Such scholarship explores a range of social, cultural and economic transformations that occur in these localities, as well as reflecting on what the forms of the creative economy in such places suggest about understandings of creativity more broadly.

Negotiating marginality

The centripetal force of creative agglomeration is such that urban centres are dominant sites in the study of creative practices. Small and/or remote, marginal places (physically and or metaphorically) are often imagined to be culturally arid. Such image problems continue patterns of the concentration of people in cities and areas conceived to be creative centres. Distance and marginality are often, however, less a physically geographical feature as much as an imaginary and discursive condition. Remoteness often becomes code for 'limited types of creative making; wariness of newcomers and new ideas; the loss of young people; limited access to business expertise, production services and training; lack of cultural stimulation and high transport costs' (Gibson, 2014, p. 4). Recent scholarship has, however, seen a growth of stories of creativity from the margins that both counteract these imaginaries of culturally emaciated areas, but also demonstrate the creative and innovative means people use to overcome the challenges of creativity in marginal locations. The degree of marginality of these places varies from suburbia (Bain, 2013) to small towns or rural areas at a distance from larger urban agglomerations, to areas considered to be marginal, remote or peripheral in national or international terms (Gibson, 2014; Harvey et al., 2012). Yet what is interesting about these 'other geographies' is both to reflect on them for their own sake, but also to explore how they develop further arguments regarding diverse creativities. Creativity in marginal or peripheral locations is often understood differently; so too are terms such as proximity (Gibson, 2014, p. 3). As Gibson and others ask, what does it mean when remoteness and smallness challenge the notion of proximity when you are a long way from 'happening' places and scenes (Gibson, 2014, p. 4)? Other questions concern the relationships between marginal places and their more central counterparts. The image problems and actual problems of remote or marginal places leads to large places exerting centripetal forces, drawing attention and people, as creative workers move for career development or pilgrimage and never return (Bennett, 1999). The local scenes that may then have cultivated the creative talent of these individuals are left struggling once more.

For others, the negotiation of marginality is possible; Leadbeater and Oakley (1999, p. 14), argue that marginal locations in relation to the flows of global culture and economic imperatives can be negotiated through local know-how and skills. Similar to the use of Etsy and the arguments made about temporary clusters, technology might enable creative workers to 'sell into much larger markets but rely on a distinctive and defensible local base' (Gibson, 2010; Power and Jansson, 2008). This can include 'local' distinctiveness becoming the source and means to 'stand out from the crowd', enabling one to make a name for oneself in the global market. There are a series of examples of how technology continues to enable distance too. As Warren and Evitt (2012) describe, for indigenous hip-hop musicians in the Torres Strait new telecommunications and technologies enable the functioning of their creative work at a distance from key urban centres. Communally provided computer equipment enables the recording of music that is produced using free software and then distributed using Myspace and YouTube, enabling these hip-hop musicians to find audiences through the internet. But, as they note, opportunities are still limited by cultural ideals of what indigenous music might be, that when combined with distance mean paid performances are rare and it is a challenge to build a mainstream audience base.

The demands of distance and the opportunities technology offers to overcome them enable an enjoyment of the pleasures of isolation of countryside and coastline life. As Thomas et al. (2010) note, for those in rural Cornwall, the coming of high-speed internet connections enabled the growth of an animation cluster on the far south-western most tip of the UK, a long day's journey to London. Central to this cluster is Spider

Eye Animation. Once based in London, they moved in 2001 to St Just Cornwall where rent prices were a fraction of their previous London costs, thus enabling them to produce more attractive tenders. Pushed out by the expensive rents and long commutes, but also drawn by the quality of life, the owners relocated to this remote area of Cornwall. As one of the poorest regions in the EU, Cornwall was eligible for funding to help develop and support a high-speed connection infrastructure. As a result, from the small wind-swept coastal area programmes are made for Disney, Universal Pictures and the UK children's channel cBeebies. The Cornish animation cluster is just one story of a group of creative workers who reconcile the pleasures of isolation with the demands of distance that can at times be overcome by technology.

As well as technology allowing challenges to be overcome, distance from key stakeholders, events and institutions encourages new forms of networking and distribution practices. As the example of the rural cluster Krowji (also in Cornwall UK) discussed by Harvey et al. (2012) demonstrates, clusters are not only urban phenomena but can also be reworked for rural areas. Here, much as design fairs provide the means to draw together the benefits of temporary and permanent clusters, so too does the Krowji complex with its cluster of studio spaces and creative services. Krowji is a far cry from an urban creative cluster, located on the outskirts of the market town of Redruth (population 12,500) in west Cornwall (UK), it is at the heart of an area once dominated by farming and mining and which is now struggling to recover from the closure of the mines and associated industries (Harvey et al., 2012).

Krowji currently houses around 95 people, working within a partially-converted former school (see Figure 8.1) including a range of creative micro-businesses, together with influential creative governance organisations.

Krowji's various tenants do not just focus their attention within the cluster, but are also responsible for developing networks that spread across the surrounding area. These might offer skills and training, or they might be focused on supporting and developing infrastructures or events, such as open studio weekends. The latter enables artists and makers across the rural area to open their houses and studios to display and sell their products to audiences who spend the weekend travelling around a mapped trail. As a result, while Krowji is a distinct cluster in the form of a building on a site, it is also the central node in a series of rural

Figure 8.1 Krowji, Cornwall

networks that enable the temporary coming together of a dispersed rural creative community. This has been crucial, as while many creative makers recount enjoying the seclusion and the local scenery, a number report feeling isolated from both markets for their work and from critical discussions of artistic practice. Thus the creative support services in Krowji have developed a series of tailor-made activities and events designed to target the specific needs of the local artists; whether this be training schemes that include travel bursaries, allowing people to travel from the islands as well as dispersed mainland locations, or discussions of the use of technology as a means to reach clients. In Krowji we see an example of how cluster policies can be altered and rescaled to suit the local needs of the diverse artistic population, to engage local heritage as well as to respond to issues with both the remoteness of the location, and the need to serve a distributed population served by poor transport networks.

It is not just creative producers that are coming to value rural areas, but interestingly galleries too. Hauser and Wirth are an international commercial art dealership, with galleries in Zurich, London, New York and LA. Founded in 1992 they represent some of the world's best-known contemporary artists and modern masters. In 2014, after over 20 years of development, they took the perhaps unusual step for a city based international art dealers of opening a rural gallery space in Bruton, Somerset (Figure 8.2). A rural gallery in south-west England, 1.5 hours drive from London, the space brings together a commitment to contemporary art and architecture (shared with their urban galleries) with concerns for work made in response to local community and environment.

The gallery, shop and café complex, which also includes a residency space for artists, is focused on the redevelopment of a series of grade II listed farm buildings, Durslade farm. The core of their work at the gallery is an artist-in-residency programme that brings international artists, such as Mark Wallinger (an English sculptor) and Pipilotti Rist (a Swiss video artist) to Bruton. Working in residence for up to a year, the artists explore their local surroundings, integrating with the local community and often making work inspired by the local environment. The art that is produced

Figure 8.2 Hauser and Wirth, Bruton, Somerset

as a result is internationally orientated with a very local flavour. It reflects, on the one hand, a sense of trends in contemporary art, whilst at the same time perhaps also reflecting the change in the profile and imaginary of rural areas. Rist, for example worked on the video-piece 'Mercy Garden' that uses images and film from the local area in her characteristic luscious and erotic video work (Hawkins, 2015b). The gallery is very commercially viable in this area; close to London, it is a favoured location for those moving out of the city to raise families or for those with weekend retreats from busy city lives. Further it taps into the growing wealth and lifestyle capital of those living in the area.

What is also clear about these marginal creativities is how they reorient how we think about creativity. So we see in these discussions of marginal creativity a sense that creativity does more than fulfil an economic role. For Warren and Evitt (2012) the hip-hop cultures of the young aboriginal people were an important means of self-determination and political expression. For Thomas et al., (2010) Spider Eye Animation were motivated by personal as well as business desires. For Hauser and Wirth to relocate in Somerset was to enable artists to become part of a rural community to make art-work. Mayes (2010), writing of postcard making in rural Ravensthorpe, Western Australia notes the complex role and place of this creative practice in this marginal location. Small-scale semi-industrial production of sets of local postcards, based on local photographs and drawings, while part of a cultural economy are perhaps equally well understood to be part of serving community needs rather than being driven by economic imperatives. As Mayes (2010) notes, in this very remote area community is important, settlers need to provide themselves with social and cultural infrastructure and meet their own social and cultural needs. The postcards emerge through the local interviews Mayes conducts as less of a novel creative activity and more as developing an understanding of 'creativity as a quotidian activity in the service of solving daily problems' (p. 19). These problems include the need to keep busy as well as engaging and developing a sense of community in this marginal place. The local community here is the context for cultural production, but more than this we see creative production as a way of enhancing local participation in community. As Mayes (2010) notes this 'local work is suggestive of the means of creativity to enhance interaction rather than (just) interaction as a means to enhance creativity'. In short, creative producers and consumers have a whole suite of ways to enable them to negotiate the challenges of marginality whilst also reaping its benefits.

MARGINAL CREATIVITIES 2: VERNACULAR CREATIVITIES – CREATIVITY ON THE MARGINS

Given the centrality of the creative economy to narratives of creativity, vernacular creativities might themselves be considered marginal. Indeed, vernacular creativities often layer together multiple forms of marginality such as location and practices driven by other than economic motives. Oftentimes, these creativities are located in suburban areas, those locations often overlooked in discussions of the creative city, dismissed either as uncreative (see Box 8.1), or just not considered at all. Such marginal urban spaces have long been attractive sites of creative practitioners, offering not just inspiration, but also affordable living and working spaces (Bain, 2013). Furthermore, vernacular creativities are also those clearly driven by motives other than the economic, and practised by those who might not be considered to be members of the 'creative class'.

Box 8.1 SUBURBAN CREATIVITY

David Gilbert

The Butterfly Tattoo studio stands in a very ordinary row of shops in the very ordinary West London suburb of Hanwell (Figure 8.3). Butterfly has a team of artists, who will create a work of art on your arm, or leg, or right across your back in a range of styles, including 'realistic', 'Japanese', and 'new school'. There are many things that can be said about the geographies of creativity at Butterfly, most obviously about the recent rise of the body itself as a site for art, and the expansion of tattoo art beyond the subcultural and avant-garde into the mainstream. However, it is easy to overlook the significance of its location. The shops are surrounded by Edwardian terraces, inter-war and post-war semi-detached houses, and a few rather grander properties, particularly in the

---

Figure 8.3 Butterfly tattoo studio (Photo: author's own)

parts of the suburb that residents and estate agents like to call 'Golden Manor' and 'Poets' Corner' (after the road names, Shakespeare and Milton, rather than any tendency for locals to write verse). Before the 1960s, London tattoo parlours were most likely to be found around the docks and other working-class areas,

Continued

with perhaps a few specialists in Soho; later in the twentieth century such body art might be found in and around the alternative consumption spaces of Camden or Kensington Market. In twenty-first century Britain, the tattooing boom has a new geography, often suburban or small town, like many other creative activities driven out of more central spaces. In London, particularly, the intensity of the 'financialisation' of space has led to the loss of central 'urban interstices', relatively affordable marginal spaces vital for creative practice, and a suburbanisation of creativity (Pike and Pollard, 2010, Gilbert, 2013).

The idea of suburban creativity cuts against some very entrenched tropes and expectations, notably of suburbia as literally 'sub' urban, dependent on the city and somehow less than it. Suburbs, certainly in the forms that developed in the UK, North America and Australasia from the late nineteenth century, have been seen as spaces of passivity and conformity. For Lewis Mumford the modern suburb was 'anti-city'; in the suburb, 'life ceased to be a drama, full of unexpected challenges and tensions and dramas: it became a bland ritual of competitive spending' (1961, p. 494). This sense of containment and stifling domesticity was reinforced by critical commentaries, ranging from conservative attacks on the trivializing effects of mass society, through Marxian critiques of commodity fetishistism, to feminist analyses of the patriarchal entrapment of women in the stultifying domestic order of the suburbs (Gilbert, 2010). In all of these, and in more recent writing on creative cities, clusters and quarters, there is an implicit version of Ernest Burgess's famous Chicago School model: concentric zones of creativity, with a central creative district of high official culture, then an edgy, urban zone of creative transition in the post-industrial inner city, then beyond a cultural desert of endless sprawl, at best the site of passive cultural consumption.

This view of the uncreative nature of suburbia is being challenged on two fronts. Firstly, recent work on the lives and locations of artists and other creative professionals reveals a much more fluid geography, as well as the de-centralizing effects of high rents and property prices. Where once, the marginal space of artistic practice was a relatively central loft or old commercial space, studio spaces are increasingly outside the urban core. In her recent study of creative artists in Toronto, Alison Bain indicates the growing presence of creative workers in suburbia, the growth of artistic networks beyond the city, and the role of 'intermediate landscapes' as a source for artistic inspiration (Bain, 2013). A longer line of work in English cultural studies has indicated the importance of suburban origins and spaces for popular music. English music from the Kinks through punk, electronica and Brit-Pop, to the Croydon Dubstep of the The Streets and beyond, has had an ambivalent relationship with the 'sound of the suburbs' (Frith, 1997; Huq, 2013). In the USA too, the home of the garage-band, suburbia has been a contradictory home and inspiration for musical creativity, more often than not as the subject of songs of escape or ennui.

Secondly, suburban creativity has been emphasised in significant revisionist approaches that attempt to disentangle the complexities of suburban culture from the characteristics ascribed to it by academics and social commentators. In his study of *The Intellectuals and the Masses* (1992), John Carey argued that characterisations of suburban life as conformist and stultifying were often little more than snobbish prejudices about the growing power and prosperity of a new middle class. A pioneering 1981 study of the English semi-detached house, *Dunroamin,* challenged its common architectural castigation as cheaply-built, ugly and mundane; viewed from the perspective of homeowners, these were liberating spaces of self-expression and self-development, particularly through the creative practices of interior design, home improvement and gardening. (Oliver et al., 1981) The more recent revisionist interpretation of consumption practices as potentially active creative engagements, alongside the recognition of 'vernacular creativity' has shifted approaches to the suburban world. Practices such as cooking or gardening are no longer seen as markers of suburban conformity but as potential expressions of everyday creative expertise (Barker, 2009).

Continued

The view of suburbs as sterile, cultural wastelands drew upon on pervasive trope of bland, homogeneity; to quote Mumford again, 'a multitude of uniform, unidentifiable houses, lined up inflexibly, at uniform distances on uniform roads, in a treeless communal waste, inhabited by people of the same class' (1961, 486). This dominant image of suburbia as socially and racially homogeneous, perhaps a viable generalisation for mid-twentieth century America, now does no justice to its complex geographies of social, cultural and ethnic diversity. One feature of the twenty-first century suburb is its position in transnational networks of migration and culture (Wei Li, 2006). Again, a conventional social geography that sees suburbs as creatively secondary to the melting pots of the inner city is anachronistic. These are spaces of creative hybridity, but also places that have transnational connections that increasingly by-pass the sterile commodified spaces of global cities, or the delimited bubbles of well-known cultural quarters, disturbing conventional expectations of central and marginal spaces. In very ordinary Hanwell, Butterfly's body artists are English, Polish and Bulgarian, the studio has a sister branch in a Black Sea resort, and designs and techniques are discussed and shared internationally.

The potential of vernacular creativities to offer richly complex accounts of marginal creative practices that challenge dominant accounts of creativity is very clear in the alternative mappings of creativity in Wollongong that are offered by the work of Andrew Warren and Chris Gibson (2011). Focusing on the youth-custom-car design scene in Wollongong, a blue-collar industrial Australian city once dominated by coal and steel industries, they examine how such narratives of vernacular creativity provide a valuable counter-point to the normative positioning of creativity in the regeneration of regional economic futures. Seeking to recognise alternative, everyday forms of creativity that might be missed in an audit of creative occupations, and recognizing the value of these creative practices in the social and cultural production of place, Warren and Gibson explored custom-car cultures as locally embedded forms of cultural production, a form of vernacular creativity with is 'own skills, networks, circuits and spaces of production' (2011, p. 2706).

As with other cities around the world, creativity was seen by Wollongong as a way of saving the city from its industrial decline. And, as with other cities, there was a tension between the industrial history of the city and its turn to sanctioned forms of creativity with their associated class and taste distinctions. Noting the gendered and classed-based nature of custom-car culture, Warren and Gibson assert the value of such apparently 'rough' creative practices in relation to those forms of creativity more normally sanctioned as part of creative city narratives. They also observe the importance of appreciating such 'other' forms of creativity in a city with a declining youth population where young people are often ignored and overlooked and their unemployment rates are high. They lay the foundations for the consideration of custom-car cultures as a legitimate, economically viable creative vehicular industry.

Developing an ethnography with fourteen individuals, largely from blue-collar labour intensive industries – carpenters, electronics, mechanics and spray painters – they explored the networks of rich social capital of automotive knowledge and fabrication skills deployed to 'produce ostentatious, unique and eye catching rides' (p. 2712). Identifying ten components of production from paintwork to wheels, engines, brakes and sound systems, Warren and Gibson were able to map these activities across the city. The result was an alternative map of creativity to that produced when the 'official' statistics on creativity and creative occupations were used. In place of a focus on the urban core and trendy areas, their alternative mapping of creativity focused on industrial estates and suburbs.

Creativity in the process of custom-car design often brings together experience from the designer's blue-collar jobs with aesthetic sensibilities. After obtaining a car (valued for its potential rather than its purchased state), often through social networks, different aspects become the focal point. For some, driven by interest, skill or money, styling is the focus, for others it is performance, which often means achieving a loud, fast ride. Creativity thus becomes a focus of skills and personal interests, refined through daily practice, and brought together with the skill sets of others in the group. As Gibson and Warren emphasise, such forms of creativity are often reliant on oral forms of knowledge production, stem from personal passions, are reliant on generosity and sharing as well as quests to perform often idiosyncratic elements of personal identity. As such, these practices might also be linked to other practices of restoration, repurposing and modification that might otherwise have escaped reflection as part of the creativity literatures. We might think, for example, of the reconditioning or modifying of 'finds' obtained from car boot sales or charity shops (Gregson and Crewe, 2003) or the work of enthusiasts (Delyser, 2014; Delyser and Greenstein, 2015). Such practices might not seem marginal, but clearly take on such a positioning when viewed in relation to the dominant narratives of the creative city, and the commercialised practices of a creative class.

MARGINAL CREATIVITIES 3: SUBCULTURAL CREATIVITIES – NEGOTIATING LIFE ON THE MARGINS

The margins are not just geographic locations, but also cultural positions, a location in relation to wider society that some people take up by choice, whereas others are forced into. For both groups, creative practices can form an important part of the negotiation of such sub or counter-cultural identities, enabling the self-stylings of bodies as well as the improvisational ways of being in the world. In Box 8.2 Murphy discusses the creativity of counter-cultural dwelling practices through the lens of creative methods. In the body of the discussion that follows, the diverse creative practices of Indonesian street children offer them the means to negotiate their precarious lives and offer some sense of collective security and identity.

Box 8.2 A PHOTOGRAPHIC EXPLORATION OF LIVING OUTSIDE SOCIETY

Ben Murphy

In a remote mountainous region of south-east Spain individuals with mixed counter-cultural ideologies gather from many other parts of the world to live outside mainstream society in transient intentional communities. Some adhere to the tradition of American and British hippy culture established in the 1960s and 1970s in places such as Drop City, Colorado and New Buffalo, New Mexico, while others come from the more recent anti-establishment youth movements Punk, Rave, DIY and radical environmentalist groups. As well as people with specific beliefs, there are also those with no fixed affiliations or who are unable or unwilling to live in the confines of conventional society. If they abide by the unwritten rules and codes of behaviour within the communities they are accepted as outsiders here. Neo-nomadic tribal identities are reinforced through co-existence in loosely structured, self-regulating, intentional communities, which alienate themselves from conventional modes of living through their choice and construction of habitat. Makeshift dwellings and the environments they occupy can be seen as deliberate symbols of rejectionist ideologies. The aim of my research is to explore these identities using highly considered photographic practice, which can be interpreted through notions of home, place and space, as expressed in the work of Blunt and Dowling (2006) and Miller (2005) among others. In this way I want to demonstrate that photography, when applied with a distinct method and understood through an informed position, is a valuable and effective medium with which to consider how dwelling and habitat are implicitly linked to and can indicate identity and thus progress understanding of dwelling in place.

A case study methodology has been employed utilizing a systematic approach to image making. This includes large format camera equipment, the use of film, images devoid of people, specific lighting conditions and the documentation of many counter-cultural dwellings. The research process seeks then to identify what German artists Bernd and Hilla Becher termed 'typologies' of buildings through contemplation of similarities and differences in certain types of structures. A particular aesthetic quality is determined by consideration of light, subject, detail, composition, colour and materials. The intention of this approach is to set up in the photographs an unmanipulated stage where the aesthetic qualities of the images and the subject can have a complex interplay that stimulates responses to the questions of

Continued

Figure 8.4 Untitled (Photo: Ben Murphy)

the research. I suggest that one way of looking at the work could be defined by what Barrett (1986) refers to as 'Explanatory' photographs, 'Descriptive' photographs and 'Aesthetically Evaluative' photographs.

Making photographic studies of their self-constructed dwellings and environments using a large format analogue plate camera, usually in diffused lighting conditions, the resulting images offer a heightened sense of reality with a clarity of fine details. These images then are capable of revealing information about the dwellings, environment and habitations of these people, which might otherwise be overlooked. On close inspection these pictures expose the tensions, conflicts and ambiguities between the counter-culture, the host society and the landscape.

In the context of producing valid research, photography has potential to offer multiple layered connotations in ways that words cannot; however, so as to bring focus to my questions and avoid

Figure 8.5 Untitled (Photo: Ben Murphy)

misinterpretation, different processes including the juxtaposition of participant text from those inside and outside these communities allows for alternative readings of the images, giving voice to the subjects and offering further insight and understanding. The work will be edited and made into a book of photographs with text and the accompanying thesis will use the images as the principle reference material from which arguments will be drawn out. Gillian Rose's theory of compositional analysis and content analysis (2007) will be used as a basic method of image analysis. Using theories of phenomenology as expressed in the work of Husserl and Heidegger, among others, I will substantiate the argument that through my practice it can be seen that places and spaces contain complex meanings and significances. I will argue that objects are imbued with meaning through their histories, interactions and placement within space by people. The research draws on theories concerned with place, space and identity set

Continued

out in the work of Massey (2005), Cresswell (2014) and others to argue that relationships to place and space are important in how we understand and express ourselves. Social histories from nineteenth century British and American intentional communities and more recent histories of western counter-cultures from the mid-to-late twentieth century allow me to contextualise the subject matter of my practice.

For those living marginal lives, creative practices – whether forms of ingenuity, of self-expression or skilled practices such as music making – are ways of surviving on the street. In her discussion of the Street children of Yogyakarta, Indonesia Harriot Beazley (2002, 2003a, 2003b) discusses the 'repertoire of strategies' they develop in order to survive. Subcultural practices become the means to understand street children's agency and to counteract those commentators who present them either as victims, as cunning criminals or as lacking in agency all together.

The presence of street children in Indonesia is seen as a problem, most often couched in terms less of moral issues, but rather in terms of a challenge to the state's development philosophy, and the ideological construction of the ideal family, home and child, which is used for social control (Beazley, 2003a). Street children in Indonesia are the result of the country's economic growth strategy, aimed at integrating Indonesia into the global economy. But they have also evolved from the discomfort of homes, violence, hostile or absent parents, as well as the attraction of street children's subcultures (ibid.). Subcultural practices often form the basis for attempts to resolve collectively experienced problems arising from contradictions in the social structure, alienation in society and harassment by the law. As subcultural theories make clear, socialisation by subcultures can help people to redefine negative self-concepts by offering a collective identity and a reference group from which to develop a new individual identity and thus face the outside world (Hebdige, 1979). Subcultures in this case are less a lifestyle choice than a strategy for collective and individual survival, creative strategies become the means to resist and overcome social and spatial oppression.

Beazley's discussions pivot around the formation of the subculture 'Tikyan', meaning little but enough. For these street children making do and getting by is not an artistic strategy, romanticised through its picturing on the street, but rather is a way of living. In attempts to create solidarity in the midst of oppression, homeless street children have created their own distinct social worlds, urban subcultures with their own systems of values beliefs and hierarchies (Beazley, 2003b). As Hebdige (1979, p. 81) writes of subcultures more generally, each subcultural 'instance' represents a solution to particular problems and contradictions. In the case of these street children, variously shining shoes, scavenging for goods to recycle, busking or selling handicrafts on public transport, finding places to sleep and building shelters, or dressing and marking their bodies, creative practices offer the means to appropriate public spaces as subcultural spaces and a means to identify themselves as part of something larger than themselves (Beazley, 2003a).

Understanding the creative practices of this subculture is to appreciate a cross-section of intersecting practices. On the one hand, creative practices are an important source of income, as street children busk with guitars or sell handicrafts to tourists. Within the group, busking is one of the highest work roles and confers a lot of subcultural capital. Those younger children, who shoe-shine, often try to save money for a guitar so they can learn and play (Beazley, 2003b). On the other hand, there are myriad ways in which the children's ingenuity and creativity has come to create the group as a force. For example, they have developed their own street language and vocabulary that develops a sense of solidarity. As Beazley (2003b) notes 'bahasa senang' (the happy language) develops a private vocabulary for events, activities and objects and in doing so reinforces a sense of belonging and excludes outsiders who cannot understand.

As with many subcultural practices, the body becomes the site of creative engagement and identity formation (Hebdige, 1979). As well as displaying adult male working-class habits as ways of building subcultural identity, for example smoking, getting drunk, taking drugs, indulging in free sex, the children will also develop some rather more symbolic grounded aesthetics (Beazley, 2002; Willis, 1998). The body is as Willis (1978, p. 10) asserts 'the source of productive and communicative active' a modifiable surface upon which the group ascribe their subcultural identities. As Beazley (2002) notes the children she studied cultivated their own style by searching for and appropriating objects and appearances through hair-style, tattooing, make-up, posture and dress. Clothes, despite the fact that they often had to be scavenged for, were a key component in identity construction and Beazley understands these through the lens of modern practices of bricolage, adopting and adapting ordinary objects and subverting their meanings. This might involve particular items, such as religion or dress, or it might involve raiding and rehabilitating styles from other subjects, such as punk. She describes how the children created their own punk style with Mohawks, safety-pinning and spray-painting their clothes. Others turned to Rastafarian culture, celebrating Bob Marley and Marijuana culture through adapting various colours and styles associated with this group. Hair, a particular site of cultural control in Indonesia, was an important site of rebellion for the children. Short and neat is the usual style, those with long hair risk being arrested, street children experimented with Mohawks and dreadlocks, or cut zig-zags in their hair. Others shaved it completely as a sign of solidarity with those bodies who have been 'cleaned up' by the police; arrests would often result in heads being shaved as a means of stigmatizing the arrested individual. Further, body modification, tattoos and piercings, are seen in Indonesia as a sign of subversion, of deviation from the ideal body, through their multiple decorations street children turn their bodies into sites of subversive practice.

For those living precarious lives on the margins of society, creative practices, whether formal or vernacular can offer an important means of negotiating this precarious, marginal status. Offering the means both of employment, of improvising living spaces in challenging conditions, as well as fostering importance, collective identities that are not only central to resistant actions but are often important ways to help keep safe when living on the streets.

MARGINAL CREATIVITIES 4: MARGINAL METHODS FOR MARGINAL PLACES AND PEOPLE

Marginal places and landscapes have long formed a rich subject and site for artists, novelists, poets and others; whether it be the imaginaries of the *terrain vague*, nature writers' interest in marshy places and abandoned coastlines, or the contemporary fascination with urban and industrial ruins. The cultural imaginary of the margins is often marked by a sense of transition, of contingency and, whilst in some cases margins are sinister, troubling places, for many they are also sites of possibility. One of the most popular sets of marginal sites for geographers to occupy of late has been the abandoned urban wastelands, derelict sites and industrial ruins left by shifts in industrial activity, and now increasingly rapidly colonised by private developers. These sites have formed fertile territories for geographers interested in exploring both their specificity but also for a range of methodological experiments (Garrett, 2014). The history of such ruin lust is long, from its association with ruined and failed histories, to its melancholic dwellings on transience and the ephemeral. For geographers, ruins have a place within those picturesque landscape paintings, wherein the real conditions of living in the countryside were turned away from, in favour of an aesthetic appreciation of tumble-down buildings and overgrown barns (Cosgrove and Daniels, 1988; Barrell, 1983). More recently, the topological variety of ruin sites has abounded within geography, from derelict infrastructures, to industrial buildings, dwellings and even entire islands (Edensor, 2005a; DeSilvey, 2007; Lorimer and MacDonald, 2002). Capturing the common valorisations of these ruins Edensor writes (2005a, 846)

> ruins foreground the value of inarticulacy, for disparate fragments, juxtapositions, traces, involuntary memories, uncanny impressions, and peculiar atmospheres cannot be woven into an eloquent narrative. Stories can only be contingently assembled out of a jumble of disconnected things, occurrences, and sensations.

For others, ruins are a site of resistance, but for most it is this general sense of the productivity and possibilities of these marginal sites that are important. Tim Edensor (2005a, 2005b), for example, explored the possibilities of ruins as contingent, playful spaces using creative writing practices and photographic essays. De Silvey (2007) experimented with arts-inspired curatorial methods and creative writing as a means to engage with the assemblage of human-non-human objects found at the abandoned site she was working at. Dixon et al. took a rather different tack in their exploration of the future ruins of Hashima Island. Developing a creative writing and performance project around Hashima Island, off the coast of Nagasaki Japan, they explore less the past of this poisoned, abandoned mining island, and more its monstrous becomings. They sought the means to create a written response to this ruined island that grasped its complex materialities and spatiotemporalities. Marginal spaces and their particular characteristics have clearly become sites for geographers to conduct creative experiments and, as Rupert Griffiths outlines in Box 8.3 below have also offered the sites for thinking through what such creative experiments might mean for geographers more generally.

Box 8.3 PICTURING MARGINAL PLACES

Rupert Griffiths

Recent years have seen an increasing interest in exploring the margins between disciplines. In particular, there has been an interest in what creative practices can bring to geography, as methodological toolkits, as mode of enquiries, as subject of research and as means of dissemination (see Hawkins, 2013, 2015a). Equally, and not unrelated to this, there has been an interest in marginal spaces, places and landscapes within geography and urban studies, with notions such as *terrain vague*, loose space, drosscape, friche and indeterminate or unplanned space gaining currency. Similarly, there has been much work around discourses of entropy, ruin, enchantment and exploration, geographies of waste urban natures and hybrid geographies.

Margins are by definition hybrid, an admix of one apparently homogeneous entity and another. The notion of marginality or hybridity does not, however, equate to being peripheral. The margin can instead be thought of as a key site for innovation and creativity, where a bricolage of objects and ideas, methods and positions jostle together, a porous territory where unexpected patterns of coherence can emerge. The margin is in this sense both disciplinary and anti-disciplinary, a region of exploration, emergent possibility, creative production and, of course, risk.

My own work is driven by the possibilities offered by this and makes use of both academic and artistic modes of enquiry, looking for example at the relationship between creative practice, the creative subject and the urban margin. I argue that the urban margins are not specific spatial or temporal regions in the city, but rather that they emerge through processes that entwine materiality, the body, creative practice and identity. The urban margin is not a place, space or landscape in its own right, but rather it is co-produced by multifarious materialities, human and non-human agents.

Some material conditions – wastelands, industrial ruins, for example – always have a sense of marginality clinging to them. Is this marginality out there in the landscape or does it exist between the creative subject and material landscape? Another way to look at this is to think of both propositions as equally valid, each trying to constantly assert itself. Here notions of distinction and indistinction co-exist; a distinct perceiving subject in a distinct material landscape is one moment plausible, the next absurd. A line between subject and landscape is both drawn and erased, over and over, smudged and faltering.

Creative practice can make this distinct sense of subjective identity, a distance between self and the landscape, and equally dissolve it. The distance between the documentary photographer and the documentary photograph is an example of this (see Figure 8.6). Equally, creative practice can develop an experience of dissolution between the subject and the material landscape through the imaginaries of transgression, dirt, decay or contamination and the acts of climbing, digging, and generally getting one's hands dirty. The margin can be considered as a place where identity is in tension, driven by conflicting desires to make assertions about self and landscape and to dissolve into an a-subjective in-distinction with blind matter. The photographer Stephen Gill is a good example of an artist whose work oscillates between these two positions.

Practices such as photography and film making are for these reasons valuable not only as a means of collecting or documenting research material, but equally as a means of accessing a self-reflexive understanding of the role subjectivity plays in research. Auto-ethnographic accounts of creative practice, for example, can be valuable in augmenting the researchers' understanding of the relationship between self and materiality. This can form a useful

Continued

Figure 8.6 Untitled (Photo: Rupert Griffiths [2015])

sensory, experiential and embodied register that strengthens one's critical capacity when dealing with research that engages with wider relationships between the body and the city. The notion of the margin is important here because it gives one the freedom, as hinted at earlier, to be both disciplinary and anti-disciplinary. It enables one to be able to explore and develop a research question in an open-ended way, whilst simultaneously being able to locate it within a strong critical framework. The margin is a region between disciplines, a method, a stance and a mode of dissemination that brings creativity and criticality together. It aims precisely at the slippage between subject and materiality that is important to many aspects of cultural geography.

The margin is thus a common denominator across multidisciplinary understandings of identity, space, place and landscape. It is an area of research that seeks out the dappled ground between disciplines and that equally seeks out that same ground 'out there' in the fabric of the world, a social, cultural and material milieu. These two areas are mutually entwined and inseparable and it is this desire, willingness and capacity to critically engage with both disciplinary and material/cultural margins that offers a fascinating and unique area of study.

MARGINAL CREATIVITIES

Margins are much mythologised spaces, the ambiguity and tensions that sit at the heart of these spaces plays through in how they are lived, practised and imagined. Margins are the spaces from which dominant norms are challenged, where rules and regulations are often relaxed and where modes of informal life and work are often made possible. As well as romanticised spaces, they are also precarious sites, where creativity can be a life skill, a necessary part of survival rather than a professional practice. Conceptual margins are, as these diverse discussions have shown, sites at which dominant discourses – in this case of the creative city – are negotiated, as creativity becomes something done elsewhere, by other people and in other ways, other than through the creative class producing and consuming in the sleek, chic spaces of the cultural quarter.

9

NATION: NEGOTIATING NATIONAL AND GLOBAL CREATIVITIES

July 1997, newly elected Prime Minster of the UK Labour Party Tony Blair threw a party, not a party of politicians but one of celebrities; of the super stars of Britpop – Blur and Oasis – the darlings of the Young British Art World, and a who's who of Britain's literary and movie cultures. The accompanying soundtrack D: REAM's 'Things can only get better' symbolised not only the hopes of a nation, but also of a creative sector that had been enrolled as part of the new young government's marketing ploy. Not since the 1960s had Britain felt so cool, on the back of the rise of its music, literature and art around the world and the growing strength of its creative economy more generally, the government firmly employed creativity as part of its national economic policy, as well as within a global rebranding exercise this was the era of 'Cool Britannia'.

The nation has been, and remains, central in discussions of creativity. On the one hand, the nation has become a key site for creative policy negotiation, the lens through which policy is formed that then comes to be negotiated around the world. On the other hand, creativity is part of the development and maintenance of the 'imagined community' of the nation (Anderson, 1982), such that creative practices come to be part of the creation and promotion of national identities and brands within the country, but also around the world. This chapter will draw out these different understandings of creative nations, in each case examining the tensions between national creativity and its global circulation and uptake. Discussion begins by exploring how creative practices and policies produce and reproduce nations as imagined communities through engaging their population, it then turns, to explore how creative exports – whether policy, paintings, films or music – come to be sites of negotiation around the world. What becomes clear from these discussions is the potency of creativity as a practice of nation-making, as part of the circulation and promotion of national values around the world, and thus as a key site of global negotiation. Discussion begins with an exploration of creative policy, its formation on a national scale and its circulation around the world, both in terms of creative economy policy, as well as the wider uptake of creativity within agendas for global development. The chapter then turns to think about the role of creativity in forging imagined communities at home and abroad, exploring how creative practices are formative of national identities, but also how they are closely enrolled in the colonial practices of soft power, thus finding creative practices to have an important role in global geopolitical practices.

CREATIVE NATIONS ONE: GLOBAL NEGOTIATIONS

The place of creativity in national policy has shifted over the centuries, but what remains consistent is the tension between economic and socio-cultural rationales within such policies. Creativity's social and cultural rationales can be dated from before the French Revolution of the eighteenth century when it became part of a monarchical tradition to support the arts. Down the centuries cultural policy has supported arts institutions and arts for arts' sake as part of national investment in being a 'civilized' nation, as well as from an understanding that arts and culture were good for social wellbeing (Duncan, 1995). Raymond Williams in an essay 'Politics and Policies: The Case of the Arts Council' notes four drivers behind the founding of what is now Arts Council England in 1946; state patronage of the fine arts; pump priming; an intervention in the market; and an expanding and changing popular culture (Williams, 1989, p. 143). The original impetus behind not only Arts Council England, but also the National Endowment for the Arts (USA) and the Australia Council were all based on ideas of 'public good' art for arts' sake, or social welfare (Caust, 2003). These social welfare narratives also drive much of the early expression of creativity within development policy, where interestingly it still has a place (Flew, 2013; Stupples, 2015).

In contrast to these earlier logics, the more recent era is one in which creativity finds its most high profile place in the political portfolio under the auspices of economic development and, in particular, urban, and increasingly rural, regeneration. Creativity has diversified away from an increasingly hollowed-out profile of arts funding, to find a home not only in economic policy but also as entrepreneurial solutions to service provision and as a cost-effective means to engage with questions of social well-being and societal coherence. The latter set of solutions have seen artists describing themselves in the role of 'social band-aids' and critics finding the delivery of social services and community cohesion policies to be landing at the feet of artists rather than being the responsibility of more expensive forms of social policy (Jones and Warren, 2015).

Creativity now appears in a range of ways in national and regional policies around the world, as well as in global development strategies, adding new levels of credence to the power given to creativity as a force for economic and social regeneration. To explore this further, discussion is going to examine a range of global creativities, starting with 'Cool Britannia' and the UK's export of creative industries policy in the early years of the new Millennium. This will be followed by an exploration of how such policy does or does not travel around the globe. To attend to creative economies in the Global South is to examine these uneven and variously successful policy exports from the Global North, as well as to consider local variations on the form and composition of Global South creative economies (Barrowclough and Kozul-Wright, 2008; Kong and O'Connor, 2009; Murphy, 2006). In addition to the issues related to calculating the proportion of income from the creative industries more generally, in developing countries a higher proportion of creative activity is masked by the largely informal nature of the sector. Alongside examining cultural policy in different cultural contexts, the form of creativity that exists in the global development agenda will also be reflected upon. What becomes clear from these examples is how problematic 'travelling' policy is and how an appreciation of diverse understandings of creativity is crucial to understanding and practising successful creative policy making around the world.

Cool Britannia and travelling policy

Cultural policy is commonly dated from before the French Revolution, yet it is most often narrativised in its current form as dating from the late

1990s under the UK Labour Government (Garnham, 2005). Interestingly, geography sits central to this work. In the late 1990s the Department for Cultural Media and Sport commissioned a 'Mapping Document', aiming to 'map' creative activities around Britain and assess their value and worth (Mould, 2015). This mapping document, later revised in 2001, was one of the early tasks of the 'Creative Industries TaskForce' set up by Tony Blair as Prime Minister. Another early task of this group was to assess the 'Hollywood model' of cultural production, seen as an iconic example of how direct and indirect economic benefits could be leveraged from creative industries, both creating economic value and branding a place. This was to find its incarnation in the Labour Government's subsequent 'Cool Britannia' and 'Creative Britain' strategies that aimed to combine economic policy and urban regeneration with a branding strategy for the whole nation.

A number of scholars have gone up and down and back and forth with the shift that New Labour's activities developed in terminology from Cultural to Creative (e.g. Flew, 2012; Garnham, 2005; O'Connor, 2007). A common telling of the story begins with the Greater London Council (GLC) as a site where the 1970s brought the first integration of culture and economy in policy. The UK's marketisation of creative activities was lagging behind the developments the US had seen in the monetisation and politicisation of cultural production. Another common telling of the story explores the negative impact of culture as a politically dubious concept in relation to creativity that was seen in a more positive light. The main reason for this positivity being creativity's seeming ability to capture an expanded sense of what creative activities might consist of in relation to the information and knowledge economy, as well as its close proximity to ideas of innovation. In short for many, 'creativity' allowed the all-important association with the economy whilst also enabling other parts of culture to still be celebrated.

The first Creative Industries Task Force Mapping Document, published in 1998, identified the creative industries as 'those industries which have their origin in individual creativity, skill and talent and which have a potential for wealth and job creation through the generation and exploitation of intellectual property'[1]. It went so far as to determine thirteen subsectors, including film, software, arts and antiques, music, publishing and advertising. Within the sector these divisions and their titles caused many problems, not least the diversity of sectors (from software and crafts, to arts and antiques). Core to all sectors lay, however, the 'potential for wealth and job creation', values that became core not only in Britain's creative industries policy but in creative policy around the world.

If Britain's Labour Government wanted to export the image of a Cool Britannia it was, perhaps ironically, the creative industries policy itself that was one of the most effective pieces of British marketing they could have created (Wang, 2004). Indeed, 'few could have predicted that the creative industries model would itself become a successful export' (Ross, 2007, p. 13). The extraordinary mobility of creativity policy has become a common focus of discussion, as Kong and O'Connor (2009) note of Asian cities – from Hong Kong to Singapore, to Shanghai and Guangzhou in China, as well as Australia and New Zealand, much of these cities recent thinking on creativity 'has derived from the European and North American Policy Landscape' (2009, p. 1). Whilst, as they observe, there is much promise in the 'creative industry, creative city' agenda, its application is 'fraught with ambiguities, tensions and translations' (ibid.). Indeed as Pratt (2009) confirms, trying to transfer these policies requires a much better understanding of the operation of these industries and their relationship with the rest of the economy than is currently in existence (p.10).

China has proved an interesting case in point here as discussions in Box 9.1 extends. The 'creative industries came to mainland China in late 2004', with its UK credentials firmly on show (Keane, 2009, p, 431). China, always strong in cultural terms having adopted soft power as an important feature of its policy (see later discussion), has also been ranked third globally (to the UK and USA) in terms of cultural export rankings (UNESCO 2005, cited in Kearne, 2009). Yet the policies had far from a simple arrival and transition, for while attractive to city planners and entrepreneurs, there was widespread scepticism regarding their uptake in the country more generally. As Wang (2004) notes, 'the thorniest question triggered by the paradigm of creative industries is that of 'creativity'. How do we begin to envision a parallel discussion in a country where creative imagination is subjugated to active state surveillance? (Wang, 2004, p. 13). Further, as Keane (2009) notes, while the idea might be fashionable there were a series of concerns over whether there was any real potential for change. Not least because creativity was seen an inherently western idea and there is a controversial sense that 'creativity is compromised in China due to official management' (Keane, 2009, p. 437). As the discussions in Box 9.1 demonstrate, ideas in China around creativity both mesh with, but also challenge ideas of creativity that circulate in the Western world. Indeed, O'Connor and Xin (2006) raise the question as to whether 'China can have creativity and innovation without social, cultural and political change'.

Box 9.1 MADE IN CHINA/ CREATED IN CHINA

China should focus its attention on a new century. From creative industries to creative economy then to creative society. Contemporary China should be Creative China; from manufacturing to creative work, from 'made in China' to 'created in China'.

(Ministry of Cultural Representation, 2004,
cited in Lindtner, 2015)

Creativity has long been a source of debate in China, with some suggesting that China had little creativity of its own and, as the world's factory, was doomed to be a site for production and manufacturing, rather than innovation and creation. Indeed, China has been unfavourably compared to Silicon Valley, with 'Made in China' standing in for 'Created in California'. Many, however, contest this western view of creativity, with its tendency to overlook the rich creativity that does exist in China. This discussion is going to visit two creative sites in Shenzhen China to examine the ways these debates have played out.

Site one: Dafen oil painting village

Dafen Village in Shenzhen China, an urban village, contains a concentration of over eight thousand painters, who collectively create millions of hand-painted oil paintings each year for export around the world, worth over 300 million yuan (Figure 9.1). Copying great masters as well as photographs of people's dogs and children, and creating millions of paintings for hotel rooms, offices and banks, the painters live and work in the narrow streets

---

Continued

Figure 9.1 Paintings in the oil painting village in Dafen (Photo:
author's own)

and alleys of this area. Dafen's painters are mostly members of
China's huge floating population. Rural migrants they are, under
Chinese law, registered in their rural homes for everything from
education to medical care, thus they remain very precarious. As

Wong (2010), outlines the Government will often award urban registration as prizes in copying competitions in the village. As Li et al. (2014) explore, the site can be considered a cluster, or in the words of Becker, an 'art world', the small urban village is now devoted to painting, and includes the whole production line, from shops selling materials and canvas, through to frames, glazers and export and shipping agents who will pack and distribute the finished works around the world.

The vision of this area of 'art industry' around the world has been that of a 'painting factory', an oil painting production line producing 'Van Gogh on Demand' (Wong, 2014). The imaginary of this industry (with the full negative force of this word in the art world) is one that sees Western artists' copyright violated through the reproduction of these works. For local officials in the area, however, this is also a democratisation of the art world for the global consumer. In this argument and more general views on the village and its copying practices, we find as Wong suggests an unfortunate resurrection of 'an age old Eurocentric idea . . . that China is a totalitarian society made up of automatons who make and consume copies, whereas the West is made up of liberal and free-thinking individuals who create and collect original things' (2010, p. 27). The point of this discussion is not, following Wong (2014, 2010), to simply reverse this dualism, but rather to demonstrate the complexity of creativity. As Wong notes, for some in Dafen the copies are technically better than the originals, for others this is an imported western form, worth less than the creativity of the traditional-ink works produced in the village. Whilst beyond Dafen there is a clear division of labour between the works of these painters and the more conceptual work of contemporary artists.

Site 2: Chaihuo Makerspace, OCT Lofts, Shenzhen

Chaihuo Makerspace is one of the growing numbers of hacker or makerspaces around China. The first Chinese example of these

Continued

Figure 9.2 Chaihuo makerspace, Shenzhen, China (Photo: author's own)

'innovation houses' (as they have been nicknamed by the Chinese Government), opened in Shanghai in 2010, and now they number many hundreds across China and around the world. Often thought to begin in counter-cultural movements in the US and Europe, these makerspaces are shared studio spaces that bring people together to share open software and hardware, as well as knowledge and ideas. A typical studio, which people pay to be a member of, includes tools such as laser cutters, 3D printers, microcontroller kits; educational workshops are hosted to teach people to use these tools. For Chris Anderson (2012, cited in Lindtner, 2015) the former editor in chief of *Wired* magazine, this 'contemporary maker movement is driving forward the third industrial revolution, a generation of technology producers that build on internet and web 2.0 techniques to make innovative productions', and in doing so remake the spaces and practices of industrial production. Together the evolution of the global hacker-space and makerspace movements have created a maker culture that revolves around 'technological and social practices of creative play, peer production, commitment to open source principles and a curiosity about the inner workings of technology'.

In the context of China, the world's largest manufacturer, such maker culture has a complex and interesting role to play. The 'innovation houses', supported by government funding, were officially part of a platform for supporting popular science work and innovation, but more broadly the funding was understood as an endorsement of China's emerging maker culture. Why would China support these DIY approaches to making technologies and free and open knowledge exchange? How does making culture manifest in China where 'making' in a DIY sense collides with China's image as the world's largest manufacturer? Lindtner (2015) makes clear that both Chinese politicians and the maker movement want to remake China, but in different ways. This is not a straightforward story of knowledge and tool transfer from west to east; rather she highlights different ideas of creativity and innovation and the negotiation, remaking and appropriation that occurs in these settings.

Shenzhen is already the home of a huge amount of often over-looked creativity and innovation with respect to its electronics industry. Infamous as the home of controversial labour practices, including the Foxconn factory which makes products for Apple, the city is also the home of a rich set of skills in copy-cat or imitation electronics. Such counterfeit electronics are based on factories developing open source strategies to bring down production costs. They have informally organised a peer-to-peer database for sharing hardware design schematic and bills of materials used in producing productions (Lindtner, 2015, p. 11). From this base has grown a Shanzhai culture (meaning mountain village and home of those who oppose and evade corrupted authority), a new form of innovation based on open source manufacturing and continuous remaking of existing products. Sourcing components from sites like Hauqiangbei electronics market (Figure 9.3)

Continued

Figure 9.3 Huaqiangbei electronics market, Shenzhen, China
(Photo: author's own)

the world's largest electronics market, people are not only coun-
terfeiting products like Apple watches and iPhones, but also
creating updated versions with improved features such as dual
sim card slots.

> As Shenzhen becomes a home for start-ups, many of these are based on the merging of the Shanzhai culture with the maker-spaces, as well as helping fund promising projects through development via crowd funding sites. For some these makerspaces, which blur work and leisure, social-cultural and economic benefits, and which attract DIY and Tech enthusiasts as well as those for whom this is a livelihood, are an important incubator space for a creative and innovative China.

The 'traffic and transferability of policy ideas between places' (Luckman et al., 2009, p.70) has become an increasingly common topic in the discussion of the creative industries. As Luckman et al. (2009) note, there are messy processes that occur when 'putatively global travelling policy discourses are translated by researchers for local actors (each with their own desires and concerns)' (p.70). Indeed what can tend to be overlooked are the 'contradictions, problems and limitations' of creativity as a universal policy tool (Rantisi et al., 2006; Scott, 2005). Increasingly, however, people are starting to query how transferable these creativity ideas are (Gibson and Klocker, 2004; Wang, 2004). Luckman et al. (2009) suggest that this interest in transference relates to the scepticism surrounding the meteoric rise of interest in creativity and the commercialised knowledge production system that surrounds it (p. 72). Indeed as they note, having explored the case study of Darwin Australia, models of cultural development imported from elsewhere are only partially transferable at best. This is not to suggest that these policies should be jettisoned. Indeed they conclude that whilst there is a need to 'remain attentive to endogenous conditions as well as wider social and political concerns' (p. 81), creative policy enables policy makers to engage new economic possibilities and encourages them to expand discussion about city futures in an inclusionary way.

Concerned to ask critical questions of the efficacy of mobile creativity policies, Peck (2012) notes how creativity typifies a new generation of 'vehicular ideas' in urban policy; it has been 'sutured to a mobile policy frame that has evidently been enabling, sustaining and normalizing a culturally tinged form of neoliberal urbanism' (p. 465). Creativity policies for Peck are thriving in the vacuum generated by neoliberal scare politics, where cities have assumed responsibilities for economic and social development and welfare without having the power to effectively engage issues in this field. As he notes, creativity seems to have become a key urban development imaginary, but it has spawned a 'decidedly unimaginative round of urban policy makeovers'. His surveys of widespread creativity policies find them to be 'remarkably similar in form and presentation ... as characteristically soft interventions they tend to be reminiscent of familiar types of economic development' (p. 472). Performing the creativity script is, Peck notes, often easy for cities, as the script 'purposefully distils and recycles readily observable features of winning cities in its circular accounts of what it takes to succeed' (2009, 2012, p. 473). Taking the case study of Amsterdam, and focusing on the impact of creative city guru Richard Florida's visit to the city, Peck concludes,

> creativity discourses, as they touched down in Amsterdam seemed to have carried with them the allure of apparently governing in fundamentally new ways, with new stakeholders and new strategic objectives ... whilst at the same time changing very little.
>
> (2012, p. 272)

Drawing on empirical material, he concludes creative industries policy makers are often going through the motions. Creativity offers a convenient narrative within which to fit a wide range of actions in the name of economic growth and inclusive urban development; often, however, instead of new goals existing policy is recycled and pet projects given new impetus. Peck explains, vehicular policy ideas are 'constructed for travel' (p. 480). Whilst often having a transitory existence they are, he suggests, best understood as wide ranging 'enabling technologies', their open ended ambiguities helping them to move between different policy making sites to engage new or existing initiatives in a range of locations around the world.

Creativity for development

> Now the time has come to fully integrate culture into the global development agenda, through clear targets and indicators, as an

overarching principle of all development policies, as an enabler of sustainability, as driver of growth.

(Irina Bokova, Director General, UNESCO, 2013)

The role and potential of creativity for development has been increasingly realised within and without the operations of large organisations like the United Nations, as well as in the activities of NGOs. Creative practices became recognised as part of the 'cultural turn' in development that was the result of the diversification and democratisation of development thinking in the 1970s and 1980s (Stupples, 2015). So important did it become that the decade from 1988 was known as the UN decade for cultural development. With the evolution of a more human-centred approach to development in the mid-1990s, funding for the arts increased as part of the support for cultural diversity and local specificity in the face of dominant western narratives of development. Interestingly, to explore the place of creativity in development is to find a positioning that reflects both the older policy location of creativity in terms of social welfare and wider social benefits, as well as to find the contemporary location of creativity firmly in economic terms, this is well-illustrated by the Banglanatak project discussed in Box 9.2.

The UNESCO Culture for Development Indicators (CDIs) include measures of the impact of culture for social development primarily by way of assessing participation in cultural life, levels of interpersonal trust, and tolerance. Further, these indicators show an increasing high percentage of developing country's GDP coming from creative industries. This includes 5% for Ecuador, similar to the contribution from their agricultural sector, 3.4% in Columbia, and 1.5% in Ghana and Cambodia.[2] A review of UN member state activities demonstrated that across strategic partnership framework plans, the place of 'culture' has doubled over the past ten years, with culture and creativity aims incorporated as part of a number of priority areas. These areas include social and economic development, human rights and governance, the communication of development messages, as well as programmes to safeguard cultural heritage. In addition, arts take up a place in discussions of social inclusion and the promotion of political stability, using arts for example in post-conflict peace-building efforts (e.g. see the discussion of dance in post-conflict Cambodia in Box 9.4). There is also a growing awareness of the possibilities of art to offer critical public spaces, where the arts offer a form of creative resistance and social critique, often in the context of otherwise oppressive regimes. As Nagar notes, for example, the rise of people's theatre, especially women's theatre in the so-called South, where

> activists and non-governmental organizations (NGOs) have deployed theatre as a vehicle to promote an alternative vision of development, a vision in which struggles over economic and political rights of the marginalized are viewed as inseparable from the development of awareness, of imagination, of a culture of the mind.
>
> (2002, p. 56)

Across discussion of the mobilisation of creativity for development are a series of reoccurring questions. Firstly, what is the current situation and how can existing cultural resources (e.g. landscapes, traditions, rituals, cultural events) be used effectively? Secondly, what is the potential here, and specifically could the creative economy really provide a viable solution to engage with current serious development problems and to create new opportunities for development? Thirdly, what is proof of success? Fourthly, how can we guard against the sanitisation of creativity in order to make it safe for investors? There are also a series of questions regarding cultural exploitation, negotiation of issues of authenticity, the co-option of cultural expression for existing interests, preservation versus development and the possibilities of missing the chance to develop and enhance dialogue, debate and forms of development for marginalised people. Such issues need to be considered to prevent a reduction in creativity's value when it is solely associated with western exported ideas of creativity.

Box 9.2 BANGLANATAK: CREATIVITY FOR ECONOMIC AND SOCIAL DEVELOPMENT

The organisation Banglanatak is a social enterprise working across India with a mission to foster the rights of women, children and indigenous communities, and to do so by way of a range of creative practices (Bhattacharya, 2011).[3] This includes using

creative forms for communication, such as theatre for development, as well as developing community led creative industries
programmes to grow the market base for cultural heritage
products. There are three key facets to their programme:

1 Street theatre-based communication for mother and child health

India has a long tradition of folk performances, but as Nagar
notes what distinguishes modern street theatre from previous
modes is its political overtones (2002, p. 59). Banglanatak's
productions are funded by The Bill and Melinda Gates Foundation
and have seen over a 1000 shows put on over eight districts in
India, targeting, in particular, 'media dark' areas where information is hard to get across. Three phases have been developed,
birth preparation (emergency preparation, registration of preg-
nant women and benefits of institutional delivery), nutrition
(prompt breast feeding, hygiene and complementary feeding)
and family planning (birth spacing, need for family planning
and contraception). The plays were performed by 24 local
teams who were trained through central workshops to help
standardise content, the actors were accompanied by trained
youth volunteers to help generate discussion after the shows and
gauge audience feedback. The performances were supported
where possible with mass media tools, like TV and radio to
promote the performances and their messages. Scholars of Indian
street theatre recognised the possibilities of its dual aim to both
entertain, but also to serve as a cultural intervention that can work
directly at the level of people's consciousness (Nagar, 2002;
Sadasivam, 2000).

2 Art for life

The Art for Life strand of the Banglanatak programme describes
itself on the website as focusing on 'ways of integrating culture
into development as a force for inclusive social-economic

Continued

development'. Their aim is to safeguard cultural heritage as means of providing local practitioners with viable livelihoods. Since 2005 they have worked with 5000 folk-dancers, singers, musicians and painters and helped them to innovate models for transforming cultural capital into economic assets. With traditional folk art forms dying out due to a lack of practice and opportunities for performance, the programme has trained young artists in traditional dance; it has also created documents and archives through sound recordings of folk singing. Working with communities in West Bengal, the group has increased the average monthly salary of local artists by 700% from 500 Rs to 3500 Rs a month. Forty per cent of the artist community has now taken arts practice as their primary income. This revival has been especially empowering for female artists, who with increasing incomes have found greater respect locally. The programme is especially happy that they have enabled over 300 girls and young women to make a living from painting. This has also helped family stability as local artists no longer migrate for day labour, but are able to stay and work in their villages.

3 Craft hubs

The third element of the programme echoes a popular spatial form for supporting the creative economy around the world – the hub. In partnership with the Department of Micro and Small Scale Enterprise and Textiles, the Government of West Bengal and UNESCO, Banglanatak has developed 10 rural craft hubs in Bengal. The objective was to support hundreds of families skilled in traditional crafts, such as mask and toy making and the textile technique kantha, to produce crafts for the national and international craft market. As a rule the craft sector is understood to embody a rich cultural heritage, and to offer a chance for the development of environmentally sustainable livelihoods with low start-up costs.

In 2013 the organisation developed an appraisal study, from which they selected a number of crafts to develop into craft hubs. The first phase involved engaging the craft communities, skills development and the making of marketable products and promotion; this was about helping to strengthen the business acumen of the makers. The next phase has involved youth training programmes to help ensure the hubs were promoted as tourist destinations and winter festivals were planned to further promote the hubs internally and externally. They are also working with the hubs to help crafts people access the various entitlements the government offers to crafts people more generally. Many of the people they work with count as vulnerable, 50% of their craft workers are women and many are also scheduled tribes, casts and religious minorities.

Whilst arts and culture have been funded in the Global South as part of development assistance since the mid-1990s, 'critical academic engagement with this cultural space remains incipient' (Stupples, 2015, p.1). This lack of attention has been put down to the lack of fit between the values associated with arts and culture and managerial approaches to development that focus either on the economy, or on basic needs and capabilities. As Stupples argues, the emergence of the creative economy as a driver of economic and social growth offers a 'more secure toe-hold in development discourse and practice' (p.3) than did previous more culturally focused arguments. As she goes onto argue, however, by legitimating action in the arts by foregrounding development outcomes driven by the economy, other less neat understandings of what creativity is and does risk being overlooked. What becomes clear through these discussions of national and global creative politics is that multiple ideas of creativity circulate with the import and export of these policy ideas and their adoption and adaptation in local contexts.

CREATIVE NATIONS TWO: CREATING THE NATION

The potential of creative practices to help forge nations has long been understood by geographers studying everything from festivals and exhibitions to paintings. Stephen Daniels' landmark volume, *Fields of Vision* (1993) explored how a series of eighteenth and nineteenth century paintings from the England and the USA, make and remake national identity. He explores J.M.W. Turner's picturing of London from the edges of the city, the imaginary of St Pauls and the other landmarks of the city of London in the wake of the Blitz bombings, as well as John Constable's picturing of rural England. Exploring the production, commissioning, consumption, reproduction and circulation of a range of these images was central to understanding how they came to produce and reproduce nations, circulating their ideals and practices around the world, for good or ill.

In 1982 Benedict Anderson, a scholar of international studies, proposed that we might consider nations 'imagined communities'. This was not to suggest that they were unreal, false or fake entities, but rather that the community of the nation is so large and spread over such a geographic area that their connections are imagined rather than created through face-to-face contact. Creative practices become an important part of the form-ation of these imagined communities, a means to connect and unite people around common ideas. For geographers, it was not just Anderson's ideas that proved central to understanding the importance of creative practices in the production and reproduction of nations, but also the work of literary scholar Edward Said (Box 9.3).

Box 9.3 EDWARD SAID: ORIENTALISM

Edward W. Said, a Palestinian-American literary scholar changed the face of discussions about representation and geographical imagination with his ground-breaking book *Orientalism* (1978). Studying 2000 years of literary texts, Said noted how the western domination of the Middle East and East Asia had led to a writing of Asia's pasts and the construction of its modern identities from a perspective that takes Europe as the norm from which the exotic Orient deviates. Oftentimes this involved exoticizing Arab cultures, but at the same time seeing them as backward and uncivilised in other ways.

For Said, Orientalism evolved during the European enlight-enment and was concretised during the subsequent colonisation of the Arab world by Europeans. The characteristics defined in Orientalism cast Europeans as superior and thus offered a

Figure 9.4 The Babylonian marriage market, 1875, Edwin Long,
© Royal Holloway, University of London

rationale for their colonizing presence, constituting the West as support to the East, and thus able to intervene and rescue the latter. As Said notes, with the West's thinking of Middle Eastern society as static and underdeveloped comes the sense that oriental cultures can be studied, depicted and reproduced by the more superior culture. It was not only that the communities of UK and Europe saw and bought into these stereotypical views but also, as Said notes, that many of the ruling elites in the Middle East played a part in producing and reproducing exoticised visions of the Orient such as Edwin Long's painting above.

The impact of Said's work is hard to underestimate, it is crucial to sets of ideas that explore how representations go to work in the world, instigating forms of oppression and reproducing power relations. As Said (1978, p. 57) argues, 'it is Europe that articulates the Orient; this articulation is the prerogative, not of a puppet master, but of a genuine creator, whose life-giving power represents, animates and constitutes the otherwise silent and dangerous space beyond familiar boundaries'. A sinister mobilisation of creative practices indeed.

Importantly, it is not just through the production and consumption of representations that imagined communities of nations are produced and reproduced, but also through actual practices of creative making. Creighton (2001), for example, explores how Japanese silk weaving workshops form individual and collective identities. Traditional Japanese handicrafts including calligraphy, painting and weaving have become an important part of the symbolism of Japanese identity. Indeed, as discussions of craft in Japan show, learning and developing amateur handicraft skills are seen as a way to develop personal collective and cultural identities (Creighton, 2001). Conducting an auto-ethnography of a silk weaving course, Creighton observes how the process of learning about silk weaving, as well as doing the actual weaving, form individual and collective identities. The two-week courses saw participants learn processes from preparing the silk to be woven through to actually weaving. As Creighton recounts, at the beginning of the course they were taught how to raise silk worms, how to boil cocoons, how to remove dead worms from cocoons, how to spin the silk threads and dye them using grasses collected from the nearby mountains. Learning the whole process reflected an important belief that to understand weaving is to understand all these stages prior to the loom. This process involved developing an understanding of the natural environment, a respect for the natural material and living creatures that were enabling the process. Once the silk was prepared, the setting up of the looms draws to the fore collective labour as the group set up 3000 warp threads, each needing to be individually tied onto the threads left still on the loom. Each woman in the group would both weave their cloth on a collective framework, but also be able to develop their own individual pattern. As Creighton recounts there was an emotional process at the end of the course when the collective cloth was cut into the women's individual sections.

In these workshops silk weaving is understood as part of Japanese cultural heritage and is used to offset the threat modern Japanese feel of a loss of culture. Weaving workshops can thus be understood as part of a collective search for identity that expresses in nostalgia. In the face of what is seen as an identity crisis amongst the Japanese brought on by internationalisation, there is a sense that uniquely Japanese cultural trends should be celebrated and preserved. As Creighton recounts, the workshop engaged these issues in a number of ways; through their location in a very rural part of Japan and the requirement that the participants live very simple lives when in residence. Travelling to the remote region becomes a symbol of travelling back to the true heart of Japan. Further, the whole process of silk weaving exemplified the principal at the heart of the craft movement in Japan, that is, 'the beauty of craft emerges because craft allows individuals to rely on each other and on a tradition. Thus almost everyone has the capacity to create works of great beauty, and this does not require individual creative genius' (quoted in Creighton, 2001). As participants in the workshop recount, in learning to weave one is not just learning a craft, but one is learning a Japanese craft and learning what it means to be Japanese and the value of being Japanese in the process. Reflecting on the discussions of the body in chapter 1, what discussions make clear is the combination of the physical doing and the materialities of making, together with iconography and national identity. If the weaving workshops created a temporary community in order to create a sense of Japanese collective identity, a series of other examples offer ways to think further about how creative and arts practices might work to connect and reinforce existing communities.

Where perhaps the greatest concentration of scholarship has occurred on creativity and national relations has been within the cultural turn in geopolitics. Here the focus has fallen on the study of popular cultural sources; video games, comic books, films and a range of other creative outputs. Most recently we see geopolitical scholars turning to art works as empirical objects. If, to begin with art was understood as representative of geopolitical orders, more recently attention has turned to studies of creative practices as a part of the arsenal of soft power. Indeed, we can see the rise of the 'creativity' script within international statecraft in terms of both the mobilisation of professionally produced work as part of state-craft, as well as in the embrace of vernacular creativities as participatory forms of geopolitical intervention.

Soft power was first identified as a form of power (alongside economic and military power) by Joseph Nye (1990, 2004). For Nye, soft power is a power of attraction rather than coercion, it arises he says from 'the attractiveness of a country's culture, political ideas and policies' (Nye, 2004, p. 4). He notes the long use of soft power, including Roosevelt's Four Freedoms for Europe, or the draw of American music and culture for those behind the Iron Curtain. On revisiting the concept in 2004, Nye notes it's entering into political parlance, but he reflects on how events have squandered it. He gives the example of the US war on Iraq in 2003, in which he claims soft power was lost, it might have been a dazzling display of military might but what is now needed, he argues, is a winning of the peace through soft power. As Nye (2004) notes, smart power is, in fact, neither hard nor soft, it is both. Soft power can be clearly seen to have a place in geopolitical thinking, where critical geopolitics has long turned to cultural forms to examine the ways in which we might understand geopolitical orders. As Ó Tuathail and Dalby (1998, p. 5) note,

> while its [geopolitics] conventionally recognized 'moment' is in the dramatic practices of state leaders (going to war, launching an invasion, demonstrating military force etc.), these practices and the much more mundane practices that make up the conduct of international politics are constituted, sustained and given meaning by multifarious representational practices throughout cultures.

As such geopolitical scholars have developed studies of a range of different cultural forms including Captain America comic books (Dittmer, 2012), James Bond films (Dodds, 2003) as well as a growing body of work on diverse arts practices (Ingram, 2011), including dance as discussed in Box 9.4. Whereas some have focused on exploring how the content of the cultural productions promote different forms of geopolitical regimes, for others it is important to appreciate the role of governmental forces in supporting and promoting certain cultural forms as means of soft power.

Box 9.4 POST-CONFLICT TENSIONS IN CONTEMPORARY CAMBODIAN DANCE

Amanda Rogers

My recent research explores how the performing arts represent war and how they are used, recovered, remembered and revived in the wake of conflict. I've been investigating this in the context of Cambodia and the legacies of the Khmer Rouge (1975–1979). Current work in political geography highlights that 'artistic practice is not just a form of resistance, refusal or critique but an index of, and contributor to, political and spatial transformation'

(Ingram, 2011, p. 218). Creative practices have therefore been seen as changing our ways of thinking, feeling and responding to extreme geopolitical events. In this regard, geography mirrors research in theatre and performance studies that examines how performance can raise political and ethical questions about war and investigates the work creative activities do in war zones and occupied territories (Thompson, 2009). In my own research I have taken the perspective that conflicts are difficult to separate temporally and spatially. That is to say, when we talk about 'post-conflict' situations, this does not necessarily mean a state of peace after a period of war, but the permeation of violent events and their legacies throughout social, political and, of course, creative worlds (Baillie, 2013). I am therefore interested in how a particular mode of performance – classical Cambodian dance – is tied to ongoing political negotiations around the role of creativity in contemporary Cambodian society.

Classical dance is synonymous with Cambodian national culture and identity. Traditionally performed solely for the monarchy, in the mid-twentieth century the image of the classical Cambodian dancer was used as part of nation-building efforts at home and abroad. However, during the Khmer Rouge around 90% of Cambodia's artists were executed because they represented royalty and an urban educated elite. After the genocide, dancers and musicians regrouped to collectively remember different dance movements, forms and stories. This process was an immense challenge; in classical dance alone there are over a thousand movements, each with a specific meaning, organised into patterns to tell stories. Given that the dances were, and still are, taught orally, visually or through touch, there was also no textual documentation of their composition.

Until relatively recently, the main impulse in Cambodian dance has therefore been on revival and reconstruction in order to document as many dances as possible before surviving masters die. However, in a developing (and corrupt) country, arts and culture lack priority compared to other issues. Much of the revival and

Continued

adaptation of dance occurs through non-governmental organisations (NGOs) with professional dance companies being funded
by private individuals and foundations in America, or, given
Cambodia's post-colonial legacies, France. Cambodian dance is
therefore vulnerable to transnational economic forces and geopolitical pressures. However, the NGO sector has opened up opportunities for creative experimentation, the funding of Cambodian
dance to a degree that would not be possible domestically, and
the construction of a new, seductive narrative about Cambodian
identity to promote abroad, one where the vibrancy of dance and
creativity replaces the image of Cambodia as a war-torn, ravaged
country. Some NGOs, and therefore some artists, benefit more
than others from these activities, but all work with the Cambodian
Government's Ministry of Culture and Fine Arts, whilst navigating
the tensions created by engaging with the state.

The NGO sector has particularly spearheaded the development
of contemporary Cambodian dance. This has entailed adapting
traditional movements and stories as well as promoting new
modes of creative expression by combining classical forms with
European or American contemporary dance. For instance, a
number of dancers use traditional folk stories to reflect upon the
Khmer Rouge era and to critically comment upon the ongoing violence in Cambodian society. Alternatively, dancers may draw upon
personal or family memories about the genocide and express them
through a contemporary dance piece. Such developments have not
always been well received in the past, with artists being censored,
investigated by the police and accused of degrading Cambodian
identity. Even when pieces do not address the genocide, contemporary dance is inflected by its legacies because this performance
form creates debates around meaning, authenticity and identity.
The classical dancing body is synonymous with nationality and yet
was almost completely erased from existence, making it *the* political site for articulating what – and who – is Cambodian.

A number of dynamics further complicate how dance represents Cambodian national and cultural identity, as the production
of community always creates conflict, tension, and unevenness

(England, 2011). For example, professional dancers train at government schools but some learn additional dances, including those officially and culturally considered 'taboo', such as male dancers learning female roles. This occurs outside official channels through traditional systems of patronage and learning where certain dances are only taught to privileged, gifted, or persistent students. The politics surrounding what dances are reclaimed and taught, and therefore what dances represent Cambodia, is also fraught, as not all elder masters have been equally recognised. Given the legacies of the Khmer Rouge, some masters also remain afraid of revealing what they know. Not everyone's knowledge is therefore valued and access to dance expertise remains uneven. This is particularly the case for 'survivor' classical dancers who have moved into a more contemporary field, experiencing numerous personal and political challenges as a result. In a post-conflict society, artists therefore constantly negotiate the meaning and value of dance in relation to the past, the state, the NGO sector, individual creative desires, and broader economic and political forces.

Importantly, and in contrast to other scholarship, much geopolitical work has focused on so-called popular geopolitics. In a lively series of articles Dodds explores how we might think of James Bond films as addressing the changing geopolitics of the Cold War and post-Cold War era (2003, 2005, 2008). While Power and Crampton (2005, p. 193) remind us, Hollywood movies provide 'a language and imagery' as well as 'reference points and ways of en-framing popular understandings of the radically changing geopolitical world'. Taking seriously the idea that 'the cinema becomes a space where 'commonsense' ideas about global politics and history are (re)produced and where stories about what is acceptable behaviour from states and individuals are naturalized and legitimated' (Lacy, 2003). To understand the value of this we should look to the political economies of the production of films. Indeed the Bush administration situated film as important not only for moral building, but also for building and promoting a very particular narrative of the US's role in the aftermath of September 11 and the onset of the war on terror (Dodds, p. 2008). Films such as *Behind Enemy Lines*, set in 1990s Bosnia, were understood as a largely self-serving portrayal of American military experiences, with the US and US service personnel in this and other films depicted as sources of goodness and moral courage. This is not a new phenomenon, as scholars have shown, under Stalin film was used to portray the struggles and triumph of the Soviet Union and the 1917 revolution. Likewise, numerous studies have explored how Hollywood is a close collaborator with the Pentagon and other agencies providing anti-communist materials. It has not just been films that have come in for attention, more recently interest has turned to video games (Shaw and Warf, 2009; Shaw, 2010) and music (see Box 9.5).

Box 9.5 THE GLOBAL RISE OF HALLYU: KOREAN WAVE AND 'GANGNAM STYLE'

Hallyu is the Korean term used to describe the 'Korean Wave' of cultural productions – music, films and video games out of South Korea that have experienced a rapid recent rise in popularity around the world. In the wake of the recession in the Korean economy in 1997 Korean Wave enabled creativity to become a pillar of the economy, diversifying away from industry and manufacturing. The export revenue from the cultural industries increased from US\$ 658 million in 2001 to US \$ 4.3 billion in 2011 (Kim and Kwon, 2014). The largest area of growth within the creative sector was Korean pop music, or K-Pop, which grew from US\$ 8 million in 2000 to US \$ 196 million in 2011. A year later, in 2012 the song 'Gangnam Style' generated US\$ 8 million alone and in 2013 pop culture exports were worth \$5 billion, a figure the government wants to double by 2017.

A range of scholars have explored how the popularity of K-Pop and television dramas, act as agents bridging popular culture with other cultural content, in other words creative goods become the means to attract people to the country. Hong (2014) writes for example of how 'Winter Sonata' a drama series attracted many Japanese tourists to Korea. For others, there is a clear sense in which cultural products, whether music, television or fashion form the means for Koreans, at home and abroad to strengthen identity (Jung and Shim 2014).

The growth of Korean Wave around the world is often understood as indivisible from technological developments that enable the streaming and downloading of media and music files around the world. The popularity of Gangnam Style was enabled by the video-sharing site YouTube, where the song became the most watched video ever, breaking previous viewing records.

The growth of Korean Wave can be seen as a result of government support of popular culture (Berg, 2015; Cho, 2011; Kwon and Kim, 2014). Strict control and censorship under the authoritarian military government gave way to deregulation and the forces of neoliberalism that both supported investments in the creative economy as well as enabling public consumption and participation in cultural activities. For example, the government offers funding, preferential loan rates and tax incentives to the creative sector. In 2005 a $1 billion investment fund was created to support the music industry (Kang, 2014, cited in Kwon and Kim (2014)) indicating the importance of this sector and hopes for its role in the country's future. As well as financial support, the government also supported the industry through the establishment of a Creative Content Agency (KOCCA) to enable the support and expansion of the industry nationally and globally. They also supported the Korea Copyright Commission, offering the industry legal regulation to enable the development of the sector as a sustainable endeavour (Berg, 2015). It is clear, however, that governmental support for the creative industries in Korea comes from more than an interest in it as an economic sector (Huat and Iwabuchi, 2008; Hong, 2014). Indeed, the growth and competitiveness of Korean popular culture has been important in forming a cultural block within Asia, as well as offering a distinct place to Korea within Japanese and Chinese cultures. Indeed, former president Kim Dae-jung predicted that popular culture would play a role in the future unification of North and South Korea (Hong 2014). So successful has the strategy been that in 2014 Japan announced a $500 million fund to cultivate a 'Cool Japan'.

Reflecting wider discussions within geography, many explorations of soft power have focused on how the combination of the representational and the non-representational works in everything from music to films and video games; in the latter case, situating video game worlds in the context of ever-tightening intersections of the military entertainment complex, a name given to the increased cooperation between the entertainment industry and the military.

Video games are produced by and productive of complex geographies that intersect spaces, practices and embodied play within the home with the practices of war and the spaces of its imperial architectures. The intimate geographies of video gaming collapse the worlds within and beyond the game, bringing together war zones on the screens and bedrooms and sitting rooms that are the site of their playing (Ash, 2015; Shaw, 2010; Shaw and Sharp, 2013). The attitudes cultivated in these domestic spaces-cum-theatres of war are in turn taken beyond them into the street, the playground and the practices and choices made in daily life. Writing about the game *Call of Duty: Modern Warfare* 2, a game released in 2009 that set a new worldwide sales record of $550 million over five days, Shaw (2010) notes these reconfigured geographies. He explores how the role of 'computer games as powerful tools for learning, socialization and training' in the military becomes extended into private spaces. The result is that the 'intimate and everyday spaces of the home computer are now instrumental to the post 9/11 cultural shift in the war on terror' (Gregory, 2008, cited in Shaw, 2010, p. 790). This was to see video game users as 'playful warriors' in a 'war deprived of its substance – a virtual war fought behind computer screens, a war experienced by its participations as a video game, a war with no casualties' (Žižek, 2002 p. 37, cited in Shaw, 2010). This particular instalment of *Call of Duty* follows the US Army's conflict with a 'new Russian ultranationalist terrorist organization' and the action takes place across a range of territories from Afghanistan to DC. Playing here is understood as a 'co-mingling of self and world' as an 'always a political moment, locked within representational logics (whether military, colonial or racist and so on)' (Shaw, 2010, p. 791). To play is to be locked into a 'military entertainment complex' which authorises orientalist representations through which wider everyday colonial presents are produced and reproduced (Gregory, 2004).

Exploring the intersections of the military and entertainment industries has led some to conclude there is a cultural turn in the US Army's war on terror. The screen becomes a 'virtual space and a site in which US soldiers train through interaction with animated enemies' as well as a site where 'millions of teenagers participate daily in the war on terror in their bedrooms' (Shaw 2010, p. 801). The intersections of the military entertainment complex around video games are complex. In 1996 the Marine Corp adopted the game *Doom II – Hell on Earth* into *Marine Doom* for training US soldiers. The relationship is said to have reached its zenith in the *Full Spectrum Warrior* an X-Box game developed at the University of Southern California, Institute for Creative Technologies. The institute is also contracted to the US Army to research and produce visual simulations and video game worlds to aid soldier training. The $45 million partnership was set up to explore what would happen if leading technologists in artificial intelligence and graphics and immersion joined forces with the creative talents of Hollywood and the game industry (ICT 2009a, cited in Shaw, 2010, p. 794). Furthermore, from ad hoc repairs made to tanks to the development of the sophisticated interfaces of drones, we witness a collapse of the video game industry and technologies and interfaces of warfare, wherein hardware and software – whether it be game controllers or screen layouts – come to feature in advanced weaponry.

Interestingly, in these enfoldings of video game worlds, we see the same production of the representations of the ruling elite, the same sense of power and ideology that we saw in landscape art. The worlds of video games tell the narratives designed in by their creators, in the case of *Call of Duty*, a very colonial norm of the sort that resonates with Gregory's (2004) discussions of the colonial present. As Shaw notes, these war games often play out over a 'simplified Islamic world in which cultural and ethical differences are flattened' (2010, p. 796). In an animated Orientalism, The Middle East becomes populated by a series of signifiers producing and reproducing the oriental imagination; these include headscarves, turbans, camels, belly dancers, deserts, harems and bazaars. As Shaw (2010) observes, it is not just a spatial other that is represented but also a violent other, cities are portrayed as maze-like, and everyday spaces are portrayed as constant sites of conflict. As such, the military entertainment complex imagines Middle East cities as perpetual war zones inhabited not by families, but by terrorist guerrillas, a collective and anonymised force to be fought. Just as those ideologies of eighteenth-century picturesque served a purpose for the elite, so the imaginary worlds of video games also serve a purpose. For Shaw they are 'banal technologies that distribute carefully crafted military aesthetics . . . play is a political practices locked within a violent imperial topos' (2010, p. 799). As well then as practically training soldiers, these games are also training the soldiers to come and are 'generating support amongst the civilian population for the increasing use of American Military power' (ibid.). Rather than, or as well as simply a recruitment tool, the games are also seen as formative of cultural consent for military action; more than 30% of 16–24 year olds had a more positive impression of the US Army after playing America's Army, a game funded by Pentagon, who support servers storing game data to the tune of over $6 million/year.

To understand the place-making and citizen-shaping power of the military- entertainment complex formed from games and films, we have to reflect not just on representations but also affect. If early studies of films focused on representation, analysing plot lines, people and places for their symbolism, then more recently scholarship has recognised how power registers corporeally through the capacity of bodies to affect and be affected by other bodies. As Connolly (2002, p. xiii) suggests, popular media such as film provide important ways in which 'cultural life mixes into the composition of body/brain processes'. Film therefore comes to be seen (among many other things) as a particular technique through which to amplify the contagious force of affect. This is a contagion that flows,

> across bodies as well as across conversations, as when anger, revenge, or inspiration is communicated across individuals or constituencies by the timbre of our voices, looks, hits, caresses, gestures, the bunching of muscles in the neck, and flushes of the skin. Such contagion flows through face-to-face meetings, academic classes, family dinners, public assemblies, TV speeches, sitcoms, soap operas, and films.
>
> (Connolly, 2002, p. 75)

This requires that we not only explore how video games and films might produce and reproduce, or disrupt particular ideas, but also how they might engage us through emotion and affect. This, as Connolly (2002, p. 14–15) suggests, involves a shift away 'from appraising the ideological politics, narrative form, and cultural message [of film] to exploring the relations between narrative flow and specific techniques of delivery'.

Digital creative communities

Alongside studies of existing cultural geopolitical forms, we also see, in a move similar to the broadening of creativity more broadly, a growth of vernacular creative geopolitics, or what could, after creative economic policy, be termed the 'creativity' script within statecraft. In work that queries the popularity of popular geopolitics, Pinkerton and Benwell (2014) explore how creative practices understood in relation to 'statecraft' might work to unsettle and complicate previously tidy geopolitical categories of the 'popular', the 'formal' and the 'practical'. Exploring the place of social media in citizen statecraft in the ongoing discussions around the Falkland Islands, they suggest that in such vernacular creative practices they find a 'flourishing of new modes of international dialogue between communities in dispute' (p. 12). As they explore, Twitter, Facebook, YouTube and other online fora and blogs are becoming sites at which distinctions between the 'intellectuals of statecraft' and 'everyday citizenry' are being blurred. Connecting and disconnecting with 'traditional' media in complex ways, these spaces are the new political arenas where debates about the island's sovereignty are playing out between politicians, marketing agencies, veterans and the Falkland Islanders, Argentines and British people. This is to make popular geopolitics, really much more popular than in its previous incarnation where it had been the preserve of the producers and consumers of the cultural elite. Building on wider theories of digital democracy (e.g. Burgess, 2006, p. 202) that concern how online social media enables people from non-elite social contexts to 'exercise more creativity and agency' through their participation in media cultures, the growth of such digital democracies becomes a way for channels of contact to be opened.

Throughout their discussion, Benwell and Pinkerton explore a number of examples of how creative practices become tactics to respond to what is considered to be an asymmetrical military and geopolitical relationship with the UK. They thus explore how creative statecraft might be seen as a 'weapon' of the weak, examining how a controversial advertisement produced by the Argentine Government was described by that government as an alternative to using military might to reinforce the island's sovereignty claims. They also explore how citizens have been using the capacities of social media, including organizing a flash rally through Twitter and Facebook. Here against the acknowledged backdrop of artist and other cultural elites as able to 'contest, invert and reframe political meaning' (Ingram, 2011, p. 221), they explore how everyday people, ordinary citizens enabled by technology might do the same. There is clearly scope for further work on what digital media might enable for digital democracy, how it might give voice to marginalised or overlooked peoples on a stage that is accessible across the world.

CONCLUSIONS: MAKING NATIONS

As these contrasting examples of the intersection of nations and creativity, whether through policy or practice (or the coming together of the two) demonstrate there are a range of ways we might think of creative practices as producing and reproducing nations. This includes creative policy that focuses on national identity, and travels, or rather does not, as well as a range of creative practices. This might include high-end practices, such as painting, as well as more popular practices such as comics, films or video games. It might also, as this discussion has shown, be developed under the manipulation of governments or the elite, seeking to produce and reproduce their own identities or it might be the digital democratic practices of citizens who use creative digital practices in order to develop their own narratives and discourses. As such, creative practices offer a means by which those non-elites can work to shape and influence national narratives. What is also very telling is how in these discussions of the coming together of creativity and nationhood, different ideas of creativity are often at play. Creativity appears here as an economic, social and cultural force, with these different elements all coming together and moving apart in different situations, sometimes existing simultaneously, sometimes with the economic in the ascendency, sometimes the social. What is also clear is that creativity is important both for the internal functioning of nations, their creation as policy spaces and as imagined communities, but also in the geopolitical processes of global engagement and negotiation. Creativity, as such, must be understood as both an embodied practice and a geopolitical force, a means to create and promote nations at home and abroad.

NOTES

1   Creative Industries Mapping Documents 1998, [ONLINE] Available from: https://www.gov.uk/government/publications/creative-industries-mapping-documents-1998 [Accessed 15/8/2016].
2   United Nations Creative Economy Report 2013: Special Edition - Widening Local Development Pathways, [ONLINE] Available from: http://www.unesco.org/new/en/culture/themes/creativity/creative-economy-report-2013-special-edition/ [Accessed 25/7/2015].
3   http://www.banglanatak.com [Accessed 25/7/2015].

10

LANDSCAPE AND ENVIRONMENT

Dialling the mobile phone number, what you hear at the other end is a series of cracks, drips and creaks. Echoing down the phone line the dripping seems endless, every now and then there is a sudden splash, the dripping continues. In 2007 artist Katie Paterson installed an underwater microphone (or hydrophone) in Jökulsárlón, an outlet glacial lagoon of Vatnajökull – the largest glacier in Europe. The microphone was connected to an amplifier, which was in turn connected to the mobile phone the audience dialled into from anywhere in the world. The piece *Vatnajökull (the sound of)* created a real-time encounter between the caller-listener and an ecosystem under stress as a result of global warming (Kanngieser, 2014).

What do you think about when you think about climate change? Rising sea levels, stranded polar bears, cracking glaciers and melting ice sheets, extreme flooding as well as droughts and wildfires? Global climate change is by now recognised as a fact. What creative practices might do to inform us about or to help combat it remains, however, an open question. This chapter will situate a discussion of the possibilities of creative practice in relation to global environmental change in the context of longer-term intersections of creativity and geographical concerns with landscape and environment.

The evolution of the concept of landscape within geography (and so more broadly) has been driven, it can be argued, by engagements with creative practices. This includes both studies of the creative practices of others such as nineteenth-century landscape artists, as well as the creative geographical research and writing practices deployed in recent studies of landscape and environment. Geographical understandings of the environment, including the contentious designation of the Anthropocene, with its recognition of the profound effects of human-driven environmental change, has led scientists from many camps to recognise the need for interdisciplinary engagements with these so called 'wicked' problems. While such interdisciplinary endeavours are proceeding apace, with calls of support echoing through the pages of journals such as *Nature and Science*, these tend to focus on social science, overlooking the potential role of arts and humanities and, in particular, creative practices, in engaging with issues of global environmental change.

This chapter will focus on three points of intersection between creativity and landscape and environment. It will begin by thinking about how creative practices have long been central to how geographers know and engage the world around them, focusing on the arts of exploration and the geopoetics of eighteenth-century polymath Alexander Von Humboldt. It will then move onto reflect on the cultural geographies of landscape, foregrounding the manifold ways that creativity has been central to evolving conceptualisations of landscape and its politics and poetics. The chapter will close with a set of reflections on the emerging place of creative practices in engaging with the issues raised by the current phase of global environmental change and, in particular, the challenges raised by the Anthropocene designation of this particular era of human-driven global environmental change.

INTERSECTION ONE: PICTURING LANDSCAPE AND ENVIRONMENT – THE ARTS OF EXPLORATION

Creative practices have long been understood to 'picture' landscapes and environments, often in very literal terms. Indeed, artists' early intersections with geographical knowledge concern their role as offering 'packets of information' about a place (Balm, 2000). When Captain James Cook set sail on the Endeavour in 1769, bound for the South Seas, he was accompanied by two artists, Sydney Parkinson, and John Buchanan (Smith, 1992). Whilst draughtsmen had long been in the employ of Navies across Europe it was with Cook's voyages that the artist on-board ship came of age. Analysis of the artist's role and value has principally understood this as the situation of art in the service of science and Empire (Smith, 1988, 1992). The images produced during these and other voyages, and during the later interior explorations, circulated the world. In the course of their circulations these visions of places and peoples far distant were as important for their role in shaping Europe's geographical imagination as they were for their place in science. Such new knowledge was also to contribute to the shaping of the discipline of geography itself. For Stoddart, the year 1769 is key in disciplinary history, he observes that Captain Cook's entry into the South Pacific formally marked the end of a 'barren didacticism of capes and bays' and the inauguration of a discipline that produces and seeks to answer questions concerning 'man and environment within regions' (1987, p. 331, cited in Hawkins, 2013a, p. 132). It is interesting that at this crucial point in the evolution of geographical knowledge making, creative practices played such a key role in the production and circulation of geographical knowledge.

The early value of images over and against words has been understood to lie in their mimetic qualities, their proffering of 'packets of information' about people and environments (Balm, 2000). Such an understanding tracks across centuries of thinking about the place of art within geographical knowledge making. In one recent revival of these arguments, Balm (2000) turned to art as information as a means to chart a passage through what he perceived as the 'interdisciplinary gulf' between geography and art. Turning away from 'celebratory fine artists' that he considers has led geographers to promote, 'technique and aesthetic conventions over content', Balm asserted the need to attend more closely to art's 'expeditionary optic'. By this he meant those art works made by artists travelling between 1760 and 1860, whose empirical naturalist work was driven by the need to provide accurate information, rather than by aesthetic conventions (Smith, 1988, 1992). In his assertion Balm unwittingly echoes, but reverses, art historians and historians of science for whom this era was cast as a 'winter of the imagination'. This was an era of art making not to be celebrated but to be denounced for its focus on mimetic practices, reflecting the role art took up within Enlightenment sciences' armoury of tools and technologies. As a result, art became another means by which scholars of the era could gather, sort, categorise, and so fix of a whole host of objects, species and spaces (Smith, 1988).

Under this view a range of artistic products – from quickly captured landscape sketches, to considered drawings and paintings and even large oils – made both in the field and on-board ship, were considered as principally developed under the sway of regimes of collection and in service of science and surveillance.

This is not quite, however, the neat story of art as data that it might initially be perceived as. Indeed, even during the reign of empirical naturalism we see the artists and scientists enrolled in these projects resisting such a subordination of art to science, and with it the particular legacy of the arts as principally offering 'objective, self-effacing and precision engineered' data (Daniels, 2011a, p. 185). Indeed, looking closely at examples of those paintings produced on-board ship or the work of Alexander Von Humboldt and his followers, we see an empowering place for art within geographical history, and one that works away at the boundaries that sit at the heart of reductive discussion of 'arts' and 'sciences.'

Visual cultures of exploration

During the age of exploration the picturings of landscapes and places in the course of voyages such as Cook's were closely related to ways of knowing the world more generally. Thus these visual cultures of exploration can be situated within the shift from the valorisation of knowledge based on teleological or metaphysical expositions (e.g. beliefs in divine powers), to one based in facts derived from close observations and empirical verification. In this context, artistic practices were seen as part of the emergence of a scientific gaze, this was a methodology that demanded close and faithful observation of natural objects, second-hand vision was unacceptable, being there – in the landscape – was key (Smith, 1988, 1992). In such a role, artists were understood as 'more perfect' describers, as being able to 'make drawings and paintings as may be proper to give a more perfect idea ... than can be formed by written descriptions alone' (cited in Smith, 1988, p. 15). In the words of the naturalist and draughtsman William Burchell art was 'a means of exhibiting nature and conveying information' (cited in Driver and Martins, 2005). But such exhibitions of nature were not to be understood wholly through a mimetic gaze, indeed this was a 'scientific' gaze thoroughly inflected by aesthetic conventions and practices, rather than one that neutralised and overrode them.

Post-colonial writer Mary Louise Pratt (1992) has written of the 'Imperial Eye', the particular masculine euro-centric vision that was cast over so much of the 'New World' during these exploratory missions. As such, images were produced from a very particular point of view, and for a particular often European audience through the lens of European modes of visuality. This is illustrated in the work of ship-board artists who went with Cook on his voyages, and very clearly in the landscape painting of William Hodges, who accompanied Cook on his second voyage to the South Seas in 1772. As a pupil of Richard Wilson, one of the key artists of the Romantic Era, Hodges was well versed in the aesthetic stylings of the European conventions, especially those of the sublime and the picturesque. As such, his images frame landscapes less through an 'innocent eye' – recording purely what it sees and encounters in the field – but rather these are landscapes visioned through aesthetic conventions. The lens Hodges brought to South Seas species and landscapes was a thoroughly European one. He became known for his large canvases, such as *Tahiti Revisited* (1776). The latter is a classic Hodges painting, a concatenation of what he saw on the island and western aesthetic ideas that come together to shape the bodies and poses of the women bathing in the foreground, the gentle curve of the trees around the side of the lake, and the mountains disappearing in the hazy background. This image is at once a record, but also a fiction, a scene seen through European aesthetic lenses, and a crucial part of the formation of a European imaginary of the tropical South Seas. The practice of topography (recording the landscape) 'had always been given a humble place at the bottom of the academic table but here was an attempt to elevate exotic topography to the high places reserved for the ideal landscapes of Claude, the heroic landscapes of Poussin, and the picturesque landscapes of Salvator Rosa' (Smith, 1988, p. 6). Such 'Tropical Visions' (Driver and Martins, 2005), which Goethe once termed 'half truths' (Goethe, cited by Dettelbach, 2009, p. 34), should not be simply divorced from scientific observations, but rather can be understood as a blend of the documentary and the aesthetically ideal. Indeed, Hodges worked to capture the light and meteorological phenomena he encountered on his visits to the tropics. His engagements were shaped by his close working relationship with naturalists, astronomers and meteorologists on-board ship. This was an era from which emerged a 'way of seeing and knowing in which the tradition of landscape art was fused with a new spirit of observation informed by the experience of voyaging around the world in the company of naval surveyors, meteorologists and astronomers' (Greppi, cited in Driver and Martins, 2005, p. 24). Understood such, landscape is both a site of scientific encounter and an increasingly important artistic subject. Of course the advent of photography, with its associations with realism, was to dramatically refigure the role of the painting and sketching of places in the context of scientific inquiry and the production of geographical knowledge (Ryan, 2013).

Rather than engaging these tropical landscapes only in terms of the 'West' projecting its sense of cultural difference on the 'rest', it is important to remain aware of the power and dynamics of these landscape encounters. As a result, these tropical landscapes are not simply screens onto which things are projected but are 'a living space of encounter and exchange' in which artists would engage landscapes and local people as well as scientific specimens (Said, 1978; Driver and Martins, 2005, p. 5). These painted 'packets of information' were often sent back to England mid-voyage in order that they might be displayed, or put into circulation as prints or slides. In so doing, the western geographical imagination of the South Seas was formed as audiences came to consume painted images of places they could never hope to see. Such tropical visions were early examples of the sorts of images produced by the gaze of the 'imperial eye' that Pratt (1992) describes, they can also be linked to the later exploration of the orientalism described by Said (1978, see Box 9.3) which was perhaps less to be understood as reproducing worlds, as it was to be seen to be producing them.

The geopoetics of Alexander Von Humboldt

No discussion of creative geographies and landscapes and environments would be complete without a consideration of the work of Enlightenment polymath Alexander Von Humboldt. Humboldt's geo-aesthetics have had a long and lasting influence on geographers' deployment of art and aesthetics and related understandings of the environment (Buttimer, 2001; Hawkins and Straughan, 2015). His import can be traced in the evolution of branches of geography as different as geomorphology and the geographical imagination (Dixon et al., 2011a). More recently, Humboldt's scholarship backstopped the development of humanistic critiques of a scientized geography and continues to shape geography's revisioning of the aesthetic, whether this be human geography's exploration of the sensuous, embodied environmental engagement or, as Box 10.1 explores in more detail, physical geographers' recognition of the role of enchantment and the arts in critical approaches to their methods and scholarship (Baker and Twindale, 1991; Hawkins and Straughan, 2015). For scholars across the discipline the resources that Humboldt offers for exploring models of ecological engagement are premised on aesthetics cast, not in opposition to cognition or to systematic observations of world processes, but rather offers a mode of aesthetics that is a building block for scientific knowledge (Bunkse, 1981).

Box 10.1 ART-FULL GEOMORPHOLOGY

Over the last few decades we have witnessed a 're-enchantment' of geomorphology. As well as engaging audiences, these efforts have generated discussions as to how it is that geomorphologists should and could communicate their knowledge especially in an era of climate change (Baker and Twidale, 1991). The growing series of collaborations between geomorphologists and artists is beginning to come to the fore with the establishment of a working group of the British Geomorphological Society devoted to the arts. Some of these geomorphologist/arts collaborations are based around a collaborative exploration of 'inspirational landscapes', for others there is a concern with engaging sensory experiences of landscape in the research and communication of geomorphological knowledge (Dixon et al., 2011). This creative turn for geomorphology, like for geography more generally, can also be considered a 're-turn', tracking a history through Humboltian science into the geomorphology of G.K. Gilbert and into the present day.

The delinking of aesthetic sensibilities from scientific inquiry is often dated to the Enlightenment shift in the making of the modern-day geographic discipline. Exploring such a history makes it clear that there is a hugely expanded field of artistic inquiry that repudiated the rarefication of artworks. Indeed, we might date the

Continued

latter in the UK at least to the founding of the Royal Academy of Arts London in 1769. Prior to this and indeed continuing from this time was an expanded field of arts that encompassed maps and charts, geological cross-sections, botanic images as well as landscape paintings (Dixon et al., 2011). What emerges is a rich imaginary of art that is a lot more than a mimetic representation of landforms, including a rich aesthetic sense of art, with aesthetics here also understood to mean concerns with multi-sensuous encounters between body and world. As such, aesthetics came to work across and between practices of sketching and painting but also fieldwork practices more broadly. In the work of G.K. Gilbert (1843–1918) heralded as the 'father' of a modern-day scientific process-orientated geomorphology, we can trace a 'felt' engagement with landforms that is an important part of how we come to know them. He developed a method that involved the 'interplay of observation and deductive reasoning wherein data were accumulated within a robust and logical theoretical framework' (Chorely, 2000, 569, cited in Dixon et al., 2011, p. 237). Gilbert was working in a field of geomorphology rich in visual culture including expanded geologic cross-sections, and a range of field sketches and photographs, produced by organisations such as the Smithsonian. Gilbert was a senior geologist in the US geological survey in 1879, and produced studies of Lake Bonneville and also speculated as to the role of meteor impacts in shaping lunar landscapes. Using his full complement of senses, Gilbert was able to begin to understand the landscape and its formation. He sketched and drew his landforms, describing things as he drew, he was delineating Blue Gate and Tununk Shales, writing 'they are beautifully laminated and are remarkably homogenous. It's only in fresh escarpments that the lamination is seen, the weathered surface presenting a structureless clay' (1877, p. 5). Often his visual acuity is supplemented by touch and a sense of the texture of rocks, their softness and manners of weathering, as such his field sketches emerge from his visual imaginings of encounter between body and landscape. Field sketching is thus a way of rendering the landscape structurally intelligible, a form of topographic art that foregrounds an accuracy of delineation, but also combined in other images with a mode of working akin to view painting. In the way that Gilbert's notebooks have been understood we see a movement from 'faithful description to idealized hypotheses'.

Moving forward to the mid-twentieth century, we see an emergence of a body of scholarship that seeks to use paintings, sketches and other artistic media as data proxies (e.g. Gemtou, 2011). For geomorphologist Jean Grove verbal descriptions were not to be taken at face value, but the paintings of Mt Blanc massif 'is so accurate that full dependence may be placed on them as a source of evidence' (Grove, 1966, 135, cited in Dixon et al., 2011, p. 241). Thus we see paintings and photographs coming to form sources for the reconstruction of past glacier positions, and even famous images such as Munch's *Scream* coming to be understood as a means to date volcanic eruptions. Gemtou's extensive study uses the changing colour of skies in paintings of European cities as a way to date the changing volume of ash in the atmosphere in the wake of Pinatubo. There is, of course, an issue here, given the faith that is put in accurate representation of nature in all its colour and morphology, a faith that denies the processes of artistic making, and that varying skill levels and the fallible nature of paint and colour.

If we move forward to contemporary geomorphology we see a range of reasons for geomorphologists to turn to artists (Dixon et al., 2011). We can think, for example, of the collaboration of glacial geomorphologist Peter Knight and artist Miriam Burke, examining how it is they notice and understand landscape differently. Methodological exchange in the field and in the cold lab culminated in the exhibition 'Know this place for the first time'. Amongst the works developed were Burke's *measuring spoon* a sculptural reflection on the ad hoc nature Knight makes of the teaspoon for scale in his photographs (Figure 10.1). Another work *Albedo*, refers to the concept of albedo that concerns the reflectivity of the thing, a glacier for example, or even the planet. White things have a high albedo as they are very reflective, black things absorb

Continued

Figure 10.1 Measuring Spoon (Photo: Miriam Burke)

energy and radiation and thus have a low albedo. In the context
of climate change, a whiter planet, with areas of snow or ice,
will reflect more of the sun's energy back out to space. As snow
and ice melts, giving way to dark areas of open ocean or forest,
more of the energy is absorbed and the world is warmed. *Albedo*,
as Burke describes, is her attempt to cool the world with small
wax casts of her finger-tips placed in the natural world. Writing
about this work, Burke explores the mechanisms of art as, in the
words of Elizabeth Grosz a 'slowing down chaos', connecting
fleshy fingers with inter-planetary forces, and drawing to the fore

concerns with our bodily encounters and intimate and not-so-intimate connections with the world. In another context Knight notes,

> Despite progress in remote sensing and analytical techniques, our reconstructions of past glaciations remain tentative, our understand of modern glacial processes incomplete and our modelling of their future unreliable. In nineteenth century art, glaciers represented romance, mystery and unassailable majesty. In 21st century science their position is perhaps similar but what art calls 'mystery' science calls 'uncertainty'.
>
> (Knight 2004, p. 385)

Risk and uncertainty such as Knight explores also inspire the collaborative work of volcanologist Carina Fearnly and designer Nelly Ben Hayoun. In *The Other Volcano* (Figure 10.2) Ben Hayoun

Figure 10.2 The Other Volcano (Photo: Nelly Ben Hayoun)

Continued

developed a piece of work that communicates the experiences of living with a volcanic landscape, both the wonder but also the fear. She experiments with how this can be harnessed as a means to enable sustained reflection upon how volcanoes and their risks are to be communicated (Dixon et al., 2011). Ben Hayoun's semi-domesticated volcanoes sit in sitting rooms and galleries and randomly erupt dust and goo (sugar, gun-powers and nitrate). Ben Hayoun's and Burke's work draw out the potential of art to move beyond the accurate conveying of scientific information to communicate the ambiguities and hesitancies of earth sciences to explore these landscapes as human spaces with consequences for life.

For historians of science, Humboldt has become a key figure in the exploration of ongoing tensions around the aesthetic and the empirical. His observations on the 'sensibilities' of the field scientist, and casting of arts and humanities into an almost empirical role have made him a key figure in debates that continue to reverberate around the discipline, especially in the context of creative geographies (Dettelbach, 2009, Hawkins, 2013a). We see this clearly in his landscape profiles, and in particular the tableau physique of the Ecuadorian Volcano Mt Chimborazo (1805). Humboldt's profile of Mt Chimborazo, at the time assumed to be the world's highest mountain, creates a pictorial science that both conducts analysis and educates the viewer. For Humboldt, travelling with his naturalist friend Amié Bonpland, Chimborazo was of interest as, ascending the mountain, it was possible to experience a then unsurpassed range of different environments and plant species. In addition to marking the climatic bands Humboldt also filled in, albeit in a rather jumbled-looking manner, the names of different plant species he encountered and provided data on 13 different scientific variables he measured during his journey; from temperature and pressure, to the height of the snowline. For Humboldt such a scientific appreciation of landscape could only be gained through first-hand experience and, as such, was closely bound into an aesthetic – sensed – appreciation. Alongside his 'records' of scientific data are his more poetic writings of the azure of the sky, the shape of the hills, and the forms of the clouds.

Within Humboldt's aesthetic 'sensibility and precise measurement cannot be separated' (Buttimer, 2001, p. 477). Humboldt has been noted to be the first 'geovisualiser', a name he gained for this systematic study of the face of the earth which circulated through 'a series of diagrams foregrounding the tabulation and cross-referencing of various forces, thus enabling comparisons to be made and rudimentary forms of multivariate analysis to be conducted' (Jackson & Romanowski, 2009). For Humboldt the Earth's surface was not just to be recorded, but also to be sensed by the ideal form of the gentleman explorer. This gentleman explorer was one with a discerning aesthetic taste and cultivated sensibilities, enlivening him to the beauty of the holistic form of the landscape. He was also in possession of the financial means and scientific knowledge that would enable him to enhance his 'gaze' upon the landscape through tools and devices, including magnetic compasses, thermometers, electrometers, and so on.

It was not only in his own 'method' that we find Humboldt's scientific vision coming together with aesthetic and creative sensibilities and practices. Indeed, landscape painters of the era, perhaps most famously Frederic Edwin Church, reproduced and collaged Humboldt's field sketches to produce large canvases such as the oil painting *Heart of the Andes* (1859). Due to its vast size and detail, on being exhibited around the world the painting had a huge impact, 'when confronted with its immensity, the crowd becomes hushed. Women feel faint' (Poole, 1998, p. 107, cited in Dixon et al., 2011, p. 231). This is far from the image of science as shaped by a bloodless rigour, but rather situates Humboldt's science within the literally dizzying effects of sublime nature. The presentation of the abundance of nature on a vast scale enabled the apprehension of sensuous encounters with the landscape as whole and the communication of this reality to others. Yet, the agglomeration of vegetation types and landscapes such images presented, physically impossible in the natural landscape, were rather a distance from Humboldt's careful adherence to systematic measurement. As the key humanistic geographer Anne Buttimer remarks,

> there are aesthetic and experience-based facets of [Humboldt's] overall vision which enable the work to transcend tensions between objectivity and subjectivity, macro-scale survey and micro-scale theatre, scientific explanation and artistic representation.
>
> (Buttimer, 2001, p. 105)

As such, it is by no means straightforward to suggest that such art was simply in the service of science and Empire, for these ideas of landscape and environment merged these ways of knowing together in complex and interesting ways. Indeed, Humboldt's work demonstrates the cyclical relationship between the sensory experience of the landscape and the making of scientific knowledge about these landscapes and environments. What we see when we look to the history of landscape and environment is the key role that creative practices have played in human environmental knowledge and relations. As such, we might consider that these relations form a strong basis from which to explore more contemporary ways that landscape and environment can be known and engaged through creativity and arts practices.

INTERSECTION TWO: THE POLITICS AND POETICS OF LANDSCAPE AND ENVIRONMENT

The place of creative practices in both knowing landscape and environment and in helping develop the conceptualisations of the same was confirmed by Marxist Geographers writing about seventeenth and eighteenth-century landscape paintings, landscape gardens and landed estates (including Cosgrove, 1984; Daniels, 1993). For them, such painted and shaped landscape aesthetics are powerful, but are not to be trusted as a simple 'picturing of place'. The reason for this distrust lay in what was understood as the 'veil' or duplicitous aesthetic, a painted surface that was drawn over the 'reality' of living in these landscapes. Two classic examples of the English Landscape, and the subject of much geographical scholarship, will illustrate these ideas (Cosgrove and Daniels, 1988; Daniels, 1993). Firstly, Thomas Gainsborough's iconic painting *Mr and Mrs Andrews* (1750) depicting a rich couple viewing their estate, and secondly, Constable's classic nineteenth-century painting of the Suffolk rural idyll *The Hay Wain* (1821). Both present a rich, often urban-based art-consuming elite with idyllic visions of country life and beautified scenes of hardship and neglect enabling them to overlook conditions of rural poverty. In both cases the 'realities' of the painted surface were studied as aesthetic codifications: forces working in concert with capitalism, reflecting and reproducing the beliefs and desires of the political elite (Daniels, 1989, 1993). These aesthetic veilings did both symbolic and actual violence to landscapes and lives, masking what art-theorist John Barrell (1983) called 'the dark side' of rural life.

Shaping geographical studies of these ideologies of landscape is a concern with the links between these painted forms and their material force in the world; in other words, an appreciation of these aesthetic codifications as more than symbolic reproductions of the power of the ruling elite (Daniels, 1989). Geographical scholarship examines how the painted and landscaped forms of these aesthetic codifications produce and reproduce power as they circulate in the material form of images and imaginaries. Studies explore, for example, how the physical landscape is shaped, how the type and form of vegetation is controlled and even how structures are built and maintained (Daniels, 1993). Discussions of the picturesque form note, for example, how workers' cottages were kept in a 'charmingly' rundown state by eighteenth-century landed estate owners who sought to emulate these aesthetics. Further, vegetation was planted and cropped in a particular manner to ensure that the 'variegated delights' of the picturesque where adhered to, no matter whether this was efficacious for those dwelling in and working the landscape. Later land-owners would turn to the aesthetic conventions codified in the practices of great landscape gardeners such as Humphry Repton and Capability Brown, shaping their grounds according to the standards laid out in these gardeners' plans (Daniels, 1999). This might include the building of a Ha-Ha, a ditch that could give a land owner uninterrupted views out over their estate, without letting animals graze too close to the house and gardens. On estates in industrial northern England large amounts of earth were moved and vegetation cropped to enable the estate owners to view the mills and forges on which their wealth was built (Daniels, 1993). To gaze on landscape paintings or to walk through landscaped gardens is to experience sets of uneven power relations painted, built and grown into the surfaces of canvas and landscape. What has been painted and formed in the landscape are spaces authored by and designed for the pleasures and the reinforcement of the powers of a ruling elite.

Importantly, such aesthetic codifications were not just rooted in place but were mobile. These English landscape aesthetics, for example, circulated through the actions of Empire, creating and recreating landscapes and lives in their image across the globe (Daniels, 1993). Of course, these aesthetics were not unchanged by their mobility and indeed the local remakings of these aesthetics at a distance have been of considerable interest for geographers (Daniels, 1993). So we find many accounts of the attempts to build landed estates on plantations in ways that were akin to European-based examples, rivers were dammed and rechannelised, trees were uprooted and new species planted, earth was sculpted into land-scape forms (Seymour et al., 1995, 1998). This was not of course only a one-way circulation of aesthetics. Many different forms of aesthetics from around the world, especially through colonialism, made their way back to England and into people's homes and gardens (Seymour et al., 1998). Such material culture and plants were adopted and adapted to suit the size and scale of English properties (Tachibana and Watkins, 2010).

If this work on the politics of landscape representations and forms tied together landscape, environment and creativity, then this legacy has continued in the latest set of discussions of non-representational, or embodied and affective understandings of landscape and environment. Such accounts foreground the phenomenological and post-phenomenological experiences of the landscape often through processes of walking, running or climbing in the landscape (Wylie, 2005; Crouch, 2010). Within this context arts and creative practices have offered rich means for geographers to explore, engage and represent these sensuous experiences of landscapes and the affective intensities of their environmental encounters. Crouch and Toogood (1999), for example, write of how the embodied experiences of the Cornish coastline (south west England), whether experienced through walking or flying shaped the abstract art of twentieth-century painter Peter Lanyon. In contrast to the symbolism of the nineteenth-century landscape painting, they are interested in how expressive gestures with brush and paint enable the conveying of a feeling of being in the landscape. Whilst for John Wylie and Hayden Lorimer a geopoetic writing style enables them to attune both themselves and their reader to their embodied experiences of walking and running through the landscape. They write of the affective experiences of being lost, of the way the materiality of the landscape goes to work on bodies, as well as the encounters with non-human others as they explore the world around them.

If the focus of such geopoetics is to convey an embodied experience of landscape to others, then for other geographers creative practices might also enable a certain form of coming to know the environment, in other words, knowledge happens through creative production in and of itself (Hawkins, 2015a, Madge, 2014). Hawkins (2015a), for example, uses drawing theory and practice to examine how the practices of creative doing might form the means to engage with place. As Hawkins (2015a) emphasises, being good at drawing is not necessarily as important as the process of attending to our surroundings that making marks on paper cultivates. Much writing on drawing explores the difference between the communicative dimension of the drawing as a finished thing, and drawing as a private process based around the event of looking. In other words, this is to get beyond a focus on drawing as offering mimetic representations and instead to open up understandings of drawing as enabling an attentiveness to embodied encounters with landscapes and environments. For Pink (2001, p.60) visual methods are mediating practices wherein, 'mediation is a process that allows us to attain richer and fuller translations of bodily experience and materiality that are located, multi-textured, reflexive, sensory and polysemious'. Following drawing theory, it can be argued that to draw is to discover, to be led to see, to be drawn into an intimate relationship with the object. Such ideas are seductive, but there is a danger that in such thinking we reproduce drawing as an easy process during which we are folded into an intimate relationship with the world. Of course, drawing, especially if we are not very good at it, might be a deeply frustrating process where you are unable to make the marks you want, where nothing goes on the paper in the way you hope. As Hawkins (2015a) notes, such frustrations are not to render drawing invalid as a method, rather drawing becomes important for what is apprehended of place through practice, rather than what was represented in the finished piece. In this, drawings were less for providing 'packets of information' or 'representing nuggets of experience', but rather appreciating the very process of image making itself as a means to come to know.

INTERSECTION THREE: CREATIVE ENVIRONMENTAL ENGAGEMENTS IN THE AGE OF THE ANTHROPOCENE

the only way to approach such a period [the Anthropocene] in which uncertainty is high and one cannot predict what the future holds, is not to predict, but to experiment and act inventively and exuberantly via diverse adventures in living.

(Dumanoski 2009, 213 quoting C.S. Holling)

The designation of the current geological epoch the 'Anthropocene', in recognition of the importance of humans as a force of environmental change, has raised questions regarding how we research and practice in this age. How, for example, we take account of the entanglement of humans and non-humans, how we manage questions of uncertainty and how we take practical steps to deal with current environmental challenges. For those within the science community, global environmental change poses challenges around how to appreciate the multiple perspectives needed to understand the environment, but also how to close gaps between understanding in the academic community and action in the world. Increasingly, in the face of such 'wicked' problems, the value of interdisciplinarity is recognised, more often, however, in terms of social sciences, than for the contribution that arts and humanities might make. Noticeably within diverse efforts to respond to these challenges, what often comes to the fore are practices of experimentalism and creativity, often coded in terms of 'anticipatory interventions', 'active experimentations', the assembly of 'hybrid research collectives' or calls for 'experimental socio-ecological practices' (Lorimer, 2010, Gibson-Graham and Rolevink, 2009). These expanded senses of experimentation redistribute the sites, spaces, efforts and subjects of environmental science beyond the academy and the laboratory to take in interdisciplinary working teams, to rethink expertise by recognizing the value of knowledges other than those of 'experts' (including those that are resolutely more-than-human, or non-human), and to make space for seeing and hearing myriad agencies constituting the collective ecosystem. As Gibson-Graham and Rolevink (2009) suggest, 'we need new ways of living and researching in the age of the Anthropocene' that are 'open to what can be learnt and to recognise our role in bringing new realities into being'. For some geographers, experimenting with creative research methods might offer one response to these questions. This discussion will explore three key dimensions of how we might think about the possibilities of creative methods with respect to tackling the challenges of global environmental change. Firstly, exploring how creative practices might help us to rethink our geographical imaginations; secondly, how creative practices might enable critical engagements with environmental science; and thirdly, how creative practices attune us to the environment differently, and so connect us differently to it.

Rethinking our geographical imaginations

Western imaginaries of environmental change often summon up distant geographies of marooned polar bears and melting glaciers calving great chunks of ice into the sea. What is more, climate change some-times seems to be an abstract science of tipping points, thresholds and planetary boundaries, the effects of exceeding these being felt by other people in other places and at other times. One of the things artists can do well is bring climate change 'back home' for Western audi-ences, enabling us to sense its effects on our own local places and environments.

Knitting is perhaps an unlikely vehicle for connecting people with local climate change. Yet the project Birdyarns (2011, Figure 10.3) conceived by artist Deidre Nelson and supported by Cape Farewell does just that. Nelson worked with communities on Mull and the surrounding Scottish Islands, to develop a simple pattern that enables amateur and experienced knitters alike to produce woolly Arctic Terns, a species once commonly seen in the islands. In the production of these felted avian forms and their eventual display, firstly on Tobermory Harbour and eventually at galleries around Scotland, knitting became a way to tell the story of the Tern's fate under climate change. In the gallery a scientific soundtrack accompanies the knitted flock narrating how the Arctic Tern, one of the longest-haul migratory species, is

Figure 10.3 Birdyarns (Photo: Deidre Nelson)

increasingly failing to mate and their offspring are starving as shifting ocean circulation patterns break up breeding grounds and threaten food stocks.

Yet *Birdyarns* is more complex than just bringing climate change back home; it offers an example of how creative making processes offer the means to forge relations between humans and non-humans (Hawkins et al. 2015, see also the discussion of taxidermy in Box 10.2). *Birdyarns* is a good example of a socially-engaged arts project, creating a global community of knitters that focused on the Woolly Wednesdays knitting group on the island of Mull, where the project started. However, *Birdyarns* extended the social relations it catalysed to the non-human social. As Hawkins et al. (2015) note, our environmental imaginaries might benefit not only from being brought closer to home, but also from a recognition of the expanded ecologies and material conditions that constitute the environment (Gabrys and Yusoff, 2011). Most obviously, *Birdyarns* encourages us to appreciate these ecologies through the spoken-word expert soundtrack, written by an ornithologist and narrated by a local wildlife photographer that accompanied the terns. *Birdyarns* registers non-human collectives in other ways too, ways that are more about the practices and materialities of art-in-the-making rather than in its consumption. We see, for example, engagements with the matter of the world formed through the processes of knitting itself. So, the making of the birds connected the fleshy corporealities of the knitters with the woolly bodies of the birds they produced. As Hawkins et al. (2015) recount, working from a pattern, and with wool from the island's sheep, the shaping of birds' woolly bodies involved repetitive practices of manipulating wooden needles and spun yarn in variously experienced hands. This developed interactions-in-the-making, connecting the fleshy bodies of the birds, their knitted form and the corporealities of the knitters. As Hawkins et al. (2015) note, knitted birds become a prompt to engage with environmental change less as matters of scientific fact and more as forces affecting more-than-human collectives, 'in place of climate change as an "out there", distributed across a global imaginary of distant places, *Birdyarns*' climate change is located and materialised in particular ways'. *Birdyarns*, together with Straughan's account of the non-human relations cultivated through taxidermy (Box 10.2), suggest that making practices, whether knitting or other forms of craft practices, have the potential to bring us into close relation with the non-human.

Box 10.2 CRAFTING TAXIDERMY: GEOGRAPHIES OF SKIN

Elizabeth Straughan

The skin is an organ understood physiologically to be composed of three layers: the epidermis, the dermis and the subcutaneous or fat layer. It is then, a three-dimensional organ, which houses nerve endings, as well as hair follicles and sweat glands, and as such it is the locus for the sense of touch. In this research I took the skin as my object of analysis, considering it as a space where the body meets environments and technologies. I was interested in resulting material effects to this organ which can be manipulated and altered, as well as the role touch plays in such meetings with regards to the physiology and physicality of this sense and its metaphorical connotations, which align it with feeling and emotion.

As one case study through which to consider skin in this way, I chose to explore taxidermy, a practice whereby skin is removed and rearranged with the use of tools. This research project was never intended to be creative. Rather, at first I was influenced by the feminist argument for the use of 'supplication'. This is informed by an understanding that the interviewer, 'seek[s] reciprocal relationships based on empathy and mutual respect . . . often sharing their knowledge with those they research' (England, 1994, p. 243). Alongside this, I was concerned with the question of how to craft and ask interview questions about the tactility and haptics of a practice with which I had never engaged. And so, I took up taxidermy.

My 'learning' of this practice initiated with a one- and-half-day course with a local taxidermist in Wales, who on the course's completion, sent me home with a bag full of frozen moles (see Figure 10.4). On arrival back at Aberystwyth University, I sought a space in which to continue my 'learning'. After enquiring into some potential practice spaces in both the University's Geography and Biology departments, I eventually found a home at the town's Scout hut and later the University Art department. This was a

Continued

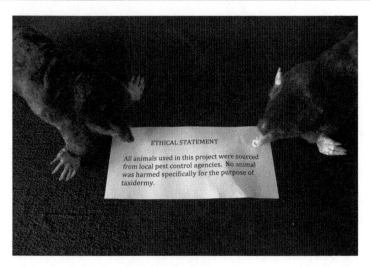

Figure 10.4 Ethical statement collection

process which raised interesting questions about certain spaces in which taxidermy practice was deemed 'out of place'.

At first I practised on my own, spending hours bent over a desk in the Scout hut documenting my progress using film and jotting notes in a research diary. This produced some interesting comparative comments on the different moles and their materiality regarding, for example, their size, the damage done by the trap, the levels of decomposition, how I work with these issues and my improving skill. Keen to get input from other 'learners' I began to ask friends to taxidermy the moles with me, a move that proved fruitful in gaining interesting documentary film footage. That is, I had found solitary practice of taxidermy produced uneventful camera footage that did not always parallel my field notes, which relayed, for example, dealings with decomposition and frustrations with slicing through skin.

In contrast, consider the following transcript that documents a particular moment in a taxidermy practice session, which took place in the Art department, where, due to a shortage of mole bodies, fellow learner and artist Dawn Olive and I took turns in the work of taxidermy. At this stage I work the mole while Dawn takes photographs:

Me: Ok this one must have started rotting before they froze it.
Dawn: Oh really?
[Once the incision is made I work quite quickly prising the carcass from the skin, as it has started decomposing the carcass seems to have less hold on the skin and falls away fairly easily. I cut through the two back legs and after 10 minutes am now ready to pull the skin over the top of the carcass.]
Me: Oh my god Dawn.
Dawn: Is it Ming?
[I pull skin back over the back.]
Me: This is disgusting. It's all rotting.
Dawn: Oh my word, that isn't anything like the ones we did. [Dawn takes some photos over my shoulder] Oh fucking hell, I just got a whiff of that. [Dawn walks out of the shot.]
Stinky!
Me: Oh my god! [I dust some chalk over the skin and get back to skinning.]
Dawn: Go on you can do it, you can do it!
Me: At least it is making me work quite quickly!
Dawn: Yeah!
Me: It's just, it's really hard, there is like, hardly any definition really, and it's hard to get the tissue tension to cut off right, so I am cutting bits of muscle off when normally it is really easy not to.
(24/08/2010)

Engaging with others in the process of 'learning' to do taxidermy, I found generated discussion on the process, highlighted body language and actions, drew attention to senses other than touch and outlined the context and evolution of events as they happened.

As a result of my practice and acknowledgment of the capacity of 'lively' or 'vibrant' dead matter to have affect (Bennet, 2010), I

Continued

began to consider a different ethics, that of '[d]ifference-in-relation' (Whatmore, 2002, p.159). This is an ethics, Harrison states, which reveals a, 'concern with ways of living, folding and becoming with 'more-than-human' others' (2006, p.132). As such in my practice and interviews, I began to also consider the effect of dead animals on the living human taxidermist. Establishing a community of learners was then beneficial in gaining a richer sense of this practice.

Birdyarns is just one example of how creative practices, participatory or otherwise enable the relocation of our imaginaries of climate change very firmly on our back doorsteps and in our back gardens. Indeed, Birdyarns, joins projects like Junaita Scheffer-Muller's Climate Hope Garden, a collaborative art–science project that used three different climate change scenarios for Switzerland in 2015 to grow the sorts of plant species that might normally be found in Swiss back yards and household gardens. The result of these and many other projects that do similar work on our environmental imaginaries is to relocate environmental change from a place faraway to one closer to home. Further, the intimacies and proximities sculpted through creative making practices can enable a realisation of the connectedness of global environmental systems and our own connectedness to non-human others.

CREATIVE EXPERIMENTS WITH ENVIRONMENTAL SCIENCE

Critical discussions of environmental science have recently emphasised the 'dangerous maths of climate science' that underpins techno-scientific interpretations of climate change and thus so many of the practices and discourses of climate security. An increasingly large proportion of art about climate change is made by artists working in collaboration with scientists. A number of organisations and funding bodies facilitate these collaborations, enabling artists to take up residencies in scientific laboratories, giving them access to equipment, techniques and data of environmental science. Oftentimes, the result is art that has not only evolved in response to these scientific tools, but that is also critical of the terms upon which scientific knowledge is made.

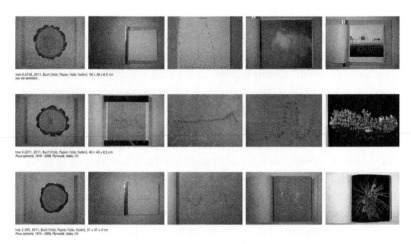

Figure 10.5 You are Variations (Photo: Christina della Giustina)

'What might the trees feel and sound like?' is a question that Swiss artist Christina della Giustina asked the environmental scientists she worked with during her residency at WSL, the Swiss Federal Laboratory for research into Forest, Landscape and Snow, based near Zurich. At WSL scientists use century-long data records concerning atmospheric composition and vegetation health, together with live data streams and field recordings, to develop contemporary predictions of environmental change. Accompanying them into the field and working alongside them in the lab, della Giustina explored how these scientists 'know' the trees they study. Her ongoing project *You Are Variations* (see Figure 10.5) encourages myriad ways of knowing these 'specimens'. For example, she experimented with a range of different registers of visualisation. Valuing hand drawing as much as high-resolution images, she draws together science and art to encourage those who studied the trees to see them from different perspectives. She also worked with a group of scientists to develop a computer programme and musical score that encourages sciences and other audiences to reflect on what trees might sound like. Incorporating the sonification of live data streams with improvised orchestral compositions, she experimented with creating installations that rendered in sound the processes of gaseous exchange central to tree growth. Her intentions throughout these sensory experiments with science have been to encourage scientists to reflect on the processes by which they produce knowledge about the environment.

As well as critiquing the practices of science, art projects also present an alternative, enabling not only a critique of expertise but also empowering and making possible different forms of knowledge making. Scientists, it is often assumed are the climate experts, they are the ones with the information about causes and effects, the strategies for intervention and mitigation. But yet, amateur knowledge has a long history as part of the making of climate knowledge; historically, for example, community records and artistic representations have had a key role in environmental reconstructions (Dixon et al., 2011). More recently, artists have combined participatory practices with citizen science to remake the terms of the production of science. Unseating the singular image of the climate expert, they open out a broader sense of who understands and can intervene within our changing environments.

We might consider a range of participatory projects that do this sort of work (Hawkins et al., 2015). Consider, for example, the project *Wrecked: On the Intertidal Zone* (2015) carried out on the Thames Estuary by an arts collective led by UK-based Arts Catalyst. *Wrecked* saw the development of a series of public experiments combining myriad ways of knowing the estuary, setting the practices of local bird watchers, fishermen and mud walkers amongst scientific studies and information gleaned through political and corporate wranglings. The result of these and a host of other projects is to see creative practice as unseating the scientific expert, enabling other voices to be heard and other practices to be valued in the making of scientific knowledge. For Barry and Born the ability to promote such 'public experiments' is one of the most interesting and innovative dimensions of recent interdisciplinary art–science work. We might reflect too on the proximity here of creative practices to those of citizen science, which enrol citizens as amateur scientists, data collectors and as environmental sensors.

Sensing environments: ethical engagements with the environment

> Geographers need to experiment with different ways of engaging and writing the world involving '[n]either words nor images but both and more besides.
>
> (Thrift, 2007 p. 22)

Recent geographical scholarship on sound as a research method has explored how such work might enable a range of different things, from 'highlight[ing] hidden or marginal characteristics of places and their inhabitants' to promoting ethical environmental engagements (Gallagher and Prior, 2014, p. 268; Kanngieser, 2014). Within the wider remit of sonic methods Gallagher and Prior have placed emphasis on phono-graphic methods, in other words, recording sound and 'associated prac-tices of listening, playback, performance and distribution' (p. 268). Their discussion of these practices, whilst careful to distinguish it from some-thing that is more artful, often uses the practices of sound artists as a backstop for exploring these processes. As such, for Gallagher and Prior these methods often fold together concerns with 'accuracy' and a recording of the more-than-representational. They are interested, for example, in how recorded sound might offer the chance to engage us with entangled 'human and more-than-human actors: beings and objects vibrating in the world' (p. 277). In place of ideas of 'capturing', what is on offer in such performances are 'detailed traces'. This is to align with the wider sense within geography that creative performances enable our engagement 'with intangible, imperceptible, ephemeral and affective dimensions of life' (p. 277). The point, as they note, is to bear in mind the 'contingency and spontaneity of the interaction between listener, play-back apparatus and environment (p.278).

A good example of this is the experimental audio-walk around a ruined Scottish landscape that Gallagher (2014) created, Kilmahew Audio Drift No. 1. The piece was designed to be listened to on portable MP3 players whilst walking in the landscape. Its aim is to 'reconfigure listeners' relationships to place, to open up new modes of attention and movement, and in doing so rework places, albeit in ways that may be small-scale, temporary and difficult to articulate' (p. 469). As such, these sound methods, like other creative practices not only represent place, but also participate in its remaking. The notion of 'drift' that Gallagher uses to inform the production of his work and its discussion encompasses his own movements around the Kilmahew site – a wooded glen near Glasgow – and the audience's movements around their own landscapes. Gallagher describes 'layers of environmental field recordings and voices blending and bleeding into each other, aligned so as to create tangential associations that might invite wandering of both the mind and the feet' (p. 471). These sonic elements include intersecting field recordings of abandoned buildings, the natural landscape and recorded voice inter-views to explore the past, present and future for the site.

As a result of listening, Gallagher hoped that visitors' movements through site would be effected, inviting imaginative meanderings and shifts in attention. Thus intervening in 'the production of Kilmahew as storied place and sounded space' (p. 472). Listened to in the landscapes in which they were created 'folds the sounds of a place back into that same place', creating a displacement-replacement that doubles a place back on itself, returning the sounds to it as revenants that can generate 'uncanny affects of ambiguity, haunting and hallucination' (ibid.). The result is a way of doing geography that enables an intervention into places, the production of immersive experiences through which audiences can move and, in doing so, exposing often overlooked elements of places, challenging meanings and opening out ways of knowing places that are more-than-representational.

What might this mean for our engagements with the environment? Sound theorists have long argued that sound objects bring the world into proximity with our body; for anthropologist Tim Ingold (2013) sound has been understood as an elementary affective force influencing how we relate to the world and the materials that comprise it. As Kanngieser argues, sound and sound objects 'can bring bodies into contact with unseen and otherwise difficult to encounter systems and situations' in ways that have 'significant consequences for how we relate to anthropogenic ecological crisis' (2014, p. 83). The explanation behind this often turns us to the embodied and phenomenological experience of sound that emphasises how in its very mechanics and materialities, sound cuts across matter and beings, sound as Kanngieser continues 'renders apparent that the world is not for humans. The world is rather with humans – a relation that is not without antagonism' (ibid.). As such, sound, she argues, can underpin an epistemological perplexity and vulnerability that creates connections and disconnections between bodies and worlds.

We see such a creation of environmental connections rendered in a rather different way through the work of Land or Earth artists such as Robert Smithson, Andy Goldsworthy or Ana Mendieta. The works of these artists offer us a way to think about how matter and forces might come to be central to thinking about creative practices (Hawkins, 2013; Matless and Revil, 1995; Yusoff and Gabrys, 2006). Amidst the huge volume of writings on their work we find studies of artistic use of natural materials and processes including water's ebb and flow, land's gradual or seismic shifts, the creep of soil, the decay of organic material, and the remnants of geological and glaciological processes. In such work, artistic practices and materialities come together with critical effect, formulating an aesthetics of mutability whose focus is the destabilisation of any sense of landscape as timeless, fixed and static (Housefield, 2007). Furthermore, these works demand a formulation of the artistic process that recognises its distributed nature, for it is through the workings of these natural processes that the piece gets made. The processes of the works' making are as much a function of the human intentional and unintentional processes as they are the human surrender to the natural processes as co-creators. Creativity is thus a distributed venture, enrolling unpredictable natural processes as well as the artist's processes. For Art Historian Amanda Boetzkes (2010, p. 12) such distributed creativities and these points of contact between artist, audience and earth 'mediate[s] a visceral contact with nature in order to suggest a way of interacting with it'. Such interactions, she argues, enable us to register the earth more definitively as a living, changing elemental force.

It is not only in sculptural and sonic engagements with the environment that the basis for more ethically attuned engagements and connections might be found, we might also look to poetic practices. Eco-poetry is gradually gaining attention within geography and importantly is a genre that expressly articulates itself as concerned with relationships between non-human physical/ecological geographies and sociocultural human geographies. Magrane's discussion of writing the desert in Box 10.3 offers a good introduction to what geopoetics might mean and do. For feminist theorist, geographer and award winning poet Sarah de Leeuw (2004, 2013) poetry offers the basis from which to explore and evoke connections between our own messy, fleshy bodies and the materialities of the environment. Her work offers, it could be argued, a practice-based exploration of how writing feminist eco-poetry might enable a move beyond the cerebral. In her poetry she presents organic images of bodies and nature as interpenetrating, such that a continuum is created between fleshy animal and human bodies and earthy environments. Whether these are birds in flight or tearing through flesh with talons, or glaciers carving into rivers, her work aligns these natural images with erotic imaginaries of the female body. While, such a folding together of bodies and nature is not to be undertaken naively, in De Leeuw's work it offers a powerful basis for a politics and ethics that is based in worldly connections, materialities and distributed affective forces, that once more find humans, non-humans and the matter of the environment in intimate and proximate relations.

Box 10.3 WRITING THE DESERT

Eric Magrane

of course a poem is not human
a poem is its own animal

I thought I'd get a sheet of glass, put a poem on it, and take it out in the desert. It struck me that at a certain time of day, if held correctly, the light would be just so to make the poem cast no shadow. At other times, each letter would stretch out upon the ground. I can picture groups of people fitting within the center of an O. Still other times the letters would merge and lose their shapes in each other. I imagine this would be something like a room where many conversations are taking place.

~~~

Figure 10.6 Geopoetics: Lyric Earth (Photo: Eric Magrane)

~~~

These poems are ephemeral. A certain assemblage of light, earth, and language. Collaboration with landscape. The site is distributed—from the rock to the language/referent to the earth to the documentary still image of the poem on this page. It is literally geography in the sense that it is earth-writing.

As geopoesis (earth-making) it is minor in scale, more immanent than transcendent. No longer there on the landscape. Momentary. It makes nothing happen but captures time and place, then lets them go, impermanent.

In contrast, geopoesis in a major scale is a further distance from ephemeral. Climate change and the Anthropocene are examples of geopoesis writ large. 'We have literally written ourselves into the strata and atmosphere of the earth,' as I put it elsewhere, on the Anthropocene. Here in the desert southwest of the U.S., one might think of the Glen Canyon Dam as an instance of embodied geopoesis writ large.

~~~

Symbiosis is a term for ecological relationships between species. In mutualism, both species benefit; in parasitism, one benefits and the other is harmed; in commensalism, one benefits and the other isn't affected. Commensalism may slide over time into mutualism, which may slide over time into parasitism, which may slide over time into commensalism, and so forth. In other words, relationships are not necessarily static.

~~~

Continued

Figure 10.7 Geopoetics: Nothing Separate (Photo: Eric Magrane)

~~~

The desert's palo verde tree has very small leaves as an adaptation to the dry climate. Bigger leaves would cause it to lose more moisture. The tradeoff is that it then has less leaf-space to photosynthesize. To compensate for that, it makes energy through its green bark: green tree photosynthesizing through green bark! Imagine if humans could do such a thing and make energy through their skin.

Thermodynamics, energy transfer, entropy, decomposition, successional pathways and regrowth . . . these are all aspects of creativity and of geopoetics.

~~~

In *Desert Water, Desert Light*, a poetry project in which I sand-blasted short poems onto mirrors and installed them in a Tucson gallery – so that entering the gallery you walked into a poem – I wrote:

not the mind's desert
the green desert
after rain

out here
distinctions blur

against blue sky
morning moon
desert rock
against desert rock

Within this poem I imagine philosophy coming from the desert earth. It is naïve about the elemental qualities of water and of light. Geopoetics may be a way to do something else with subject-object. It is also a means through which to consider appropriate technology.

CONCLUSION: LANDSCAPE AND ENVIRONMENT – RESEARCHING AND LIVING DIFFERENTLY

Extracting three points of intersection of creativity and landscape and environment from the complex and rich history of these relations is not a straightforward task. The episodes chosen here reflect the key points in a much longer and more complex story, but points that hopefully illuminate the possibilities of these intersections for both helping us to research landscape and environment differently, so evolving our ideas of these crucial concepts, but also, importantly for helping us to live differently too. In short, enabling us to reflect on how creative practices as part of our scholarship might enable the closure of what environmental scientists have appreciated as the gap between knowledge and action. This is to query how creative practices might enable us to know and encounter our landscape and environments differently, whether through sensory relations, an appreciation of difference or through embodied encounters from which might propagate an ethics. Not only might these practices enable us to encounter the environment differently, they also might help us remake the terms upon which we know it, reshaping the forms of scientific expertise to revalue knowledge made by other people, in other places and through other means.

The global nature of the agreement reached at COP21 in 2015 might make history in getting most of the world around the table to agree to combat climate change, but these promises, inadequate as they are, still need to be translated into action. While climate change remains the preserve of scientific experts and politicians and activists, and while the popular imaginary is dominated by stranded polar bears, confused penguins and cracking ice-bergs, action is likely to be limited. Localising climate change, developing critical–collaborative engagements with climate scientists, and enabling grass-roots amateur experimental practices and local knowledge might not seem like much but they have the power to shift imaginaries, empowering us to remake climate science from below. One thing seems clear; if we are to come close to tackling this global crisis some creative action is needed.

11

CREATIVITY'S PROMISES

'Making Other Worlds Possible' is, suggests a recent edited collection by Roelvink et al. and the alterative economies collective (2015), a complex and contradictory process. In their diverse, always inspirational, propositions for other ways of living and acting within capitalist systems, creativity does not really figure as a direct set of practices, or interventionary, transformatory actions. But yet, thinking creatively, acting experimentally is evoked in many different ways by this collective as they imagine and bring about different forms of human non-human relation and economic actions other than those of capitalism's globe talk. To read their collection, to engage their work, is to come to an awareness of the myriad ways that creativity offers the means to participate in the making and creating of other possible worlds.

Who does not want to be creative? Creativity promises much; economic regeneration, material prosperity, identity formation, well-being, connected communities, subject transformation, social cohesion, knowledge production, place shaping, environmental engagements, world-making. The ideas of creativity presented in this volume celebrate it as a diverse concept and one cut through with promises and tensions. These promises and tensions have been refracted here through a strategy of putting creativity in its place. To do so required an exploration of how creative activities are not only shaped by the places in which they are produced and consumed, but how they also make places, shape subjects and sculpt communities, environments and worlds. Such geographies enable, this book suggests, the means to help us understand the challenges of diverse creativities, to explore their demons and to engage their promises and possibilities.

By way of a conclusion, the text offers an examination of the promissory nature of creativity. Without falling into the trap of universally celebrating and fetishising creativity discussion draws together the key tensions and debates that have worked throughout the text. As this text has argued from the outset, attending to the critical geographies of creativity enables us to understand and potentially to mobilise the promise that creativity might offer us the means, individually and collectively, to live differently. Such critical geographies of creativity have, throughout this text, formed the means to find productive ways through the challenges posed by an expanded field of creativity, ways that let us explore and engage the potentialities of creativity, while acknowledging the tensions and challenges within this diversity.

To recap, the first of these critical geographies of creativity concerns creativity as a socio-spatial practice, not just a force of the mind, but a material, embodied and 'placed' practice. To explore creativity as a socio-spatial practice is to examine how creative practices are not just situated, but they are also socially-spatially productive, producing social effects (communities, identity, etc.), spaces, places and subjects. As such, we might think of creative practices' world-making potentials to lie in the whole suite of relationships between bodies, materials and matter, technologies and objects that they configure.

Secondly, the critical geographies of creativity draw our attention to the hugely variegated political and critical potential of creative practices, and the tensions between the different political deployments of creativity. So we find the co-existence of a range of points of view, encompassing the enthusiastic enrollment of creativity within the neoliberal production of urban spaces and economies (Landry, 2006; Florida, 2005) as well as a sense of avant-garde creative practices as practices of activism and resistance. We find, for example, creativity as part of a logics of practice for individuals and collectives who seek consciously or unconsciously to produce subjects, experiences and spaces that challenge neoliberal and capitalist projects (Gibson-Graham, 2008; Kanngieser, 2014; Katz, 2004). Negotiating these possibilities is far from clear, especially as creativity has often become enrolled within neoliberal agendas precisely because it is associated with this latter set of 'cool' politicalities; whether such enrollments 'de-tooth' creativity or offer opportunities for further forms of creative difference remains a contentious question (Daskalaki and Mould, 2013; Kanngieser, 2014).

Thirdly, and perhaps most importantly, underpinning the previous two sets of questions, the text appreciates creativity as a force, asking what it is that creativity does. It does so not only in terms of the production of goods and services, but also in terms of the making and forming of subjects, spaces, politics and knowledge. Asking questions around how this force manifests (is its power in practice, in representation, in affect?) who has the ability to bring it about and what work does it do, enables a querying of how creativity might make the world otherwise. It also points us towards some of the dangers that are inherent in any potential fetishisation of creativity and its possibilities.

These geographies, as they have evolved through this text, are critical in three main ways. Firstly, these geographies are of importance in the context of the economic, political and social dimensions of creative industries and practices. These might underpin our understandings of the political economies of the 'creative city', or they might concern the value of creative practices for social well-being. If creativity is to be one of the economic and social themes of our time then it is imperative that we understand the geographies of its production and consumption – whether these concern the organisation of the creative cluster or the micro-geographies of embodied creative labour practices and their affective conditionings of lived experience. Secondly, these are critical geographies of creativity because, taking leave from critical traditions within geography, the discussions that evolve enable us to engage with the tensions that cut across different bodies of work on creativity; principally, as we shall see, those concerning politics and social justice. For example, whilst creativity is well acknowledged as one of the narratives of neoliberalism circulating the globe (Luckman et al., 2009; Peck, 2005), it also remains a powerful source for those who seek, often collectively, to remake our worlds differently (Kanngieser, 2014). Finally, and perhaps most importantly, the text proffers these critical geographies as the means to engage with the relations and tensions between the diverse ideas, practices and politics of creativity. It argues that to understand the geographies of creativity is to be equipped to critically engage with creativity's diversity, to understand the tensions within this richness and to make possible a realisation, and even a mobilisation, of some promise of creativity to enable us to live differently. To explore what exactly this means, the remainder of this conclusion is going to revisit the challenges of creativity's diversity before reengaging with each of those three critical geographies in turn.

DIVERSE CREATIVITIES

Creativity is an idea with much promise, but many tensions. Core to those explored in this text has been the sense of diverse creativities that have recently come to characterise our understandings of creativity. In his exploration of the idea of creativity Tatarkiewicz (2012) notes three phases of thinking about creativity. In the first, creativity is located in the power of the divine, in the second, creativity is understood in relation to human agency and in the third, creativity becomes understood primarily in relation to the arts. This text has, perhaps, explored a fourth evolution of creativity, wherein it is once more a distributed phenomenon, taking in the arts and perhaps inheriting certain concepts from them, but not dominated by them. Instead, creativity is an expanded concept, a domain general concept that takes in both individual and collective processes and products, that is on the one hand both a property of many, as well as the refined talents of few. It is a practice found in specialist spaces of the studio and the museum, but also in the home and on the streets, it is a practice wrought from decades of skilled learning and bodily habituation, as well as an unconscious in the moment act of improvisation or just part of what it means to be in the world.

Appreciating such diversity does not come without its challenges, indeed for some such an expanded definition of creativity renders the idea useless rather than useful. For others, however, far from rendering null the analytic value and critical purchase of creativity such diversity forms the basis for the promise and possibilities of creativity. As the introduction stated, the intent of this text was to deploy an appreciation of the critical geographies of creativity, its production, consumption and circulation, as a means to engage, understand and celebrate creativity's diversity. And, what is more, to find in this diversity the root for the fulfillment of creativity's promise.

Creativity has manifested in diverse ways across this text, as economic, social, political and cultural geographies – live, work, create. We should, of course, appreciate the economic force of creative practices and their mobilisation by and through diverse political groups and motivations.

Creativity is also, of course, about professional artists, musicians and writers, whose creative products grace the specialist spaces of consumption and sales, whether this be a gallery, a concert hall or the internet. But, creativity need not exclusively be about such professional creativities, for appreciating diverse creativities is also to embrace a set of ideas that enables us to understand creativity as part of the spaces and practices of our everyday lives. Creativity appeared here as a grounded aesthetics and mundane practices such as housework, gardening and dressing, as well as in more obviously 'creative' practices of home-making, such as decorating and DIY and very clearly in the range of hobbies that are part of our daily lives. Some perspectives test the limits of these ideas by requiring us to take account of creativity as an unconscious action, part of our improvisational going about in the world as well as even wider sets of thinking about pancreationism, and process philosophies that propose that all reality is creative in some way, shape or form.

Important too in these diverse creativities and their geographies are to appreciate how creativity has long been part of geographical knowledge-making enabling us to research differently; whether this be the place of artists on-board ships, or the more recent role of creative practitioners in engaging with urban spaces and landscapes and environments, or hard-to-reach people. As such, creativity has, for centuries, offered geographers the means to research differently, to challenge and extend their ways of knowing and representing the world around them.

Across this volume a range of work from within and beyond geography has drawn out commonalities and differences between these economic, political, social and cultural understandings of creativity. At times these feel like closely linked bodies of work, with shared concepts, methods and epistemologies. At other times, however, they offered different ways of thinking about the world alighting on the same issue; whether this be what makes creative workers cluster, the nature and value of creative labour, or concerns with what creativity does. To set these different bodies of work in productive tension with one another, to draw on this variety of intellectual resources, is an important way to get beyond the recent fetishisation of creativity, but also to dismantle its demonisation. Acknowledging the diversity of creativity throughout this text is precisely not to let these very different ideas and forms of creative practices and approaches to it collapse into an undifferentiated acritical mass. Rather, it is to use these critical geographies as a lens through which to view the similarities and differences present across these ideas of creativity, and so to move us closer to the ability to explore and engage with creativity's promises, its abilities to enable us to research and to live differently.

Creativity as a socio-spatial practice

Much has been made within human geography of late of the question of scale. Efforts have been made to denaturalise scalar thinking that casts the local as the disempowered position in the context of globe talk and practices. Other approaches seek to re-territorialise such globe talk, to ensure it touches down at specific locations and through such specificities challenge dominant singular narratives. Combined with such approaches has been the building of alternative imaginaries that value local, grass roots action, that find agency and force in embodied activities rather than always seeking to scale-up such micro-geographic stories. Of course, in an era where creativity is seen as an economic panacea, where cities the world over are turning to creativity as a means of regeneration, and where soft power is becoming such a crucial part of the diplomatic arsenal, to underestimate creativity as a global force would be foolish. Indeed, as this volume has hopefully made clear, the geographies of creativity on a global scale are fascinating. Comparisons between different modes and expression of creativity around the world enable a closer look at what we mean by creativity, while creativity's problematic mobility as a policy apparatus demonstrates the importance of local contextualisation in the context of globe talk. Further, as the ongoing enrollment of arts and culture in practices of statecraft continues to casts the force of creativity in geopolitical terms, it seems there is an even clearer need to understand the geographies of creativity, its production, consumption and circulation on global terms.

Throughout this text attention has been paid to the common spatial forms of creativity, and especially those forms of the creative economy and the financial support offered to them. Perhaps dominant throughout the volume have been the creative cluster and the creative city. The first focused primarily on efficient economic production, the second more mixed use but increasingly focused on consumption by elite social groups. Such forms have been challenged of late by the rise of other geographies of creativity, but remain of central importance to the discussions of creativity. Such other geographies include both alternative spatial organisations such as networks and temporary clusters, as well as ideas of marginal or remote creativities, where the creative economy happens at a distance from urban centres. Indeed, asking questions of where creativity happens, to attune ourselves to other geographies of creativity is to assert creativities that either escape or are willfully overlooked in creative economy discussion, seen as being of little direct or indirect economic value.

To consider creativity is, however, as discussion throughout this text also made clear, to be aware of its micro-geographies, the embodied practices that entwine mind, body, materials and tools in the making of things. It is important that these perspectives not be dismissed from the telling of wider economic stories. To do so is to overlook where exactly creativity happens, its biopolitical force, both in terms of the possibilities it presents for bodies and subjects to grow and extend, but also the stresses and strains it places on minds and bodies. As the often auto-ethnographic accounts of creative practices detailed throughout the volume make clear, creative practices often involve fleshy bodies in learning to become skilled, shaping muscles but also neural pathways to develop dexterity and perfect techniques. This is not, as accounts have made clear, just about appreciating fleshy, sensing bodies and their experiences of the production and consumption of creative outputs, but is rather about appreciating bodies and minds coming together. Further, as we saw, what is used within the process becomes crucial; not only tools, but also the animate, vibrant matter that is worked with. As discussion made clear, to appreciate the volatile materials involved in making is to enable an understanding of creativity that once more decentres the dominance of the transcendental artistic genius, and instead finds agency in the materials being worked with. It is also to open up possibilities for particular relations to the world established through engagement with these volatile materials in relation with our own messy fleshy forms. As Carr and Gibson (2015) make clear, there is much potential to be found in thinking through what it is that processes of making that take account of volatile materials might make possible. There is also an imperative in understanding the political economies of creative labour, its affective and immaterial forms, especially as it becomes idealised by workers and companies alike.

Creativity and politics

Perhaps one of the tensions that has cut across this volume most clearly and that sits at the heart of some of the issues people have with diverse creativities is that of politics. For many, the rise of creativity in contemporary times is firmly interlinked with its neoliberal expression. At the heart of the popularity of this expression is creativity's rhetorical power;

> creativity as a distinctly positive, nebulous-yet-attractive, apple-pie-like phenomenon: like its stepcousin flexibility, creativity preemptively disarms critics and opponents, whose resistance implicitly mobilizes creativity's antonymic others — rigidity, philistinism, narrow mindedness, intolerance, insensitivity, conservativism, not getting it.
>
> (Peck, 2005, p. 765)

As Peck argues elsewhere 'who wants to be uncreative?' (2005, p. 113). Yet it appears that creativity smuggled in through these policies might often not be about creativity at all. For some, it is too focused on creative consumption, so festivals, galleries, and so on; for others, it is about the creation of an atmosphere that attracts a creative class, most of whom are not creative producers. For other opponents of neoliberal creativity, it can actually be detrimental to the support of creative practitioners' livelihoods. Not to mention the rejection of subversive forms of creative practices such as skateboarding or urban exploration, apart from when the co-option of such practices would enable the production of cultural capital by way of the acquisition of their sub-cultural characteristics.

The ideology of creativity also veils the darker side of creative labour and creativity driven urban redevelopment. The challenges of gentrification have been well explored, but also with the fetishisation of creativity comes the promotion of creative jobs characterised by precariousness, temporality and job insecurity. In short the creative labour being promoted by those creative quarters and creative city policies and being rolled out to other sectors beyond the creative labour force is a system characterised by precariousness, by high degrees of informality, by self-exploitation and lack of security masked by the celebration of the entrepreneurial self. It is a sector where people often hold down multiple jobs, where unpaid internships are normal and the insecurities of project-based labour and freelance work are common. Creativity policies and schemes can unfortunately come to normalize such labour characteristics.

One of the characteristics of creative policy has been its global circulation, and hence the international reproduction of ideas of creative clusters, creative cities and other agglomerative forms around the world. In a sense we can take lessons from Roelvink et al. (2015), who remind us that such globe talk is not all powerful, that we need to attend to differences on the ground, to think through locally appropriate policy and action.

For all the negative press that creativity receives from some quarters for the forms of planning it involves, the kinds of spaces it creates and the people it excludes, creativity is not only associated with governmental forms of politics. It is also closely associated with the activist left, whether this be in terms of arts works or performances that intervene in place and community, or whether this be in terms of the enrollment of aesthetic practices into activist tool-kits. We might also, of course, think about those subversive forms of creativity, such as skateboarding and parkour, which are not necessarily being carried out with a political programme in mind, but in their material practice offer a means of intervention and resistance carried out through daily, mundane practices. To look across the range of creative practices, is, as this volume has illustrated to see a rich range of urban interventions and environmental attunements amongst other possibilities. Lest, however, we get carried away by the possibilities of creativity and its variegated political forms for critical spatial practice and urban subversion and intervention, it behoves us to attend too to the situated expressions of these ideas, their capture and enrollment within various capitalist creative enterprises and their instrumentalisation by those on the look-out for the next big thing to feed the desires of the consumerist-driven creative city.

Visual arts, for example, are often famously co-opted in gentrification schemes, rendered complicit in the rising of rents and silenced in the face of the explusion of marginalised peoples. Subversive practices too have been 'captured' by capitalism. Their aura now very much in tune with forms of tactical urbanisms and hipster practices that now form the basis for edgy, apparently local and grass roots urban programmes and schemes. Such capturing by capitalism, combined with the instrumentalised enrollment of arts and culture within political arenas that include social inclusion and diversity discussions, or within development discourses, does of course risk diluting the interventionary possibilities and prospects of these art works. But it does not mean that all arts and creativity are thus de-toothed, it just suggests the need for further careful critical work in this regard. We need to attend carefully to individual examples of how might creative practices configure our cities and our environments as spaces of possibly and engagement, as sites for the making of new futures.

Creative practices hold within them the possibilities to make other worlds collectively and individually. As discussion has explored across this text, the ways in which this happens, intentionally and unintentionally is varied. Art might be understood to intervene in place; it might be understood to connect community. We might also consider how arts practices enable the collective imagining of new futures and, in doing so, remake political spaces and practices in more lively and magical ways. We could reflect on the ways that crafting together, the material practices of knitting and weaving, come to shape identities and relate us to our communities, nations and also our environments. What remains to be done is to consider in more depth how exactly it is that these practices do their political work, how in short does creativity make worlds? These are dynamics and questions that remain to be explored in more detail and that should form a central part of any future agenda for creative geographies.

Creativity as a force

What marks out geographical studies of creative practices and outputs – whether this be art works, performance practices, literary texts of musical scores – from other studies is an ongoing commitment to understanding how it is they go to work in the world. In other words, what is it that they do? The productive, and sometimes not so productive, force of creativity as a maker and shaper of places, people and knowledge has sat at the heart of the geographies developed throughout this volume.

Perhaps where creativity's force has been most clearly seen and promoted recently is in its role as an economic saviour. Florida situated creativity at the heart of the 'great reset', and the facts bear out the centrality of creativity to current forms of global capitalism. Creativity policy may be compromised as it circulated the globe and touches down around the world, but it has provided a key rallying point around which new forms of economic practice can emerge; a mouthpiece through which the voices of oftentimes marginalised groups can be heard; and a means by which policy making agendas can be pushed and viewed from another angle. This is not, of course, to suggest that creativity is inherently good. It is also associated with precarity and with forms of biopolitical violence exacted on the minds and bodies of creative labourers, as well as the destruction of communities and the spending of huge volumes of money on large-scale, exclusive white elephant schemes around the world. Creativity as an economic force is clearly not always to be applauded.

Furthermore, creativity must be considered a force in terms of making of life-worlds and subjectivities. It is interesting to note how much of the literature on creativity and the making of subjects and forming of identities, relates to subjects in what might be understood as precarious conditions. This might be those individuals in temporary precarious states at points of change in their life course, but they might also concern those for whom precarity is a way of life, and for whom creativity offers a tool box of survival strategies.

Creativity has been understood too, as a force for the production of knowledge. In the context of geographical knowledge-making, creative practices have come to offer not only objects for study, but also collaborative opportunities and new forms of research method. As such, new knowledge is produced individually and collectively that can have an important role in challenging disciplinary norms and the progress of our ideas and understandings. Such deployments do not, of course, come without their challenges. We need to take seriously the sense in which creative practices and geography are different sets of practices, they are learned differently, and they involve different skill-sets and different modes of engagement and assessment. To respect the skill and experience of other disciplines is crucial to this endeavour. This is not to limit what geographers could do, but is rather to urge caution and a reality check as geographers move increasingly towards creative methods. These are methods that can be done and enjoyed by amateurs, but should be undertaken with a realistic sense of what the desired goal is, often to learn through the creative process, rather than to produce an output fit for exhibition.

Finally, of course, the force of creativity can be ultimately thought of as opening up the chance to make new worlds possible. This involves not only imagination that enables us to think through what those worlds might be, but also sets of actions that are able to draw us individually and collectively towards changed practices and worldly attunements. These might be large projects aimed at structural changes and high-profile shifts in attitudes or behaviours, but perhaps more often they are small political acts, daily creativities, generosities and ethical engagements that bring people together. In short, and to return where we started, P and H creativity (creativity for society and creativity for the individual) are both based around novelty, the new might be a worldly new, or it might be a personal new, but in such newness, such novelty, space is made for the imagining and making – the creating – of other possible worlds.

CRITICAL CREATIVE GEOGRAPHIES: RESEARCHING AND LIVING DIFFERENTLY

Beginning from an intellectual location shaped by the desire to understand the promise of creativity, to appreciate its potentials, this volume has sought to explore how an appreciation of the geographies of creativity enables a critical account of such a promise. It has sought, through these critical geographies of creativity, to positively engage with creativity as a diverse and myriad concept, rather than seeking to move towards a singular definition of creativity. Indeed, whilst appreciating the challenges expanded understandings pose to critical acuity and precision, this volume has argued that through attending to the geographies of creativity we can remain open to its diversity and the tensions within this in productive ways. A geographical understanding of diverse creativities, far from reducing the criticality of creativity operates, in fact, in the exact opposite way. It operates by attuning us to how it is we can understand creativity's myriad promises for researching and living differently, how we can fulfill its possibilities for remaking worlds.

BIBLIOGRAPHY

Abrahamsson, C. and Abrahamsson, S. (2007) 'A Body Conveniently Known as Sterlac', *Cultural Geographies* 14 (2): 293–308.

Adams, D. and Hardman, M. (2013) 'Observing Guerrillas in the Wild: Reinterpreting Practices of Urban Guerrilla Gardening', *Urban Studies* 51(6): 1103–1119.

Adamson, G. (2009) *The Craft Reader,* Oxford: Berg.

Adamson, G. (2007) *Thinking Through Craft,* Oxford: Berg.

Adey, P. (2013) 'Air/Atmosphere of the Mega City' *Theory Culture Society* 30 (7): 291–308.

Aksoy, A. and Robins, K. (1992) 'Hollywood for the 21st Century: Global Competition for Critical Mass in Image Markets', *Cambridge Journal of Economics* 16 (1): 1–22.

Alfrey, N.J., Daniels, S. and Postle, M. (2004) *Art of the Garden: The Garden in British Art, 1800 to the Present Day,* London: Tate Publishing.

Ameel, L. and Tani, S. (2012) 'Parkour: Creating Loose Spaces?', *Geografiska Annaler: Series B, Human Geography* 94(1): 17–30.

Amin, A. and Thrift, N. (2007) 'Cultural-Economy and Cities', *Progress in Human Geography* 31(2): 143–161.

Amin, A. and Thrift, N. (2002) *Cities: Reimagining the Urban,* London: Polity Press.

Amin, A. and Thrift, N. (1992) 'Neo-Marshallian Nodes in Global Networks', *International Journal of Urban and Regional Research* 16 (4): 571–587.

Ammerman, N. (ed) (2007) *Everyday Religion,* Oxford: Oxford University Press.

Anderson, B. (1982) *Imagined Communities,* Penguin: London.

Anderson, B. and Harrison, P. (2010a) 'The Promise of Non-Representational Theories', in Anderson, B. and Harrison, P. (eds.) *Taking-Place: Non-Representational Theories and Geography,* 1–35, Farnham: Ashgate.

Anderson, B. and Harrison, P. (eds) (2010b) *Taking Place: Non-Representational Theories and Geography,* Farnham: Ashgate.

Anderson, S. (2015) *Content Curation: How to Avoid Information Overload,* London: Sage.

Arvidsson, A., Malossi, G. and Naro, S. (2010) 'Passionate Work? Labour Conditions in the Milan Fashion Industry', *Journal for Cultural Research* 4 (3): 295–309.

Ash, J. (2015) *The Interface Envelope: Gaming, Technology, Power*, New York: Bloomsbury.

Atkinson, M. (2009) 'Parkour, anarcho-environmentalism, and poiesis', *Journal of Sport & Social Issues* 33(2): 169–194.

Atkinson, P. (2009) 'Do it Yourself : Democracy and Design', *Journal of Design History* 19 (1): 1–10.

Attali, J. (1984) *Noise: The Political Economy of Music*, Minneapolis: University of Minnesota Press.

Bagaeen, S. (2007) 'Brand Dubai: The Instant City; or the Instantly Recognizable City', *International Planning Studies* 12 (2): 173–197.

Baillie, B. (2013) 'Capturing Facades: Structural Violence and the (re)construction of Vukovar's Churches', *Space and Polity* 17: 300–319.

Bain, A. (2013) *Creative Margins: Cultural Production in Canadian Suburbs*, Toronto: University of Toronto Press.

Bain, A. (2007) 'Claiming Space: Fatherhood and Artistic Practice', *Gender, Place and Culture* 14(3): 249–265.

Bain, A. (2005) 'Constructing an Artistic Identity', *Work, Employment and Society* 19 (1): 25–46.

Bain, A. (2004a) 'In/Visible Geographies: Absence, Emergence, Presence, and the Fine Art of Identity Construction', *Tijdschrift voor Economische en Sociale Geografie* 95 (4): 419–426.

Bain, A. (2004b) 'Female Artistic Identity in Place: The Studio', *Social and Cultural Geography* 5 (2): 179–193.

Bain, A. (2003) 'Constructing Contemporary Artistic Identities in Toronto Neighbourhoods', *The Canadian Geographer* 47 (3): 303–317.

Bain, A. and McLean, H. (2013) 'The Artistic Precariat', *Cambridge Journal of Regions, Economy and Society* 6(1): 93–111.

Baker, V.R. and Twidale, C.R. (1991) 'The Re-enchantment of Geomorphology', *Geomorphology* 4 (2): 73–100.

Balm, R. (2000) 'Expeditional Art: An Appraisal', *Geographical Review* 90 (4): 585–602.

Balzar, D. (2014) *Curationism: How Curating Took Over the World*, Coach House Books: London.

Banks, J.A. and Deuze, M. (2009) 'Co-creative Labour', *International Journal of Cultural Studies* 12 (5): 419–431.

Banks, M. (2007) *The Politics of Cultural Work*. Basingstoke, UK: Palgrave

Banks, M. and Hesmondhalgh, D. (2009) 'Looking For Work in Creative Industries Policy', *International Journal of Cultural Policy* 15 (4): 415–430

Barker, P. (2009) *The Freedoms of Suburbia*, London: Frances Lincoln.

Barnett, C. (2008) 'Political Affects in Public Space: Normative Blind-Spots in Non-Representational Ontologies', *Transactions of the Institute of British Geographers* 33(2): 186–200.

Barrell, J. (1983) *The Dark Side of The Landscape; The Rural Poor in English Painting, 1730–1840*, Cambridge: Cambridge University Press.

Barrett, T. (1986) 'Teaching about Photography: Types of Photographs,' *Art Education* 39 (5): 41–44.

Barrowclough, D. and Kozul-Wright, Z. (eds) (2008) *Creative Industries and Developing Countries: Voice, Choice and Economic Growth*, London: Routledge.

Bassett, K., Griffiths, R. and Smith, I. (2002) 'Cultural Industries, Cultural Clusters and The City: The Example of Natural History Film-Making in Bristol', *Geoforum* 33 (2): 165–177.

Bathelt, H., Malmberg, A. and Maskell, P. (2004) 'Clusters and knowledge: local buzz, global pipelines and the process of knowledge creation', *Progress in Human Geography* 29 (1): 31–56.

Battista, K., LaBelle, B., Penner, B., Pile, S. and Rendell, J. (2005) 'Exploring "an area of outstanding unnatural beauty": a treasure hunt around King's Cross, London', *Cultural Geographies* 12 (4): 429–462.

Beazley, H. (2002) 'Vagrants wearing make-up: Negotiating spaces on the streets of Yogakarta, Indonesia', *Urban Studies* 39 (9): 1665–1683.

Beazley, H. (2003a) 'Voices from the margins: Street children's subcultures in Indonesia', *Children's Geographies* 1 (2): 181–200.

Beazley, H. (2003b) 'The construction and protection of individual and collective identities by street children and youth in Indonesia', *Children, Youth and Environments* 13 (1): 105–133.

Becker, H.S. (1984) *Art Worlds*, Berkeley: University of California Press.

Beckett, M. (1947) 'Paris Forgets This Is 1947', *Picture Post*, 27 September: 220–224.

Beegan, S. and Atkinson, P. (2008) 'Professionalism, Amateurism and the boundaries of design', *Journal of Design History* 21 (4): 305–313.

Behlen, B. (2012) 'Does Your Highness feel like a gold person or a silver one?' Princess Margaret and Dior', *Costume* 46 (1): 55–74.

Bennett, A. (1999) 'Subcultures or neo-tribes? Rethinking the relationship between youth, style and musical taste', *Sociology* 33 (3): 599–617.

Bennett, J. (2010) *Vibrant Matter: A Political Ecology of Things*, Durham: Duke University Press.

Berg, S.H. (2015) 'Creative Cluster Evolution: The film and television industries in Seoul South Korea', *European Planning Studies*, 23 (10): 1993–2008.

Berger, J. (1984) *And Our Faces, My Heart, Brief as Photos*, London: Penguin.

Bhagat, A. and Mogal, L. (2008) *An Atlas of Radical Cartography*, LA: Journal of Aesthetics and Protest Press.

Bhattacharya, Ananya (2011) 'Developing sustainable tourism destinations offering authentic folk experiences', in Proceedings of ICOMOS 17th General Assembly, 2011-11-27 / 2011-12-02, Paris, France. Available from http://openarchive.icomos.org/1277/ [Accessed: 24/7/2015].

Biggs, I. (2010) 'Essaying Place: Deep Mapping', [online] Available from: http://www.iainbiggs.co.uk/text-deep-mapping-as-an-essaying-of-place/ [Accessed: 24/7/2015].

Bishop, C. (2006) *Participation (Documents of Contemporary Art)*, Boston: MIT Press.

Bishop, C. (2005) *Installation Art: A Critical History*, London: Tate.

Bishop, C. (2004) 'Antagonism and Relational Aesthetics', *October* 110 (fall): 51–79.

Bissell, D. (2012) 'Habit Displaced: The Disruption of Skilfull Performance', *Geographical Research* 51 (2): 120–129.

Blunt, A. and Dowling, R. (2006) *Home*, London: Routledge.

Boden, M.A. (1996) *Dimensions of Creativity*, Boston: MIT Press.

Boden, M.A. (1990) *The Creative Mind: Myths and Mechanisms*, London: Weidenfeld and Nicolson.

Boetzkes, A. (2010) *The Ethics of Earth Art*, Minnesota: University of Minnesota Press.

Bohm, D. (2004) *On Creativity*, London: Routledge.

Bohm, D. and Peet, F.D. (2000) *Science, Order and Creativity*, London: Routledge.

Bolt, B. (2011) *Heidegger Reframed*, London: I.B. Tauris.

Bolt, B. (2010) *Art Beyond Representation: The Performative Power of the Image*, London: I.B. Tauris.

Bolt, B. (2007) 'Material Thinking and the Agency of Matter', *Studies in Material Thinking* 1 (1): 1177–1180.

Bond, S., DeSilvey, C. and Ryan, J. (2013) *Visible Mending: Everyday Repairs in the South West*, Axminster: Uniform Books.

Bonehill, J. and Daniels, S. (eds) (2010) *Paul Sandby (1731–1809): Picturing Britain*, London: Royal Academy.

Bonnett, A. (1992) 'Art, Ideology, and Everyday Space: Subversive Tendencies from Dada to Postmodernism', *Environment and Planning D: Society and Space* 10 (1): 69–86.

Bonnett, A. (1989) 'Situationism, Geography, and Poststructuralism', *Environment and Planning D: Society and Space* 7 (2): 131–146.

Borden, I. (2001) *Skateboarding, Space and the City: Architecture and the Body*, Oxford: Berg.

Borén, T. and Young, C. (2013) 'Getting Creative with the 'Creative City'? Towards New Perspectives On Creativity In Urban Policy', *International Journal of Urban and Regional Research* 37 (5): 1799–1815.

Bourriaud, N. (2002) *Relational Aesthetics*, Paris: Les Presses du Reel.

Brace, C. and Johns-Putra, A. (2010) 'Recovering Inspiration in the Spaces of Creative Writing', *Transactions of the Institute of British Geographers* 35 (3): 399–413.

Brady, E. (2007) 'Aesthetic Regard for Nature in Environmental and Land Art', *Ethics, Place and Environment* 10 (3): 287–300.

Brand, P. and Dávila, J.D (2011) 'Mobility Innovation at the Urban Margins: Medellin's *Metrocables*', *City: analysis of urban trends, culture, theory, policy, action* 15 (6): 647–661.

Brickell, K. (2012) '"Mapping" and "Doing" Critical Geographies of Home', *Progress in Human Geography*, 36 (2): 225–244.

Brown, A., O'Connor, J. and Cohen, S. (2000) 'Local Music Policies within a Global Music Industry: Cultural Quarters in Manchester and Sheffield', *Geoforum* 31 (4): 437–451.

Buckley, C. (1998) 'On the Margins: Theorizing the History and Significance of Making and Designing Clothes at Home', *Journal of Design History* 11 (2): 157–171.

Buckley, C. (1990) *Potters and Paintresses: Women Designers in the Pottery Industries 1870–1955*, London: Women's Press.

Buckley, C. (1989) 'The Noblesse of the Banks: Craft Hierarchies, Gender Division and the Roles of Women Paintresses and Designers in the British Pottery Industry, 1980–1939', *Journal of Design History* 2 (4): 257–273.

Buckley, C. (1986) 'Made in Patriarchy: Towards a feminist analysis of women and design', *Design Issues* 3 (2): 3–14.

Bunkse, E.V. (1981) 'Humboldt and an Aesthetic Tradition in Geography', *Geographical Review* 71 (2): 127–146.

Buran, D. (2004) 'The Function of the Studio,' in Doherty, C. (ed) *From Studio to Situation*, London: Black Dog Press.

Buran, D. and Repensek, T. (1979) 'The Function of the Studio', *October* 10 (Autumn): 51–58.

Burgess, Jean E. (2009) 'Remediating vernacular creativity : photography and cultural citizenship in the Flickr photosharing network', in Edensor, T. Leslie, D. Millington, S. and Rantisi, N. (eds) *Spaces of Vernacular Creativity : Rethinking the Cultural Economy*, 116–126, London: Routledge.

Burgess, J. (2007) *Vernacular Creativity and New Media* [Online PhD Thesis] Availble from: http://eprints.qut.edu.au/16378/1/Jean_Burgess_Thesis.pdf [Accessed 25/7/2015].

Burgess, J. (2006) 'Hearing Ordinary Voices: Cultural Studies, Vernacular Creativity and Digital Storytelling', *Continnum: Journal of Media and Cultural Studies*, 20 (2): 201–214.

Burk, A.L. (2006) 'Beneath and Before: Continuums of Publicness in Public Art', *Social & Cultural Geography*, 7 (6): 949–964.

Butler, T. (2006) 'A Walk of Art: The Potential of the Sound Walk as Practice in Cultural Geography', *Social & Cultural Geography* 7 (6): 889–908.

Butler, T. and Lees, L. (2006) 'Super-gentrification in Barnsbury, London: globalisation and gentrifying global elites at the neighbourhood level', *Transactions of the Institute of British Geographers* 31: 467–487.

Butler, T. and Miller, G. (2005) 'Cultural Geographies in Practice: Linked: A Landmark in Sound, a Public Walk of Art', *Cultural Geographies* 12 (1): 77–88.

Buttimer, A. (2001) 'Beyond Humboldtian Science and Goethe's Way of Science: Challenges of Alexander Von Humboldt's Geography', *Erdkunde* 55 (2): 105–20.

Buttimer, A. (1976) 'Grasping the Dynamism of Lifeworld', *Annals of the Association of American Geographers* 66 (2): 277–292.

Cameron, S. and Coaffee, J. (2005) 'Art, Gentrification and Regeneration – From Artist as Pioneer to Public Arts', *European Journal of Housing Policy* 5 (1): 39–58.

Cameron, L. and Rogalsky, M. (2006) 'Conserving Rainforest 4: Aural Geographies and Ephemerality', *Social & Cultural Geography* 7 (6): 909–926.

Campbell, C. (2005) 'The Craft Consumer, Craft and Consumption in a Post-Modern Society', *Journal of Consumer Culture* 5 (1): 23–42.

Cant, S.G. and Morris, N.J. (2006) 'Geographies of Art and the Environment', *Social & Cultural Geography* 7 (6): 857–61.

Carey, J. (1992) *The Intellectuals and the Masses. Pride and Prejudice Among the Literary Intelligentsia, 1880–1939*, London: Faber and Faber.

Carpenter, E. (2010) 'Activist Tendencies in Craft', in *Concept Store no 3. Art, Activism and Recuperation*, Bristol: Arnolfini Publishing.

Carr, C. and Gibson, C. (2015) 'Geographies of Making: Rethinking Materials and Skills for Volatile Futures', *Progress in Human Geography*, 40 (3): 297–315.

Carter, N. (2011) 'Man with a Plan: Masculinity and DIY House Building in Post-war Australia', *Australasian Journal of Popular Culture* 1 (2): 165–180.

Carter, P. (2004) *Material Thinking: The Theory and Practice of Creative Research*, Melbourne: University of Melbourne.

Casey, E. (2002a) *Painting and Maps*, Minnesota: University of Minnesota Press.

Casey, E. (2002b) *Representing Place: Landscape Painting and Maps*, Minnesota: University of Minnesota Press.

Casey, E. (1998) *The Fate of Place: A Philosophical History*, Berkeley, CA: University of California Press.

Caust, J. (2003) 'Putting the 'art' back into arts policy making: how arts policy has been 'captured' by the economists and the marketers', *International Journal of Cultural Policy* 9 (1): 51–63.

Caves, R.E. (2000) *Creative Industries: Contracts between Art and Commerce*, Cambridge, MA: Harvard University Press.

Chang, T.C. (2000) 'Renaissance revisited: Singapore as a 'global city for the arts', *International Journal of Urban and Regional Research* 24 (4): 818–831.

Cho, Y. (2011) 'Desperately Seeking East Asia amidst the popularity of South Korean Pop Culture in Asia' *Cultural Studies* 25 (3): 383–404.

Christopherson, S. (2008) 'Beyond the Self-Expressive Creative Worker: An Industry Perspective on Entertainment Media', *Theory, Culture & Society* 25 (7–8): 73–95.

Christopherson, S. and Storper, M. (1986) 'The city as studio: the world as back lot. The impact of vertical disintegration on the location of the motion picture industry', *Environment and Planning D: Society and Space* 4 (3): 305–320.

Clark, N. (2010) *Inhuman Nature: Sociable Life on a Dynamic Planet*, London: Sage.

Clarke, T.J. (2006) *The Site of Death: An Experiment in Art Writing*, New Haven: Yale University Press.

Coe, N.M. (2001) 'A hybrid agglomeration? The development of a satellite-Marshallian Industrial district in Vancouver's film industry', *Urban Studies* 38 (10): 1753–1775.

Coe, N.M. (2000) 'The view from our West: embeddedness, inter-personal relations and the development of an indigenous film industry in Vancouver', *Geoforum* 31 (4): 391–407.

Cole, A. (2008) 'Distant Neighbours: the new geography of animated film production in Europe', *Regional Studies*, 41: 1–14.

Coles, A. (1999) *Site-Specificity: The Ethnographic Turn*, London: Black Dog Publishing.

Colls, R. (2012) 'Bodiestouchingbodies', *Gender, Place and Culture* 19: 175–192.

Comunian, R. and Mould, O. (2014) 'The Weakest Link: Creative industries, flagship cultural projects and regeneration', *City, Culture and Society* 5(2): 65–74.

Conlon, F. (1995): 'Dining Out in Bombay', in Carol Breckenridge (ed) *Consuming Modernity: Public Cultures in a South Asian World*, Minneapolis: University of Minnesota Press.

Connolly, W. (2002) *Neuropolitics: Thinking, Culture, Speed*, Minneapolis, Minnesota: University of Minnesota Press.

Cook, I. et al. (2001) 'Follow the Thing: Papaya', *Antipode* 36: 642–664.

Cook, I. et al. (2000) 'Social Sculpture and Connective Aesthetics: Shelley Sacks's Exchange Values', *Cultural Geographies* 7 (3): 337–343.

Cosgrove, D. (2008) *Geography and Vision. Seeing, Imagining and Representing the World*, London: I. B. Taurus.

Cosgrove, D. (2005) 'Maps, Mapping, Modernity: Art and Cartography in the Twentieth Century', *Imago Mundi* 57: 35–54.

Cosgrove, D. (1999) *Mappings*, London: Reaktion Books.

Cosgrove, D. (1996) 'Ideas and Culture: A Response to Don Mitchell', *Transactions, Institute of British Geographers* 21 (2): 574–575.

Cosgrove, D. (1985) 'Prospect, Perspective and the Evolution of the Landscape Idea', *Transactions of the Institute of British Geographers'* 10 (1): 45–62.

Cosgrove, D. (1984) *Social Formation and Symbolic Landscape*, London: Croom Helm.

Cosgrove, D. (1979) 'John Ruskin and the Geographical Imagination', *Geographical Review* 69: 43–62.

Cosgrove, D. and Daniels, S. (1988) *The Iconography of Landscape: Essays on the Symbolic Representation, Design and Use of Past Environments*, Cambridge: Cambridge University Press.

Cosgrove, D. and della Dora, V. (2008) *High Places: Cultural Geographies of Mountains and Ice*, London: I.B. Tauris.

Cosgrove, D. and Fox, W. (2010) *Photography and Flight*, London: Reaktion Books

Coverley, M. (2012) *Psychogeography*, London: Oldcastle Books.

Crampton, J.W. (2011) 'Cartography: Cartographic Calculations of Territory', *Progress in Human Geography* 35 (1): 92–103.

Crampton, J.W. (2009) 'Cartography: A Field in Tension?', *Cartographica*, 44 (1): 1–3.

Crampton, J.W. (2009) 'Cartography: Performative, Participatory, Political', *Progress in Human Geography* 33 (6): 840–848.

Crampton, J.W. (2009) 'Cartography: Maps 2.0', *Progress in Human Geography* 33 (1): 91–100.

Crampton, J.W. and Krygier, J. (2006) 'Introduction to Critical Cartography', *ACME: An International E-Journal for Critical Geographies* 4 (1): 11–33.

Crang, M. (2010a) 'The Death of Great Ships: Photography, Politics, and Waste in the Global Imaginary', *Environment and Planning A* 42: 1084–8102.

Crang, M. (2010b) 'Visual Methods and Methodologies', in Delyser, D., Herbert, S., Aitken, S., Crang, M. and McDowell, L. (eds) *The Sage Handbook of Qualitative Geography*, 208–825, London: Sage.

Crang, M. (2003) 'Qualitative Methods: Touchy, Feely, Look-See?', *Progress in Human Geography* 274: 494–504.

Crang, P. (2010) 'Cultural Geography: After a Fashion', *Cultural Geographies* 17 (20): 191–201.

Crang, P. (1994) 'It's Showtime!: On the Workplace Geographies of Display in a Restaurant in Southeast England', *Environment and Planning D: Society and Space* 12 (6): 675–704.

Creighton, M.R. (2001) 'Spinning Silk, Weaving Selves: Nostalgia, Gender, and Identity in Japanese Craft Vacations', *Japanese Studies* 21 (1): 5–29.

Cresswell, T. (2014) 'Place', in Lee, R. et al. (eds) *Sage Handbook of Human Geography*, London: Sage.

Cresswell, T. (2013a). *Place: A Short Introduction*, London: Wiley-Blackwell.

Cresswell, T. (2013b) *Soil*, London: Penned in the Margin.

Cresswell, T. (2012) 'Value, Gleaning and the Archive at Maxwell Street, Chicago', *Transactions of the Institute of British Geographers* 37: 164–876.

Cresswell, T. (2010) 'New Cultural Geography – an Unfinished Project?' *Cultural Geographies* 17 (2): 169–874.

Cresswell, T. (2006a) *On the Move: Mobility in the Western World*, London: Routledge.

Cresswell, T. (2006b) '"You Cannot Shake That Shimmie Here": Producing Mobility on the Dance Floor', *Cultural Geographies* 13 (1): 55–77.

Cresswell, T. (2003) 'Landscape and the Obliteration of Practice', in Thrift, N., Pile, S. and Anderson, K. (eds) *The Handbook of Cultural Geography*, 269–81, London: Sage.

Cresswell, T. (1996) *In Place/Out of Place: Geography Ideology and Transgression*, London: University of Minnesota Press.

Cresswell, T. and Hoskins, G. (2008) 'Place, Persistence, and Practice: Evaluating Historical Significance at Angel Island, San Francisco, and Maxwell Street, Chicago', *Annals of the Association of American Geographers* 98 (2): 392–413.

Crewe, L. (1996) 'Material Culture: Embedded Firms, Organisational Networks and Local Economic Development of Fashion Quarter', *Regional Studies* 30 (3): 257–272.

Crewe, L. and Beaverstock, J. (1998) 'Fashioning the city: Cultures of consumption in contemporary urban spaces', *Geoforum* 29 (3): 287–308.

Cropley, D.H. Croley, A.J., Kuafman, J.C. and Runco, M.A. (2010) *The Dark Side of Creativity*, Cambridge: Cambridge University Press.

Crouch, D. (2011) *Flirting with Space: Journeys into Creativity*, London: Ashgate.

Crouch, D. (2010) 'Flirting with Space: Thinking Landscape Relationally', *Cultural Geographies* 17 (1): 5–18.

Crouch, D. (2009) 'Creativity, Space and Performance: Community Gardens', in T. Edensor, Leslie, D. and Millington, S. (eds) *Spaces of Vernacular Creativity*, London: Routledge.

Crouch, D. (2001) 'Spatialities and the Feeling of Doing', *Social & Cultural Geography* 2: 61–75.

Crouch, D. and Toogood, M. (1999) 'Everyday Abstraction: Geographical Knowledge in the Art of Peter Lanyon', *Cultural Geographies* 6 (1): 72–93.

Currid, E. (2007a) *The Warhol Economy: How Fashion, Art and Music Drive*, New York City: Princeton University Press.

Currid, E. (2007b) 'The economics of a good party: social mechanics and the legitimization of art/culture', *Journal of Economics and Finance* 31 (3): 383–393.

Currid, E. (2006) 'New York as a Global Creative Hub: A Competitive Analysis Of Four Theories on World Cities', *Economic Development Quarterly* 20 (4): 330–350.

Currid, E. and Connolly, J. (2008) 'Patterns of Knowledge: The Geography of Advanced Services and the Case of Art and Culture', *Annals of the Association of American Geography* 98(2): 414–434.

Currid, E. and Williams, S. (2010a) 'The Geography of Buzz: art, Culture and the social milieu in Los Angeles and New York', *Journal of Economic Geography* 10 (3): 423–451.

Currid, E. and Williams, S. (2010b) 'Two Cities, Five Industries: Similarities and Difference Within and Between Cultural Industries in New York and Los Angeles', *Journal of Planning Education and Research* 29 (3): 322–335.

Daniels, S. (2011a) 'Geographical Imagination', *Transactions of the Institute of British Geographers* 36: 182–187.

Daniels, S. (2011b) 'Art Studio', in Agnew, J. and Livingston, D. (eds) *The Handbook of Geographical Knowledge*, 137–148, London: Sage.

Daniels, S. (2010) 'Maps of Making', *Cultural Geographies* 17 (2): 181–184.

Daniels, S. (1999) *Humphry Repton Landscape Gardening and the Geography of Georgian England*, Yale: Yale University Press.

Daniels, S. (1993) *Fields of Vision: Landscape and National Identity in England and the United States*, Princeton: Princeton University Press.

Daniels, S. (1992) 'Love and Death across an English Garden: Constable's Paintings of His Family's Flower and Kitchen Gardens', *The Huntington Library Quarterly* 55 (3): 433–457.

Daniels, S. (1989) 'Marxism, Culture, and the Duplicity of Landscape', in Thrift, N. and Peet, R. (eds) *New Models in Geography: The Political Economy Perspective* Vol. 2. 196–220, London: Unwin Hyman.

Daniels, S. (1984) 'Human Geography and the Art of David Cox', *Landscape Research* 9 (3): 14–19.

Daniels, S. and Bonehill, J. (2009) *Paul Sandby: Picturing Britain*, London: Royal Academy.

Daniels, S. and Nash, C. (2004) 'Lifepaths: Geography and Biography', *Journal of Historical Geography* 30: 449–458.

Daniels, S. and Seymour, S. (1990) 'Landscape Design and the Idea of Improvement 1730–1914', in Butlin, L.R.A. and Dodgshon, R.A. (eds) *An Historical Geography of England and Wales*, London: Academic Press.

Daniels, S., Delyser, D., Entrikin, N.J. and Richardson, D. (2012) *Envisioning Landscape: Making Worlds*, London: Routledge.

Daniels, S., Pearson, M. and Roms, H. (2010) 'Editorial: Fieldworking', *Performance Research* 15 (4): 1–5.

Darby, H.C. (1962) 'The Problem of Geographical Description', *Transactions and Papers (Institute of British Geographers)* 30: 1–14.

Daskalaki, M. and Mould, O. (2013) 'Beyond Urban Subcultures: Urban Subversions as Rhizomatic Social Formations', *International Journal of Urban and Regional Research* 37 (1): 1–18.

Datta, A. (2008) 'Building Material Differences: Material Geographies of Home(s) Among Polish Builders in London', *Transactions of the Institute of British Geographers* 33 (4): 518–531.

David, A. (2009) 'Performing for the gods? Dance and embodied ritual in British Hindu temples', *South Asian Popular Culture* 7 (3): 217–231.

Davies, G. (2010) 'Where Do Experiments End?' *Geoforum* 41 (5): 667–670.

Dear, M. Ketchum, J. Luria, S. and Richardson, D. (2012) *GeoHumanities: Art, History and Text at the Edge of Place*, New York: Routledge.

Debord, G. (1983 [1967]) *The Society of the Spectacle*. London: Rebel.

De Certeau, M. (1997) *Culture in the Plural* [trans. T. Conley], Minneapolis: University of Minnesota Press.

De Certeau, M. (1984) *The Practice of Everyday Life*, London: University of California Press.

Degen, M., DeSilvey, C. and Rose, G. (2008) 'Experiencing Visualities in Designed Urban Environments: Learning from Milton Keynes', *Environment and Planning A* 40: 1901–1920.

De Leeuw, S. (2013) *Geographies of a Lover*, Edmonton: NeWest Press.

De Leeuw, S. (2004) *Unmarked*, Edmonton: NeWest Press.

Deleuze, G. and Guattari, F. (1988) *A Thousand Plateaux*, London: Continuum.

Delyser, D. and Greenstein, P. (2015) '"Follow that Car!" Mobilities of Enthusiasm in a Rare Car's Restoration', *The Professional Geographer*, 67 (2): 255–268.

Delyser, D. (2014) 'Tracing Absence: Enduring Methods, Empirical Research and a Quest for the First Neon Sign in the USA', *Area* 46 (1): 40–49.

DeSilvey, C. (2012) Making sense of transience: an anticipatory history, *Cultural Geographies*, 19 (1): 34–54.

DeSilvey, C. (2010) 'Memory in Motion: Soundings from Milltown, Montana', *Social & Cultural Geography* 11 (5): 491–510.

DeSilvey, C. (2007a) 'Art and Archive: Memory-Work on a Montana Homestead', *Journal of Historical Geography* 33 (4): 878–900.

DeSilvey, C. (2007b) 'Salvage Memory: Constellating Material Histories on a Hardscrabble Homestead', *Cultural Geographies* 14 (3) : 401–424.

DeSilvey, C. (2006) 'Observed Decay: Telling Stories with Mutable Things', *Journal of Material Culture* 11(3): 318–338.

DeSilvey, C. (2003) 'Cultivated Histories in a Scottish Allotment Garden', *Cultural Geographies* 10 (4): 442–468.

Dettelbach, M. (2009) 'The Stimulations of Travel: Humboldt's Physiological Construction of the Tropics', in Driver, F. and Marins, L. (eds) *Tropical Visions in an Age of Empire*, Chicago: University of Chicago Press, 43–58.

Deutsche, R. (1996) *Evictions: Art and Spatial Politics*. Boston: MIT Press.

Dewsbury, J.D. (2012) 'Affective Habit Ecologies: Material dispositions and immanent inhabitations', *Performance Research – a Journal of the Performing Arts* 17: 74–89.

Dewsbury, J.D. (2010) 'Seven Injunctions: Performative, Non-Representational, and Affect-Based Research', *Handbook in Qualitative Geography*, London: Sage.

Dewsbury, J.D. and Naylor, S. (2003) 'Practising Geographical Knowledge: Fields, Bodies and Dissemination', *Area* 34: 253–60.

Dickens, C. (1998) *Our Mutual Friend*, London: Penguin.

Dickens, L. (2008a) 'Placing Post-Graffiti: The Journey of the Peckham Rock', *Cultural Geographies* 15 (4): 471–496.

Dickens, L. (2008b) '"Finders keepers": Performing the street, the gallery and the spaces in-between', *Liminalities: A Journal of Performance Studies* 4(1): 1–30.

Dillon, S. (2007) *The Palimpsest: Literature, Criticism, Theory*, London: Continuum.

Dittmer, J. (2012) *Captain America and the Nationalist Superhero: Metaphors, Narratives and Geopolitics*, Philadelphia: Temple University Press.

Dittmer, J. (2010) 'Comic Book Visualities: A Methodological Manifesto on Geography, Montage and Narration', *Transactions of the Institute of British Geographers* 35 (2): 222–236.

Dixon, D. (2009) 'Creating the Semi-Living: On Politics, Aesthetics and the More-Than-Human', *Transactions of the Institute of British Geographers* 34 (4): 411–425.

Dixon, D. (2008) 'The Blade and the Claw: Science, Art and the Creation of the Lab-Borne Monster', *Social and Cultural Geography* 9 (6): 671–692.

Dixon, D. and Cresswell, T. (2002) *Engaging Film: Geographies of Mobility and Identity*, London: Rowman and Littlefield.

Dixon, D., Hawkins, H. and Straughan, E. (2011a) 'Wonder-Full Geomorphology. Sublime Aesthetics and the Place of Art', *Progress in Physical Geography* 37 (2): 227–247.

Dixon, D., Hawkins, H. and Straughan, E. (2011b) 'When Artists Enter the Laboratory', *Science* 331 (6019): 860.

Dixon, D. and Jones, J.P. III. (2004) 'What Next? Guest Editorial', *Environment and Planning A* 36: 381–390.

Dodds, K. (2008) 'Hollywood and the Popular Geopolitics of the War on Terror', *Third World Quarterly* 29 (8): 1621–1637.

Dodds, K. (2005) 'Screening Geopolitics: James Bond and the Early Cold War Films (1962–1967)', *Geopolitics* 10 (2): 266–289.

Dodds, K. (2003) 'Licensed to Stereotype: Geopolitics, James Bond and the Spectre of Balkanism', *Geopolitics* 8 (2): 125–156.

Doherty, C. (2004) *From Studio to Situation*, London: Black Dog Publishing Limited.

Domosh, M. (1991) 'Towards a Feminist Historiography of Geography', *Transactions of the Institute of British Geographers* 16: 95–104.

Douglas, G.C.C. (2014) 'Do it yourself urban design: the social practice of infomal 'improvement' through unauthorised alteration', *City and Community* 13 (1): 5–25.

Dreher, C. (2002) 'Be creative—or die', *Salon* 6 June [online] Available from: www.salon.com [Accessed 27 January 2005].

Driver, F. (2012) 'Hidden Histories Made Visible? Reflections on a Geographical Exhibition', *Transactions of the Institute of British Geographers* 38 (3): 420–435.

Driver, F. (2003) 'On Geography as a Visual Discipline', *Antipode* 35: 227–231.

Driver, F. (2001) *Geography Militant: Cultures of Exploration and Empire*, Oxford: Wiley-Blackwell.

Driver, F. (1998) 'Scientific Exploration and the Construction of Geographical Knowledge: Hints to Travellers', *Finisterra* XXXIII (65): 21–30.

Driver, F. (1995) 'Visualizing Geography: A Journey to the Heart of the Discipline', *Progress in Human Geography* 19 (1): 123–134.

Driver, F. and Jones, L. (2009) *Hidden Histories of Exploration: Exhibiting Geographical Collections*, London: Royal Geographical Society with IBG/Royal Holloway.

Driver, F. and Martins, L. (2005) *Tropical Visions in an Age of Empire*, Chicago London: University of Chicago Press.

Driver, F., Nash, C., Predengast, K. and Swensen, I. (2002) *Landing Eight Collaborative Projects between Artists + Geographers*, London: Royal Holloway, University of London.

Drobnick, J. (2002) 'Toposmia: Art, Scent, and Interrogations of Spatiality', *Angelaki* 7 (1): 31–47.

Duffy, B.E (2015) 'The romance of work: gender and aspirational labour in the digital cultural industries', *International Journal of Cultural Policy*, 19 (4): 441–457.

Duncan, C. (1995) *Civilising Rituals: Inside Public Art Museums*, London: Routledge.

Dwyer, C. (2015) 'Reinventing Muslim space in suburbia: the Salaam Centre in north London', in S.D. Brunn (ed) *The Changing World Religion Map*, London: Springer.

Dwyer, C. (2014) 'Photographing Faith in Suburbia', *Cultural Geographies*, 22 (3): 531–538.

Dwyer, C. and Davies, G. (2007) 'Qualitative Methods III: Animating Archives, Artful Interventions and Online Environments', *Progress in Human Geography* 31 (2): 257–266.

Edensor, T. (2011) 'Entangled agencies, material networks and repair in a building assemblage: The mutable stone of St Ann's Church, Manchester', *Transactions of the Institute of British Geographers* 36 (2): 238–252.

Edensor, T. (2005a) 'The Ghosts of Industrial Ruins: Ordering and Disordering Memory in Excessive Space', *Environment and Planning D: Society and Space* 23: 829–849.

Edensor, T. (2005b) *Industrial Ruins: Space, Aesthetics and Materiality*, Oxford: Berg.

Edensor, T. and Millington, S. (2012) 'Blackpool Illuminations: revaluing local cultural production, situated creativity and working-class values', *International Journal of Cultural Policy*, 12 (2): 1–17.

Edensor, T. and Millington, S. (2009) 'Illuminations, Class Identities and the Contested Landscapes and Christmas', *Sociology* 43 (1): 103–121.

Edensor, T., Leslie, D., Millington, S. and Rantisi, N. (eds) (2009a) *Spaces of Vernacular Creativity*, London: Routledge.

Edensor, T., Leslie, D., Millington, S. and Rantisi, N. (2009b) 'Introduction: Rethinking Creativity: Critiquing the Creative Class Thesis', in Edensor, T., Leslie, D., Millington, S. and Rantisi, N. (eds) *Spaces of Vernacular Creativity Rethinking the Cultural Economy*. London: Routledge: 1–16.

Elden, S. and Hawkins, H. (2016) 'Towards an interdisciplinary visual politics', *Annals of the Association of American Geographies*, Forthcoming.

Elkins, J. (2000) *Our Beautiful, Dry and Distant Texts: Art History of Writing*, London: Routledge.

Engell, J. (2013) *The Creative Imagination: Enlightenment to Romanticism*, Cambridge, MASS: Harvard University Press.

England, K. (1994) 'Getting Personal: Reflexivity, Positionality, and Feminist Research', *The Professional Geographer* 46 (1): 80–89.

England, M. (2011) 'Community', in Del Casino, V.J., Thomas, M.E. Cloke, P. and Panelli, R. (eds) *A Companion to Social Geography*, Oxford: Wiley-Blackwell: 91–107.

Enigbokan, A. and Patchett, M. (2012) 'Speaking with Specters: Experimental Geographies in Practice', *Cultural Geographies* 19 (4): 335–346.

Entwistle, J. and Rocamora, A. (2006) 'The Field of Fashion Materialised: A Study of London Fashion Week', *Sociology* 40 (4): 735–751.

Evans, G. (2009) 'Creative Cities, Creative Spaces and Urban Policy', *Urban Studies* 46(5–6): 1003–1040.

Evans, G. (2005) 'Measure for Measure: Evaluating the Evidence of Culture's Contribution to Regeneration', *Urban Studies* 42(5/6): 959–984.

Evans, J. and Jones, P. (2008) 'Sustainable Urban Regeneration as a Shared Territory', *Environment and Planning A* 40 (6): 1416–1434.

Fenton, J. (2005) 'Space, Chance, Time: Walking Backwards through the Hours on the Left and Right Banks of Paris', *Cultural Geographies* 12 (4): 412–428.

Finnegan, R. (2007) *The Hidden Musicians: Music-Making in an English Town*, Middleton: Wesleyan University Press.

Fisher, C. (1997) '"I bought my first saw with my maternity benefit"': Craft production in west Wales and the home as the space of re (production)', in Cloke, P. and Little, J. (eds) *Contested countryside cultures: Rurality and Socio-cultural Marginalization*, London: Routledge.

Flew, T. (2013) *Creative Industries and Urban development: Creative Cities in the 21st Century*, London: Routledge.

Flew, T. (2012) *The Creative Industries: Culture and Policy*, London: Sage.

Flew, T. (2003) 'Beyond Ad Hocery: defining creative industries', in *Cultural Sites, Cultural Theory and Cultural Policy : The second international conference on cultural policy research* [Online] Available from: http://eprints.qut.edu.au/256/ [Accessed, 25/7/2015].

Florida, R. (2013) 'More Losers Than Winners in America's New Economic Geography', CityLab, 30th Jan, [Online] Available from: http://www.citylab.com/work/2013/01/more-losers-winners-americas-new-economic-geography/4465/ [Accessed 4th September, 2014].

Florida, R. (2012a) *The Rise of the Creative Class Revisited*, New York: Basic Books.

Florida, R. (2012b) 'What Critics Get Wrong About Creative Cities', CityLab, 30th May [Online] Available from: http://www.citylab.com/work/2012/05/what-critics-get-wrong-about-creative-cities/2119/ [Accessed 4th September, 2014].

Florida, R. (2010) *The Great Reset: How New Ways of Living and Working Drive Post-Crash Prosperity*, New York: Harper.

Florida, R. (2009) 'The Great Reset' [Online] Available from: http://www.the-atlantic.com/magazine/archive/2009/03/the-great-reset/307303/ [Accessed 30/7/2015].

Florida, R. (2005) *Cities and the Creative Class*, New York: Basic Books.

Florida, R. (2004) 'Revenge of the Squelchers: The Great Creative Class Debate.' [Online] Available from: CreativeClass.org [Accessed 25/7/2015].

Florida, R. (2002) *The Rise of the Creative Class: and How it's Transforming Work, Leisure, Community and Everyday Life*, New York: Basic Books.

Foster, K., and Lorimer, H. (2007) 'Some Reflections on Art-Geography as Collaboration', *Cultural Geographies* 14 (3): 425–432.

Fox, A. and Macpherson, H. (2015) *Inclusive Arts Practice and Research: A Critical Manifesto*, New York: Routledge.

Frith, S. (1997) 'The suburban sensibility in British rock and pop', In R. Silverstone (ed) *Visions of Suburbia*. Abingdon: Routledge.

Gabrys, J. and Yusoff, K. (2011) 'Art, Sciences and Climate Change: Practices and Politics at the Threshold', *Science as Culture* 21 (1): 1–24.

Gallagher, M. (2014) 'Sounding Ruins: reflections on the production of an 'audio drift' ', *Cultural Geographies* 22 (3): 467–485.

Gallagher, M. and Prior, J. (2014) 'Sonic Geographies: Exploring phonographic methods' *Progress in Human Geography* 38 (2): 267–284.

Gandy, M. (2012) *Urban Constellations*, London: Jovis.

Gandy, M. (1997) 'Contradictory Modernities: Conceptions of Nature in the Art of Joseph Beuys and Gerhard Richter', *Annals of the Association of American Geographers* 87 (4): 636–665.

Garland, W.J. (1969) 'The Ultimacy of Creativity', *Southern Journal of Philosophy* 7 (4) 361–376.

Garnett, J. and Harris, A. (2011) 'Faith in the home: Catholic spirituality and devotional materiality in East London', *Material Religion* 7(2): 299–302.

Garnham, N. (2005) 'From Cultural to creative industries: An Analysis of the implications of the "creative industries" approach to arts and media policy making in the UK', *International Journal of Cultural Policy* 11 (5): 15–29.

Garrett, B. (2016) 'Picturing Urban Subterranea: Embodied aesthetics of London's sewers, Environment and Planning A', online: doi:10.1177/0308518X16652396

Garrett, B.L. (2014) 'Undertaking recreational trespass: urban exploration and infiltration', *Transactions of the Institute of British Geographers* 39 (1): 1–13.

Garrett, B. (2013) *Explore Everything: Place-hacking the City*, London: Verso.

Garrett, B.L. (2011) 'Videographic geographies: Using digital video for geographic research', *Progress in Human Geography* 34 (4): 521–541.

Garrett, B.L. and Hawkins, H. (2013) 'And Now For Something Completely Different . . . Thinking Through Explorer Subject-Bodies: A Response to Mott and Roberts', Antipode Online. Available from: http://radicalantipode.files.wordpress.com/2013/11/garrett-and-hawkins-response.pdf [Accessed 5th September, 2014].

Gauntlett, D. (2011) *Making is Connecting*, London: Polity Press.

Gauntlett, D. (2008) *Making is Connecting: The Social Meaning of Creativity, from DIY and Knitting to YouTube and Web 2.0*, Cambridge: Polity Press.

Gelber, S.M. (1997) 'Do-it-yourself: constructing, repairing and maintaining domestic masculinity', *American Quarterly* 49: 66–112.

Gellatly, A.F. (1985) 'Historical records of glacier fluctuations in Mt Cook National Park, New Zealand: a century of change', *Geographical Journal* 151: 86–99.

Gemtou, E. (2011) 'Depictions of Sunsets as Information Sources', *Leonardo* 44 (1): 49–53.

Geoghegan, H. (2010) 'Museum Geography: Exploring Museums, Collections and Museum Practice in the UK', *Geography Compass* 4 (10): 1462–1476.

Gibson, C. (2014) *Creativity in Peripheral Places: Redefining the Creative Industries*, London: Routledge.

Gibson, C. (2010) 'Guest Editorial, Creative Geographies: tales from the margins', *Australian Geographer* 41: 1–10.

Gibson, C. and Brennan-Horley, C. (2009) 'Where is creativity in the city? Integrating qualitative and GIS Methods', *Progress in Human Geography* 41: 2595–2614.

Gibson, C. and Klocker, N. (2004) 'Academic publishing as 'creative' industry and recent discourses of creative economies: some critical reflections', *Area* 36 (4): 423–434.

Gibson, C. and Kong, L. (2005) 'Cultural Economy: A Critical Review', *Progress in Human Geography* 29 (5): 541–561.

Gibson, C., Luckman, S. and Willoughby-Smith, J. (2010) 'Creativity Without Borders?: Re-thinking Remoteness and Proximity', *Australian Geographer* 41 (1): 25–38.

Gibson, C. et al. (2010) 'Cultural Festivals and Economic Development in Nonmetropolitan Australia', *Local Cultural Development: Tensions and Opportunities* 29 (3): 280–293.

Gibson, K. (2001) 'Regional Subjection and Becoming', *Environment and Planning D: Society and Space* 19 (6): 639–667.

Gibson-Graham, J.K. (2008) 'Diverse Economies: performative practices for other worlds', *Progress in Human Geography* 32 (5): 613–632.

Giffinger, R., Hainldmaier, G. and Kramer, H. (2011) 'The role of rankings in growing city competition', *Urban Research & Practice* 3(3): 299–312.

Gilbert, D. (2013) 'A new world order? Fashion and its capitals in the twenty-first century', in Bruzzi, S. and Church Gibson, P. (eds) *Fashion Cultures Revisited: Theories, Explorations and Analysis*, London: Routledge.

Gilbert, D. (2010) 'Suburbs', in Southerton, D. (ed) *The Encyclopedia of Consumer Culture*, London: Sage.

Gilbert, D. and Breward, C. (2006) *Fashion's World Cities*, Oxford: Berg.

Gilbert, E.W. (1960) 'The Idea of the Region', *Geography* 45: 157–175.

Gill, R. and Pratt, A. (2008) 'In the social factory? Immaterial labour, precariousness, and cultural work', *Theory, Culture, and Society* 25: 1–30.

Glacken, C. (1992) *Traces on the Rhodian Shore: Nature and Culture in Western Thought from Ancient Times to the End of the Eighteenth Century*, Berkeley: University of California Press.

Godfrey, B.J. and Zhou, Y. (1999) 'Ranking World Cities: Multinational Corporations and the Global Urban Hierarchy', *Urban Geography* 20: 268–281.

Goldsmith, B. and O'Regan, T. (2003) *Cinema Cities, Media Cities: The Contemporary International Studio Complex*, Screen Industry, Culture and Policy Research Series, Sydney: Australian Film Commission.

Gómez, M.V. and González, S. (2001) 'A Reply to Beatriz Plaza's – The Guggenheim-Bilbao Museum Effect', *International Journal of Urban and Regional Research* 25 (4): 895–900.

Grabher, G. (2002) 'The Project Ecology of Advertising: Tasks, talents and teams', *Regional Studies* 36 (3): 245–262.

Grabher, G. (2001) 'Ecologies of Creativity: the village, the group, the heterarchic organisation of the advertising industry', *Environment and Planning A* 33: 351–374.

Grabher, G., Ilbert, O. and Flohr, S. (2008) 'The neglected kind: the consumer in the new knowledge ecology of innovation', *Economic Geography* 84: 253–280.

Grace, M., Gandolfo, E. and Candy, C. (2009) 'Crafting Quality of Life: Creativity and Well Being', *Journal of the Association for Research on Mothering* 11 (1): 239–250.

Graham, S. (2009) 'Cities as battlespace: the new military urbanism', *City* 13 (4): 383–402.

Green, J. and Jenkins, H. (2009) 'The Moral Economy of Web 2.0: Audience Research and Convergence Culture', in Holt, J. and Perren, A. (eds) *Media Industries: History, Theory and Method*, Malden, MA: Wiley-Blackwell, 213–225.

Gregory, D. (2004) *Colonial Present: Afghanistan, Palestine, Iraq*, Oxford: Wiley-Blackwell.

Gregory, D. (1994) *Geographical Imaginations*, Oxford: Blackwell.

Gregson, N. (2007) 'Identity, Mobility, and the Throwaway Society', *Environment and Planning D: Society and Space* 25: 682–700.

Gregson, N. (2006) *Living with Things: Ridding, Accommodation, Dwelling*, Oxford: Kingston.

Gregson, N. and Beale, V. (2004) 'Wardrobe Matter: The Sorting, Displacement and Circulation of Women's Clothing', *Geoforum* 35 (6): 689–700.

Gregson, N. and Crewe, L. (2003) *Second Hand Cultures*, Oxford: Berg.

Gregson, N., Metcalfe, A. and Crewe, L. (2009) 'Practices of object maintenance and repair: how consumers attend to consumer objects within the home', *Journal of Consumer Culture* 9: 248–272.

Gregson, N. Metcalfe, A. and Crewe, L. (2007) 'Moving Things Along: The Conduits and Practices of Divestment in Consumption', *Transactions of the Institute of British Geographers* 32: 187–200.

Gregson, N., Crang, M., Ahamed, F., Akhtar, N. and Ferdous, R. (2010) 'Following things of rubbish value: end-of-life ships, "chock-chocky" furniture and the Bangladeshi middle class consumer', *Geoforum*. 41 (6): 846–854.

Grindon, G. (2013) 'Revolutionary Romanticism: Henri Lefebvre's Revolution as Festival', *Third Text* 27 (2): 208–220.

Grodach, C. (2012) 'Before and after the creative city: the politics of urban cultural policy in Austin, Texas', *Journal of Urban Affairs* 34 (1): 81–97.

Grodach, C. (2010) 'Beyond Bilbao: Rethinking flagship cultural development and planning in three California cities', *Journal of Planning Education and Research* 29 (3): 353–366.

Grosz, E. (2008) *Chaos, Territory, Art*, New York: Columbia University Press.

Grosz, E. (1994) *Volatile Bodies: Toward a Corporeal Feminism*. Bloomington: Indiana University Press.

Guéry, F. and Dekeule, D. (2014) *The Productive Body*, London: Zero Books.

Guilbaut, S. (1985) *How New York Stole the Idea of Modern Art: Abstract Expressionsim, Freedom and the Cold War*, Chicago: University of Chicago Press.

Hackney, F. (2006) 'Using your hands for happiness: Home craft and make do and mend in British women's magazines in the 1920s and 1930s', *Journal of Design History* 19 (1): 23–38.

Hall, P. (1966) *The World Cities*, London: Weidenfeld & Nicolson.

Hall, T. (2007) 'Artful Cities', *Geography Compass* 6: 1376–1392.

Hall, T. (1997) 'Images of Industry in the Post-Industrial City: Raymond Mason and Birmingham', *Ecumene* 4 (1): 46–68.

Hallam, E. and Hockey, J. (2001) *Death, Memory and Material Culture*, Oxford: Berg.

Hallam, E. and Ingold, T. (2014) *Making and Growing*, Farnham: Ashgate.

Hallam, E. and Ingold, T. (2008) *Creativity and Cultural Improvisation*, London, Oxford: Berg.

Hamnett, C. and Whitelegg, D. (2007) 'Loft conversion and gentrification in London: from industrial to postindustrial land use', *Environment and Planning A* 39 (1): 106–124.

Hardt, M. and Negri, A. (2004) *Multitude: War and Democracy in the Age of Empire*, New York: Penguin.

Harley, J.B. (1990) 'Cartography, Ethics and Social Theory', *Cartographica* 27 (2): 1–23.

Harley, J.B. (1989) 'Deconstructing the Map', *Cartographica* 26 (2): 1–20.

Harley, J.B. and Zandvliet, K. (1992) 'Art, Science, and Power in Sixteenth-Century Dutch Cartography', *Cartographica: The International Journal for Geographic Information and Geovisualization* 29 (2): 10–19.

Harmon, K. (2003) *You Are Here. Personal Geographies and other Maps of the Imagination,* Princeton: Princeton Architectural Press.

Harris, L.M. and Hazen, H.D. (2006) 'Power of Maps: (Counter Mapping for Conservation)', *ACME* 4 (1): 99–130.

Harrison, P. (2006) 'Post structural theories', in Stuart C. A. and Valentine, G., (eds) *Approaches to Human Geography*, Sage: London.

Harrison, P. (2000) 'Making Sense: embodiment and the sensibilities of the everyday', *Environment and Planning D: Society and Space* 18: 497–517.

Hart, J. F. (1982) 'The Highest form of the Geographer's Art', *Annals of the Association of American Geographers* 72 (1): 1–29.

Hartwick, E. (2000) 'Towards a Geographical Politics of Consumption', *Environment and Planning A* 32: 1177–92.

Harvey, D. (2009) 'Is This Really the End of Neoliberalism?' Counter Punch, March 13–15. [Online] Available from: http://www.counterpunch.org/2009/03/13/is-this-really-the-end-of-neoliberalism/ [Accessed 3/9/2014].

Harvey, D. (2007) 'Neoliberalism as creative destruction', *The Annals of the American Academy of Political and Social Science* 610 (1): 21–44.

Harvey, D. (2003) 'The right to the city', *International Journal of Urban and Regional Research* 27(4): 939–994.

Harvey, D. (2000) *Spaces of Hope*, Berkeley: University of California Press.

Harvey, D. (1991) *The Condition of Post-Modernity*, Oxford: Wiley-Blackwell.

Harvey, D. (1989) 'From Managerialism to Entrepreneurialism: The Transformation in Urban Governance in Late Capitalism', *Geografiska Annaler. Series B, Human Geography* 71(1): 3–17.

Harvey, D. (1979) 'Monument and Myth', *Annals of the Association of American Geographers* 69 (3): 362–381.

Harvey, D., Hawkins, H. and Thomas, N. (2012) 'Thinking Creative Clusters beyond the city: People, places and networks', *Geoforum* 43 (3): 529–539.

Hawkins, H. (2015a) 'Creative Geographic Methods: Knowing, Representing, Intervening', *Cultural Geographies* 22 (2): 247–268.

Hawkins, H. (2015b) 'All it is is Light . . . Pipilotti Rist and the Feminist Languages of Light', *Senses and Society*, 10 (2): 158–178.

Hawkins, H. (2013a) *For Creative Geographies: Geography, Visual Art and the Making of Worlds*, London: Routledge.

Hawkins, H. (2013b) 'Geography and Art: An Expanding Field', *Progress in Human Geography* 37 (1): 52–71.

Hawkins, H. (2011) 'Dialogues and Doings: Sketching the relations between Geography and Art', *Geography Compass* 5 (7): 464–478.

Hawkins, H. (2010a) 'Turn your trash into . . . Rubbish, art and politics, Richard Wentworth's Geographical imagination', *Social & Cultural Geography* 11 (8): 805–827.

Hawkins, H. (2010b) 'The Argument of the Eye: Cultural Geographies of Installation Art', *Cultural Geographies* 17 (3): 1–19.

Hawkins, H. (2010c) 'Placing Art at the Royal Geographical Society', in Patel, V. and Ledda, T.C. (eds) *Creative Compass: New Mappings by International Artists*, London: Royal Geographical Society, 7–16.

Hawkins, H. and Lovejoy, A. (2009) *Insites: An Artists' Book*, Penryn: Insites Press.

Hawkins, H. and Straughan, E. (eds) (2015) *Geographical Aesthetics: Imagining Space, Staging Encounters*, Farnham: Ashgate.

Hawkins, H. and Straughan, E. (2014) 'Midas: The Sight/Sound of Touch', *Geoforum* 51 (1): 130–139.

Hawkins, H., Marston, S., Ingram, M. and Straughan, E. (2015) 'The Arts of Socio-Ecological Transformation', *Annals of the Association of American Geographers* 105 (2): 331–341.

Hebdige, D. (1979) *Subculture: the Meaning of Style*, London: Methuen.

Heidegger, M. (1977) *The Question Concerning Technology and Other Essays*, London: Garland Publications.

Henry, N. and Pinch, S. (2000) 'Spatialising knowledge: placing the knowledge community of Motor Sport Valley', *Geoforum* 31 (2): 191–208.

Hesmondhalgh, D. (2005) 'Media and cultural policy as public policy', *International Journal of Cultural Policy* 11 (1): 95–109.

Highmore, B. (2002a) *The Everyday Life Reader*, New York: Psychology Press.

Highmore, B. (2002b) *Everyday Life and Cultural Theory: An Introduction*, London: Routledge.

Hochschild, A.R. (1983). *The Managed Heart: Commercialization of Human Feeling*, Berkeley: University of California Press.

Hoggett, P. and Bishop, J. (1986) *Organizing Around Enthusiasms. Patterns of Mutual Aid in Leisure*, London: Comedia.

Holloway, J. (2012) 'The space that faith makes: towards a (hopeful) ethos of engagement' in Hopkins, P. Kong, L. and Olson, E. (eds) *Religion and Place: landscape, place and piety*, New York: Springer, 203–218.

Hong, E. (2014) *The Birth of Korean Cool*, New York: Picador.

Housefield, J. (2007) 'Sites of Time: Organic and Geologic Time in the Art of Robert Smithson and Roxy Paine', *Cultural Geographies* 14 (4): 537–561.

Hracs, B. (2009) 'Beyond Bohemia: Geographies of everyday creativity for musicians in Toronto', in Edensor, T., Leslie, D., Millington, S. and Rantisi, N. (eds) *Spaces of Vernacular Creativity: Rethinking the Cultural Economy*, London, Routledge: 75–88.

Hracs, B. and Leslie, D. (2013) 'Aesthetic labour in the creative industries: the case of independent musicians in Toronto, Canada', *Area* 46: 66–73.

Hracs, B.J., Jakob, D. and Hauge, A. (2013) 'Standing out in the crowd: the rise of exclusivity-based strategies to compete in the contemporary marketplace for music and fashion', *Environment and Planning A* 45 (5): 1144–1161.

Huat, C.B. and Iwabuchi, K. (eds) (2008) *East Asian Pop Culture: Analysing the Korean Wave*, Hong Kong: Hong Kong University Press.

Hughes, C. (2013) 'I draw the line at stringing pearls: the craftswoman's imperative and hopeful economies', *Journal of Cultural Economy* 6 (2): 153–167.

Hughes, C. (2012) 'Gender, Craft Labour and the Creative Sector', *International Journal of Cultural Policy* 18 (4): 439–454.

Huq, R. (2013) *Making Sense of Suburbia through Popular Culture*, London: Bloomsbury.

Ingold, T. (2013) *Making: Anthropology, Archaeology, Art and Architecture*, London: Routledge.

Ingold, T. (2010a) 'Bringing things to life: Creative entanglements in a world of materials', *Realities Working Papers 15, Realities*, London: ESRC National Centre for Research Methods.

Ingold, T. (2010b). 'The textility of making', *Cambridge Journal of Economics* 34 (1): 91–102.

Ingold, T. (2007) *Lines: A Brief History*, London: Routledge.

Ingold, T. (2001) *The Perception of the Environment: Essays in Livelihood, Dwelling and Skill*, London: Routledge.

Ingold, T. (1993) 'The temporality of landscape', *World Archeology* 25: 152–174.

Ingram, A. (2011) 'Making geopolitics otherwise: artistic interventions in global political space', *Geographical Journal* 177: 218–222.

Ingram, M. (2014) 'Washing Urban Water: The Diplomacy of Environmental Art in the Bronx, New York City', *Gender, Place and Culture* 21 (1): 105–122.

Irigaray, L. (2005) *An Ethics of Sexual Difference*, London: Continuum.

Iveson, K. (2013) 'Cities within the city: Do-It-Yourself Urbanism and the Right to the City', *International Journal of Urban and Regional Research* 37 (3): 941–956.

Jackson Lears, T.J. (1981) *No Place of Grace: Antimodernism and the Transformation of American Culture*, Chicago: University of Chicago Press.

Jackson, P. (2000) 'Rematerialising Social and Cultural Geography', *Social & Cultural Geography* 1: 9–14.

Jackson, P. (1989) *Maps of Meaning: Introduction to Cultural Geography*, London: Unwin Hyman.

Jackson, S. (2011) *Social Works: Performing Art, Supporting Publics*, London: Routledge.

Jackson, S.T. and Romanowski, S. (2009) 'Introduction: Alexander Von Humboldr and Aime Bonpland', in *The Geography of Plants*, Chicago: University of Chicago Press.

Jacobs, M. (1985) *The Good and Simple Life: Artist Colonies in Europe and America*, London: Phaidon.

Jansson, J. and Power, D. (2010) 'Fashioning a global city: global city brand channels in the fashion and design industries', *Regional Studies* 44 (7): 889–904.

Jay, M. (1993) *Downcast Eyes: The Denigration of Vision in Twentieth-Century French Thought*, Berkeley: University of California Press.

Jeffcutt, P. and Pratt, A.C. (eds) (2009) *Creativity and Innovation in the Cultural Economy*, London: Routledge.

Johnson, N. (1985) 'Cast in Stone: Monuments, Geography, and Nationalism', *Environment and Planning D: Society and Space* 13(1): 51–65.

Johnston, C. and Lorimer, H. (2014) 'Sensing the city', *Cultural Geographies* 21(4):673–680.

Jones, A. (1998) *Body Art, Performing the Subject*, Minneapolis: University of Minnesota Press.

Jones, A. and Stephenson, A. (1999) *Performing the Body: Performing the Text,* London, New York: Routledge.

Jones, A. and Warr, T. (2012) *The Artists Body,* London: Phaidon Press.

Jones, C.A. (2005) *Eyesight Alone: Clement Greenberg's Modernism and the Bureaucratization of the Senses,* Chicago: University of Chicago Press.

Jones, P. (2014) Performing sustainable transport: a RIDE across the city. *Cultural Geographies* 21 (2): 282–287.

Jones, P. and Warren, S. (2015) (eds) *Creative Economies, Creative Communities: Rethinking Place, Policy and Practice,* London: Ashgate.

Jonker, J. and Till, K.E. (2009) 'Mapping and excavating spectral traces in post-apartheid Cape Town', *Memory Studies* 2 (3): 303–335.

Joosse, S. and Hracs, B.J. (2015) 'Curating the quest for good food: the practices, spatial dynamics and influence of food related curation in Sweden', *Geoforum,* 64: 205–216.

Judovitz, D. (2002) 'De-Assembling Vision: Conceptual Strategies in Duchamp, Matta-Clark, Wilson', *Angelaki: Journal of Theoretical Humanities* 7 (1): 95–114.

Jung, S. and Shim, D. (2014) 'Social Distribution: K Pop fan practices', *International Journal of Cultural Studies,* 17 (5): 485–501.

Kanarinka (2006) 'Art-Machines, Body-Ovens and Map-Recipes: Entries for a Psychogeographic Dictionary', *Cartographic Perspectives* 53: 24–40.

Kanna, A. (2011) *Dubai, the City as Corporation,* Minneapolis: University of Minnesota Press.

Kanngieser, A. (2014) *Experimental Politics and the Making of Worlds,* Farnham: Ashgate.

Kanngieser, A. (2012) 'Creative Labour in Shanghai', *Subjectivity* 5 (1): 54–74.

Karsten, L., and Pel, E. (2000) 'Skateboarders exploring urban public space: Ollies, obstacles and conflicts', *Journal of Housing and the Built Environment* 15(4): 327–340.

Katz, C. (2004) *Growing up Global, Economic Restructuring and Children's Everyday Lives,* Minneapolis: University of Minnesota Press.

Kaufmann, T., De Costa. (2004) *Toward a Geography of Art,* Chicago: University of Chicago Press.

Kaye, N. (2000) *Site-Specific Art: Performance, Place, and Documentation,* London and New York: Routledge.

Keane, M.A. (2009) 'Creative Industries in China; four perspectives on social transformation', *International Journal of Cultural Policy* 15 (4): 431–434.

Keighren, I.M. (2005) 'Geosophy, Imagination, and Terrae Incognitae: Exploring the Intellectual History of John Kirtland Wright', *Journal of Historical Geography* 31 (3): 546–562.

Kester, G. (2011) *The One and the Many. Contemporary Collaborative Art in a Global Context,* Durham and London: Duke University Press.

Kester, G. (2004) *Conversation Pieces: Community and Communication in Modern Art,* Berkeley: University of California Press.

Kidder, J.L. (2013) 'Parkour, Masculinity, and the City', *Sociology of Sport Journal* 30(1): 1–23.

Kinman, E.L., and Williams, J.R. (2007) 'Cultural Geographies in Practice: Domain: Collaborating with Clay and Cartography', *Cultural Geographies* 14 (3): 433–444.

Knight, P. (2004) 'Glaciers: Art and History, Science and Uncertainty', *Interdisciplinary Science Reviews* 29 (4): 385–393.

Knox, P. (2012) 'Starchitects, starchitecture and the symbolic capital of world cities', in Derudder, B., Hoyler, M., Taylor, F. and Witlox, F. (eds) *International Handbook of Globalization and World Cities*, Cheltenham: Edward Elgar, 275–283.

Kong, L. (2014) 'From Cultural Industries to Creative Industries and Back? Towards Clarifying Theory and Rethinking Policy', *Inter-Asia Cultural Studies* 15 (4): 593–607.

Kong, L. (2009) 'Making sustainable creative/cultural space in Shanghai and Singapore', *Geographical Review* 99 (1): 1–22.

Kong, L. (2007) 'Cultural icons and urban development in Asia: Economic imperative, national identity, and global city status', *Political Geography* 26(4): 383–404.

Kong, L. (2005) 'The sociality of cultural industries: Hong Kong's cultural policy and film industry', *International Journal of Cultural Policy* 11 (1): 61–76.

Kong, L. (2001) 'Mapping "new" geographies of religion: politics and poetics in modernity' *Progress in Human Geography* 25(2): 211–233.

Kong, L. (2000a) 'Cultural policy in Singapore: Negotiating economic and socio-cultural agendas', *Geoforum* 31(4): 409–424.

Kong, L. (2000b) 'Culture, economy, policy: Trends and developments', *Geoforum* 31 (4): 385–390.

Kong, L. and O'Connor, J. (eds) (2009) *Creative Economies, Creative cities, Asian European Perspectives*, New York: Springer.

Kramer, R. (2010) 'Painting with permission: legal graffiti in New York City', *Ethnography* 11(2): 235–253.

Krätke, S. (2010) '"Creative Cities" and the Rise of the Dealer Class: A Critique of Richard Florida's Approach to Urban Theory', *International Journal of Urban and Regional Research* 34(4): 835–853.

Krauss, R. (1994) 'Richard Serra: Skulptur/Sculpture', in Richard Serra and Christoph Borckhaus *Richard Serra: Props*, Duisburg: Wilhelm Lehmbruck Museum, 28–109.

Krauss, R. (1988) *Richard Serra Sculpture*, New York: MOMA.

Krygier, J.B. (1995) 'Cartography as an Art and a Science?' *Cartographic Journal*, 32 (1): 3–10.

Kucklich, J. (2005) 'Precarious Playbour: Modders and the Digital Game Industry', *Fibreculture*, 5 (11), [Online] Accessed from: http://five.fibreculturejournal. org/fcj-025-precarious-playbour-modders-and-the-digital-games-industry/ [Accessed 15/8/2016].

Kurgen, L. (2013) *Close up at a Distance: Mapping, Technology, and Politics*, Cambridge: MIT Press.

Kwan, M.P. (2007) 'Affecting Geospatial Technologies: Toward a Feminist Politics of Emotion', *The Professional Geographer* 59 (1): 22–34.

Kwan, M.P. (2002) 'Feminist Visualization: Re-Envisioning GIS as a Method in Feminist Geographic Research', *Annals of the Association of American Geographers* 92 (4): 645–661.

Kwan, M.P. (2002a) 'Introduction: Feminist Geography and GIS', *Gender, Place & Culture* 9 (3): 261–262.

Kwan, M.P. (2002b) 'Is GIS for Women? Reflections on the Critical Discourse in the 1990s', *Gender, Place & Culture* 9 (3): 271–279.

Kwon, M. (2001) *Site Specificity in Art. The Ethnographic Turn*, London: Black Dog Press.

Kwon, M. (1997) 'One Place after Another: Notes on Site Specificity', *October* 80: 85–110.

Kwon, S.H. and Kim, J. (2014) 'The Cultural Industry Policies of the Korean Government and the Korean Wave', *International Journal of Cultural Policy*, 20 (4) 422–439.

Lacy, M.J. (2003) 'War, Cinema and Moral Anxiety', *Alternatives*, 28 (5): 611–636.

Lacy, S. (1995) *Mapping the Terrain: New Genre Public Art*, San Fransisco: Bay Press.

Lambert, R. (2012) *Digital Storytelling, Capturing Lives, Creating Community*. London: Routledge.

Landry, C. (2011) 'Beyond the Creative City', Talk given at PICNIC Festival, September, 2011. [Online] Accessed from: http://vimeo.com/30733079 [Accessed 3/9/2014].

Landry, C. (2006) *The Art of City-making*, London: Earthscan.

Landry, C. (2000) *The Creative City: A Toolkit for Urban Innovators*, London: Earthscan.

Landry, C. and Bianchini, F. (1995) *The Creative City*, London, Demos.

Last, A. (2012) 'Experimental Geographies', *Geography Compass*, 6 (12): 706–724.

Latham, A. (2003) 'Research, Performance, and Doing Human Geography: Some Reflections on the Diary-Photograph, Diary-Interview Method', *Environment and Planning A* 35 (11): 1993–2017.

Lazzarato, M. (2012) 'Immaterial Labour', [Online] Accessed from: http://www.generation-online.org/c/fcimmateriallabour3.htm [Accessed from: 23/7/2015].

Lea, J. (2009) 'Becoming skilled: the cultural and corporeal geographies of teaching and learning Thai Yoga Massage', *Geoforum* 40(3): 465–474.

Leach, B. (1940) *A Potter's Book*, London: Faber and Faber.

Leadbeater, C. and Miller, P. (2004) 'The pro-am revolution: How enthusiasts are changing our economy and society', *Demos*, [Online] Accessed from: http://www.demos.co.uk/files/proamrevolutionfinal.pdf [Accessed 25/7/2015]

Leadbeater, C. and Oakley, K. (1999) *The Independents: Britain's New Cultural Entrepreneurs*, London: Demos.

Lees, L. (2008) 'Gentrification and Social Mixing: Towards an Inclusive Urban Renaissance', *Urban Studies* 45 (12): 2449–2470.

Lees, L. (2003) 'Super-gentrification: The case of Brooklyn Heights, New York City', *Urban Studies* 40 (12): 2487–2509.

Lefebvre, H. (1996 [1967]) *The Right to the City*, in Kofman, E. and Lebas, E. (eds) *Writings on Cities*, London, Blackwell: 147–159.

Lefebvre, H. (1991) *The Production of Space*, Oxford: Blackwell.

Ley, D. (2003) 'Artists, Aestheticisation and the field of Gentrification', *Urban Studies*, 40 (12): 2527–2544.

Leyshon, A. (2014) *Reformatted. Code, Networks and the Transformation of the Music Industry*, Oxford: Oxford University Press.

Leyshon, A. (2009) 'The software slump? Digital music, the democratization of technology, and the decline of the recording studio sector within the musical economy', *Environment and Planning A* 41 (6): 1309–1331.

Leyshon, A., Matless, D. and Revil, G. (1998) *The Place of Music*, Guildford: Guild Press.

Li, S.M., Cheng, H.H. and Wang, J. (2014) 'Making a Cultural Cluster in China: A Study of Dafen Oil Painting Village, Shenzhen', *Habitat International* 41: 156–164.

Lindtner, S. (2015) 'Hacking with Chinese Characteristics: The Promise of the Maker Movement against Chinese Manufacturing Culture', *Science, Technology and Human Values*, 40 (5): 854–879.

Livingstone, D. (2003) *Putting Science in Its Place: Geographies of Scientific Knowledge*, Chicago: University of Chicago Press.

Lloyd, R. (2006) *Neo-bohemia: Art and Commerce in the Post-industrial City*, London: Routledge.

Loftus, A. (2012) *Everyday Environmentalism: Creating an Urban Political Ecology*, Minnesota: University of Minnesota Press.

Loftus, A. (2009) 'Intervening in the Environment of the Everyday', *Geoforum*, 40 (3): 326–34.

Lomax, Y. (2004) *Sounding the Event: Escapades in Dialogue and Matters of Art, Nature and Time*, London: I.B. Taurus.

Lomax, Y. (2000) *Writing the Image. An Adventure with Art and Theory*, London: I.B. Taurus.

Longhurst, R., Ho, E. and Johnston, L. (2008) 'Using "the Body" as an "Instrument of Research": Kimch'i and Pavlova', *Area* 40 (2): 208–217.

Lorimer, H. (2010) 'Forces of Nature, Forms of Life: Calibrating Ethology and Phenomenology', in Anderson, B. and Harrison, P. (eds) *Taking Place: Non-Representational Theories and Geographies*, Farnham: Ashgate, 55–78.

Lorimer, H. (2006) 'Herding Memories of Humans and Animals', *Environment and Planning D: Society and Space* 24: 497–518.

Lorimer, H. (2003) 'Telling Small Stories: Spaces of Knowledge and the practice of Geography', *Transactions of the Institute of British Geographers* 28 (2): 197–217.

Lorimer, H. and MacDonald, F. (2002) 'A Rescue Archeology, Taransay', *Cultural Geographies* 9: 95–102.

Lowenthal, D. (1961) 'Geography, Experience, and Imagination: Towards a Geographical Epistemology', *Annals of the Association of American Geographers* 51 (3): 241–260.

Lowenthal, D. and Prince, H.C. (1965) 'English Landscape Tastes', *Geographical Review* 55 (2): 186–222.

Lübbren, N. (2001) *Artists' Colonies in Europe 1870–1910,* Manchester: Manchester University Press.

Luckman, S. (2015) *Craft and the Creative Economy,* Basingstoke and New York: Palgrave Macmillan.

Luckman, S. (2013) 'The Aura of the Analogue in a Digital Age: Women's Crafts, Creative Markets and Home-Based Labour After Etsy', *Cultural Studies Review* 19 (1): 249–270.

Luckman, S., Gibson, C. and Lea, T. (2009) 'Mosquitoes in the Mix: How Transferable Is Creative City Thinking?', *Singapore Journal of Tropical Geography* 30: 70–85.

MacDonald, F. (2013) 'The Ruins or Erskine Beveridge', *Transactions of the Institute of British Geographers* 39 (4): 477–389.

MacFarlene, R. (2015) *Landmarks,* London: Penguin.

Mackenzie, A.D. (2006a) '"Against the Tide": Placing Visual Art in the Highlands and Islands, Scotland', *Social & Cultural Geography* 7 (6): 965–985.

Mackenzie, A.F.D. (2006b) 'Claims to Place: The Public Art of Sue Jane Taylor', *Gender, Place and Culture* 13 (6): 605–27.

Mackenzie, A.F.D. (2006c) 'Leinn Fhàin Am Fearann? (the Land Is Ours): Re-Claiming Land, Re-Creating Community, North Harris, Outer Hebrides, Scotland', *Environment and Planning D: Society and Space* 24 (4): 577–98.

Mackenzie, A.F.D. (2004) 'Place and the Art of Belonging', *Cultural Geographies* 11 (2): 115–137.

Mackenzie, A.F.D. (2002) 'Re-Claiming Place: The Millennium Forest, Borgie, North Sutherland, Scotland', *Environment and Planning D: Society and Space* 20 (5): 535–560.

MacKian, S. (2012) *Everyday Spiritualities,* New York: Palgrave, Macmillan.

Macpherson, H. (2010) 'Non-Representational Approaches to Body–Landscape Relations', *Geography Compass* 4 (1): 1–13.

Macpherson, H. (2008) 'Cultural Geographies in Practice: Between Landscape and Blindness: Some Paintings of an Artist with Macular Degeneration', *Cultural Geographies* 15 (2): 261–269.

Maddrell, A. et al. (2014) *Christian Pilgrimage, Landscape and Heritage: Journeying to the Sacred,* London: Routledge.

Madge, C. (2014) 'On the creative re(turn) to Geography: Poetry, politics and passion', *Area* 46 (2): 178–185.

Maffesoli, M. (1995) *The Time of the Tribes: The Decline of Individualism in Mass Society,* London: Sage.

Maitland, J. (1976) 'Creativity', *Journal of Aesthetics and Art Criticism* 34 (4): 397–409.

Malecki, E.J. (2000) 'Knowledge and regional competitiveness', *Erdkunde (54):* 334–351.

Mann, J. (2015) 'Towards a Politics of Whimsy: Yarn Bombing the City', *Area* 47 (1): 65–72.

Manning, E. (2008) 'Creative Propositions for Thoughts in Motion', *Inflexions,* 1, [Online] Accessed from: http://www.inflexions.org/n1_manninghtml.html [Accessed 15/8/2016].

Marcuse, P. (2009) 'From critical urban theory to the right to the city', *City* 12 (2–3): 195–197.

Markusen, A. (2006) 'Urban development and the politics of a creative class: evidence from a study of artists', *Environment and Planning A* 38(10): 1921–1940.

Markusen, A. and Gadwa, A. (2010) *Creative Placemaking*, Washington: National Endowment for the Arts. [Online] Available from: http://arts.gov/sites/default/files/CreativePlacemaking-Paper.pdf [Accessed 28/7/2015].

Marshall, M.A. (1920) *Principles of Economics*, London: Macmillan.

Marston, S. (2000) 'The Social Construction of Scale', *Progress in Human Geography* 24 (2): 219–242.

Marston, S. and De Leeuw, S. (2013) 'Creativity and Geography: Toward a politicized Intervention', *Geographical Review* 103 (2): iii–xxvi.

Marx, K. (1976) *Capital. Trans Ben Fowlkes*, London: Penguin.

Maskell, P., Bathelt, H. and Malmberg, A. (2006) 'Building Global Knowledge Pipelines: The Role of Temporary Clusters', Druid working paper [Online]. Available from http://openarchive.cbs.dk/bitstream/handle/10398/7883/DRUID_05_20.pdf?sequence=1 [Accessed 27/7/2015].

Massey, D. (2005) *For Space*, London: Sage.

Mathews, V. (2010) 'Aestheticizing Space: Art, Gentrification and the City', *Geography Compass* 4 (6): 660–675.

Matless, D. (2010) 'Describing Landscapes: Regional Sites', *Performance Research: A Journal of Performing Arts* 15 (4): 72–82.

Matless, D. and Revill, G. (1995) 'A Solo Ecology: The Erratic Art of Andy Goldsworthy', *Cultural Geographies* 2 (4): 423–448.

Mayer, M. (2013) 'First world urban activism', *City* 17(1): 5–19.

Mayes, R. (2010) 'Post-cards from somewhere: "marginal" cultural production creativity and community', *Australian Geographer* 41 (1): 11–23.

McAuliffe, C. (2012) 'Graffiti or street art? Negotiating the moral geographies of the creative city', *Journal of Urban Affairs* 34(2): 189–206.

McCormack, D.P. (2014) *Refrains for Moving Bodies*, Durham, NC: Duke University Press.

McCormack, D.P. (2008a) 'Geographies for Moving Bodies: Thinking, Dancing, Spaces', *Geography Compass* 2 (6): 1822–1836.

McCormack, D.P. (2008b) 'Engineering affective atmospheres on the moving geographies of the 1897 Andrée expedition', *Cultural Geographies*, 15(4): 413–430.

McDonough, T. (ed) (1994) *Guy Debord and Situationist International*, Boston: MIT Press.

McFarlane, C. (2011) 'The city as assemblage: dwelling and urban space', *Environment and Planning D* 29(4): 649–671.

McGuire, M. (2008) *Lived Religion*, Oxford: Oxford University Press.

McGuirk, J. (2014) *Radical Cities: Across Latin America in Search of a New Architecture*, London: Verso.

McNally, D. (2014) 'Art as a Technology of Connection' in Hawkins, H. and Straughan, E. (2014) *Geographical Aesthetics*, London, Routledge, 114–136.

McRobbie, A. (1998) *British Fashion Design: Rag Trade or Image Industry?* London: Routledge.

Meinig, D.W. (1983) 'Geography as an Art', *Transactions of the Institute of British Geographers* 8 (3): 314–328.

Meinig, D.W. (1979) *The Interpretation of Ordinary Landscapes: Geographical Essays*, London, New York: OUP.

Merchant, S. (2011) 'The Body and the Senses: Visual Methods, Videography and the Submarine Sensorium', *Body and Society* (17): 53–72.

Merrifield, A. (2011) *Magical Marxism*, London: Pluto Press.

Milbourne, P. (2012) 'Everyday (in)justices and ordinary environmentalisms: community gardening in disadvantaged urban neighbourhoods', *Local Environment: The International Journal of Justice and Sustainability* 17 (9): 943–957.

Miles, M. (2010) 'Representing Nature: Art and Climate Change', *Cultural Geographies* 17 (1): 19–35.

Miles, M. (2009) 'Aesthetics in a Time of Emergency', *Third Text* 23 (4): 421–433.

Miles, M. (2006) 'Geographies of Art and Environment', *Social & Cultural Geography* 7 (6): 987–993.

Miles, M. (2004) *Urban Avant-Gardes*, London: Routledge.

Miles, M. (1997) *Art Space and the City*, London: Routledge.

Millen, J. (2010) 'Romantic Creativity and the Ideal of Originality: A Contextual Analysis', *Cross-Sections* 6: 93–104.

Miller, D. (ed) (2005) *Materiality*, Durham, NC: Duke University Press.

Miller, D. (ed) (2001) *Home Possessions*, Oxford: Berg.

Miller, D. (1998) *A Theory of Shopping*, Cambridge: Polity.

Miller, D. (1997) (ed) *Material Cultures: Why Some Things Matter*, London: University College London.

Ministerio de Cultura del Ecuador (2011) *Políticas para una revolución cultural*, Quito: Ministerio de Cultura del Ecuador.

Mitchell, A.J. (2010) *Heidegger Among the Sculptors. Body, Space and the Art of Dwelling*, Redwood: Stanford University Press.

Molotch, H. (1996) 'LA as Design Product: How Art Works in a Regional Economy', in Scott, A. and Soja, E. (eds), *The City: Los Angeles and Urban Theory at the End of the Twentieth Century*, Berkeley and Los Angeles: University of California Press, 225–275.

Mommaas, H. (2004) 'Cultural clusters and Post-Industrial City: towards the remapping of Urban Cultural Policy', *Urban Studies* 41(3): 507–532.

Montgomery, J. (2008) *The New Wealth of Cities: City Dynamics and the Fifth Wave*, Farnham: Ashgate.

Montgomery, J. (2003) 'Cultural Quarters as Mechanisms for Urban Regeneration. Part 1: Conceptualising Cultural Quarters', *Planning Practice & Research* 18(4): 293–306.

Moore, M. and Prain, L. (2009) *Yarn Bombing: The Art of Crochet and Knit Graffiti*, Vancouver: Arsenal Pulp Press.

Moretti, E. (2012) *The New Geography of Jobs*, New York: Houghton Mifflin Harcourt.

Morris, A. (2005) 'The Cultural Geographies of Abstract Expressionism: Painters, Critics, Dealers and the Production of an Atlantic Art', *Social & Cultural Geography* 6 (3): 421–437.

Morris, J. (2012) 'Beyond Coping? Alternatives to consumption within a social network of Russian workers', *Ethnography* 14 (1): 85–103.

Morris, N. J., and Cant, S. G., (2006) 'Engaging with Place: Artists, Site-Specificity and the Hebden Bridge Sculpture Trail', *Social & Cultural Geography* 7 (6): 863–888.

Mott, C., and Roberts, S.M. (2014) 'Not everyone has (the) balls: Urban exploration and the persistence of masculinist geography', *Antipode* 46(1): 229–245.

Mould, O. (2015) *Urban Subversion and the Creative City*, London: Routledge.

Mould, O. (2014) 'Tactical Urbanism: The New Vernacular of the Creative City', *Geography Compass* 8 (8): 529–539.

Mould, O. (2009) 'Parkour, the city, the event', *Environment and Planning D: Society and Space* 27: 738–750.

Mould, O. and Comunian, R. (2014) 'Hung, drawn and cultural quartered: rethinking cultural quarter development policy in the UK', *European Planning Studies*, 23 (12): 2356–2369.

Mould, O. and Joel, S. (2010) 'Knowledge networks of "buzz" in London's advertising industry: a social network analysis approach', *Area* 42 (3) 281–292.

Mumford, L. (1961) *The City in History: Its Origins, Its Transformations, and Its Prospects*, New York: Harcourt, Brace & World.

Murphy, J.T. (2006) 'The sociospatial dynamics of creativity and production in Tanzanian industry: urban furniture manufacturers in a liberalizing economy', *Environment and Planning A* 38 (10): 1863–1882.

Nachum, L. and Keeble, D. (2003) 'Neo-Marshillian Clusters and Global Networks – The Linkages of Media Firms in Central London', *Long Range Planning* 36 (5) 459–480.

Nagar, R. (2002) 'Women's theatre and the redefinitions of public, private and politics in North India', *ACME An International e-journal for Critical Geography* 1 (1): 55–72.

Nash, C. (2000) 'Performative in Practice: Some Recent Work in Cultural Geography', *Progress in Human Geography* 24 (4): 653–664.

Nash, C. (1996) 'Reclaiming Vision: Looking at Landscape and the Body', *Gender, Place and Culture* 3: 149–169.

Nash, C. (1996) *Irish Geographies*, Nottingham: Djanogly Art Gallery.

Neate, H. (2012) 'Provinciality and the Art World: The Midland Group 1961–1977', *Social and Cultural Geography* 13 (3): 275–294.

Neff, G. (2012) *Venture Labour: Work and the Burden of Risk in Innovative Industries*, Cambridge Massachusetts: MIT Press.

Neff, G. Wissinger, E, and Zukin, S. (2005) 'Entreprenuerial Labor among Cultural Producers: 'Cool' Jobs in 'Hot' Industries', *Social Semiotics* 15 (3): 307–334.

Negus, K. and Pickering, M. (2004) *Creativity, Communication and Cultural Value*, London, Sage.

Nicholson, H. (2015) 'Absent Amateurs', *Research in Drama Education* 20 (3): 263–266.

Nicholson, H. (1997) 'In amateur hands: framing time and space in home-movies', *History Workshop Journal* 43: 198–213.

Nicodemus, A.G. (2013) 'Fuzzy vibrancy: creative placemaking as ascendant US cultural policy', *Cultural Trends,* 22 (304): 213–222.

Nolan, S. (2014) '"Social Urbanism" experiment breathes new life into Colombia's Medellin', The Globe and the Mail, 22nd December 2014. [Online] Accessed from: http://www.theglobeandmail.com/news/world/social-urbanism-experiment-breathes-new-life-into-colombias-medellin/article22185134/ [Accessed 25/7/2014].

Nold, C. (2009) *Emotional Cartography* [Online] Accessed from: https://scholar.google.co.uk/scholar?q=nold+biomapping&btnG=&hl=en&as_sdt=0%2C5 [Accessed: 27/7/2015].

Norcliffe, G. and Rendace, O. (2003) 'New Geographies of comic book production in North America: the new artisan, distancing and the period social economy', *Economic Geography* 79 (3): 241–263.

Nye, J.S. (2004) *Soft Power: The Means to Success in World Politics*, New York: Public Affairs, US.

Nye, J.S. (1990) 'Soft Power', *Foreign Policy* 80: 153–171.

Oakley, K. (2004) 'Not So Cool Britannia: The Role of the Creative Industries in Economic Development', *International Journal of Cultural Studies* 7(1): 67–77.

O'Brien, D. (2013) *Cultural Policy: Management, Value and Modernity in the Creative Industries*, London: Routledge.

Ocejo, R.E. (2015) 'Cathedrals of Craft: Workplaces for the New Elite Service, Retail, and Manual Labor Jobs', *Metropoloitiques*, [Online] Accessed from: https://www.metropolitiques.eu/IMG/pdf/met-ocejo3.pdf, [Accessed, 19/07/2015].

Ocejo, R.E. (2014a) *Upscaling Downtown: From Bowery Saloons to Cocktail Bars in New York City*, Princeton: Princeton University Press.

Ocejo, R.E. (2014b) 'Show the Animal: Constructing and Communicating New Elite Food Tastes at Upscale Butcher Shops', *Poetics* 47: 106–121.

Ocejo, R.E. (2012) 'At your Service: the meanings and practices of contemporary bartenders', *European Journal of Cultural Studies* 15 (5): 642–658.

Ocejo, R.E. (2010) 'What'll it be? Cocktail bartenders and the redefinition of service in the creative economy', *City, Culture and Society* 1 (4): 179–184.

O'Connor, E. (2007) 'Embodied knowledge in glassblowing: the experience of meaning and the struggle towards proficiency', *The Sociological Review,* 55: 126–141.

O'Connor, E. (2006) 'Glassblowing Tools: Extending the body towards Practical Knowledge and Informing a Social World', *Qualitative Sociology* 29 (2): 177–193.

O'Connor, E. (2005) 'Embodied Knowledge: Meaning and the Struggle Towards Proficiency in Glassblowing', *Ethnography* 6 (2): 183–204.

O'Connor, J. (2007) 'The cultural and creative industries: a review of the literature', London: Arts Council England.

O'Connor, J. and Xin, G. (2006) 'A new Modernity? The arrival of 'creative industries in China', *International Journal of Cultural Studies* 9 (3): 271–283.

O'Donnell, M. (2014) 'Six months in Biashizhou', [Online] Available from: maryannodonnell.files.wordpress.com/2014/04/czc-six-months-in-baishizhou.pdf [Accessed 26/7/2015].

Oliver, P., Davis, I. and Bentley, I. (1981) *Dunroamin: The Suburban Semi and its Enemies,* London: Barrie & Jenkins.

Ooi, C.S (2011) 'Subjugated in the Creative Industries: The Fine Arts in Singapore', *Culture Unbound: Journal of Current Cultural Research,* 3: 119–137.

Ooi, C.S. (2010) 'Political pragmatism and the creative economy: Singapore as a city for the arts', *International Journal of Cultural Policy* 16 (4): 403–417.

Ooi, C.S. (2008) 'Reimagining as a creative nation: the politics of place branding', *Place Branding and Public Diplomacy,* 4: 287–302.

O'Tuathail, G. and Dalby, S. (1998) Introduction – Rethinking geopolitics: Towards a critical geopolitics in O'Tuathail, G. and Dalby, S. *Critical Geopolitics,* London: Routledge.

Pace, T. et al. (2013) 'From Organisational to Community Creativity: Paragon Leadership and Creativity Stories at Etsy', Proceedings of CSCW. [Online] Accessed from: http://dl.acm.org/citation.cfm?id=2441892 [Accessed: 19/7/2015].

Parr, H. (2007) 'Collaborative Film-Making as Process, Method and Text in Mental Health Research', *Cultural Geographies* 14 (1): 114–138.

Parr, H. (2006) 'Mental Health, the Arts and Belongings', *Transactions of the Institute of British Geographers* 31 (2): 150–66.

Parr, H. (1998) 'Mental Health, Ethnography and the Body', *Area* 30 (1): 28–37.

Patel, V. and Cisneros-Ledda, T. (2010) *The Creative Compass: New Commissions by Agnes Poitevin-Navarre and Susan Stockwell,* London: Royal Geographical Society.

Paterson, M. (2009) 'Haptic Geographies: ethnography, haptic knowledges and sensuous dispositions', *Progress in Human Geography* 33 (6): 766–788.

Paton, D. A. (2013) 'The quarry as sculpture: the place of making', *Environment and Planning A* 45 (5): 1070–1086.

Pearson, M. (2010) *Site-Specific Performance,* London: Palgrave.

Pearson, M. (2007) *In Comes I: Performance, Memory and Landscape,* Exeter: Exeter University Press.

Pearson, M. and Shanks, M. (2001) *Theatre/Archaeology: Disciplinary Dialogues,* London: Routledge.

Peck, J. (2012) 'Recreative city: Amsterdam, vehicular ideas, and the adaptive spaces of creativity policy', *International Journal of Urban and Regional Research* 36 (3): 462–485.

Peck, J. (2009) 'The cult of urban creativity', in Keil, R. and Mahon, R. (eds) *Leviathan Undone? Towards a Political Economy of Scale*, Vancouver: University of British Columbia Press, 159–176.

Peck, J. (2005) 'Struggling with the Creative Class', *International Journal of Urban and Regional Research* 29 (4): 740–770.

Peck, J. and Tickell, A. (2002) 'Neoliberalizing space', *Antipode* 34(3): 380–404.

Peck, J., Theodore, N. and Brenner, N. (2013) 'Neoliberal urbanism redux?', *International Journal of Urban and Regional Research* 37(3): 1091–1099.

Pevesner, N. (1956) *The Englishness of English Art*, London: Penguin.

Phelan, P. (2003) *Unmarked: The Politics of Performance*, London: Routledge.

Phillips, A. (2005) 'Cultural Geographies in Practice: Walking and Looking', *Cultural Geographies* 12 (4): 507–513.

Phillips, P. (2009) 'Clotted Life and Brittle Waters', *Landscapes: The International Centre for Landscape and Language* 3 (3): 1–35.

Phillips, P. (2004) 'Doing Art and Doing Cultural Geography: The Fieldwork/Field Walking Project', *Australian Geographer* 35 (2): 151–159.

Pickerill, J. and Chatterton, P. (2006) 'Notes towards autonomous geographies: creation, resistance and self-management as survival tactics', *Progress in Human Geography* 30 (6): 730–746.

Pico, W. and Pico, A. (2011) *Cuerpo Festivo: personajes escénicos en doce fiestas populares del Ecuador*, Quito: Ministerio de Cultura del Ecuador.

Pike, A. and Pollard, J. (2010) 'Economic Geographies of Financialization', *Economic Geography* 86 (1): 29–51.

Pile, S. (2005) *Real Cities: Modernity, Space and the Phantasmagorias of City Life*, Oxford: Sage.

Pinder, D. (2011) 'Errant paths: the poetics and politics of walking', *Environment and Planning D: Society and Space* 29(4): 672–692.

Pinder, D. (2008) 'Urban Interventions: Art, Politics and Pedagogy', *International Journal of Urban and Regional Research* 32 (3): 730–36.

Pinder, D. (2007) 'Cartographies Unbound' *Cultural Geographies* 14 (3): 453–462.

Pinder, D. (2005a) *Visions of the City: Utopianism, Power and Politics in Twentieth-Century Urbanism*, Edinburgh: Edinburgh University Press.

Pinder, D. (2005b) 'Arts of Urban Exploration', *Cultural Geographies* 12 (4): 383–411.

Pinder, D. (2001) 'Ghostly Footsteps: Voices, Memories and Walks in the City', *Ecumene* 8 (1): 1–19.

Pinder, D. (1996) 'Subverting Cartography: The Situationists and Maps of the City', *Environment and Planning A* 28: 405–427.

Pink, S. (2001) *Doing Visual Ethnography: Images, Media and Representation in Research*, London: Sage.

Pinkerton, A. and Benwell, M. (2014) 'Rethinking popular geopolitics in the Falklands/Malvinas sovereignty dispute: creative diplomacy and citizen statecraft', *Political Geography* 38: 12–22

Plant, S. (1992) *The Most Radical Gesture. The Situationist International in a Postmodern Age*, London: Routledge.

Plaza, B. (2000) 'Evaluating the Influence of a Large Cultural Artifact in the Attraction of Tourism: The Guggenheim Museum Bilbao Case', *Urban Affairs Review* 36 (2): 264–274.

Plaza, B. (1999) 'The Guggenheim-Bilbao Museum Effect: A reply to Maria V. Gomez, reflective images: the case of urban regeneration in Glasgow', *International Journal of Urban and Regional Research* 23 (3): 589–592.

Plaza, B., Tironi, M. and Haarich, S. (2009) 'Bilbao's Art Scene and the "Guggenheim effect" Revisited', *European Planning Studies* 17(11): 1711–1729.

Podmore, S. (1998) '(Re) Reading the 'Loft Living' Habitus in Montreal's Inner City', *International Journal of Urban and Regional Research* 22 (2): 283–302.

Polanyi, M. (1966) *The Tacit Dimension*, London: Routledge.

Pollack, V. and Sharp, J. (2007) 'Constellations of Identity: Place-Ma(R)King Beyond Heritage', *Environment and Planning D: Society and Space* 5 (25): 1061–1078.

Pollard, J.S. (2007) 'Making Money, (Re)Making Firms; Micro-business financial networks in Birmingham's Jewellery Quarter', *Environment and Planning A* 39 (2): 378–397.

Pollock, G. (2006) 'Encountering Encounter', in Griselda Pollock and Vanessa Corbel (eds) *Encountering Eva Hesse*, London: Prestel, 13–23.

Pope, R. (2005) *Creativity: Theory, History, Practice*, New York: Psychology Press.

Porter, M.E. (2000) 'Locations, Clusters and Company Strategy' in Clarke, G.L., Feldmann, M. and Getler, M. (eds) *The Oxford Handbook of Economic Geography*, Oxford and New York: Oxford University Press, 253–274.

Porter, M.E. (1996) 'Competitive Advantage, Agglomeration Economies, and Regional Policy', *International Regional Science Review* 19: 85–90.

Postigo, H. (2009) 'America Online volunteers- Lessons from an early co-production community', *International Journal of Cultural Studies* 12 (5): 451–469.

Postle, M. (2009) 'Everything including the kitchen sink : from barry to bacon', in Giles Waterfield (ed) *The Artist's Studio*, Compton Verney: Hogart Arts, 42–63.

Postrel, V.I. (2003) *The Substance of Style: How the Rise of Aesthetic Value is Remaking Commerce, Culture and Consciousness*, New York: Harper Collins.

Powell, H. (2009) 'Time, Television, and the Decline of DIY', *Home Cultures* 6 (1): 89–107.

Power, M. and Crampton, A. (2005) 'Reel Geopolitics: Cinemator-graphing Political Space,' *Geopolitics*, 10: 193–203.

Power, D and Hallencreutz, D. (2002) 'Profiting from Creativity? The Music Industry in Stockholm, Sweden and Kingston, Jamaica', *Environmental and Planning A* 34 (10): 1833–1854.

Power, D. and Jansson, J. (2008) 'Cyclical Clusters in Global Circuits: Overlapping Spaces in Furniture Trade Fairs', *Economic Geography* 84 (4): 423–448.

Power, D. and Scott, A. (2004) *Cultural Industries and the Production of Culture*, London: Routledge.

Pratt, A.C. (2012) 'Factory, Studio, Loft: there goes the neighbourhood?' in Baum, M. and Christiaanse, K. (eds) *City as Loft: Adaptive Reuse as a Resource for Sustainable Development*, Zurich: Gta Verlag, 25–31.

Pratt, A, (2009) 'Policy Transfer and the field of the cultural and creative industries: what can be learned from Europe', in Kong, L. and O'Connor, J. (eds) *Creative Economies, Creative Cities, Asian European Perspectives* New York: Springer, 9–23.

Pratt, A. (2008) 'Creative cities: the cultural industries and the creative class', *Geografiska Annaler: Series B, Human Geography* 90(2): 107–117.

Pratt, A.C. (2006) 'Advertising and Creativity, a Governance Approach: a Case Study of Creative Agencies in London', *Environment and Planning A* 38 (10): 1883–1899.

Pratt, A.C. (2004) 'Creative Clusters: Towards the governance of the creative industries production system?', *Media International Australia* (112): 50–66.

Pratt, G. and Johnston, C. (2010) 'Nanay (Mother): a testimonial play', *Cultural Geographies* 17: 123–133.

Pratt, G. and Johnston, C. (2007) 'Turning Theatre into Law, and Other Spaces of Politics', *Cultural Geographies* 14 (1): 92–113.

Pratt, G. and San Juan, R.M. (2014) *Film and Urban Space*, Edinburgh: Edinburgh University Press.

Pratt, M.L. (1992) *Imperial Eyes: Studies in Travel Writing and Transculturation*, London: Routledge.

Pred, A. (1995) *Recognising European Modernities: A Montage of the Present*, London: Routledge.

Price, L. (2015) 'Knitting and the City', *Geography Compass* 9 (2): 81–95.

Prince, H. (1988) 'Art and Agrarian Change, 1710–1815', in Cosgrove, D. and Daniels, S. *The Iconography of Landscape*, Cambridge: Cambridge University Press, 98–118.

Pudup, M.B. (2008) 'It takes a Garden: Cultivating citizen-subjects in organized garden projects', *Geoforum* 39: 1228–1240.

Quinn, S., Walters, T. and Whiteoak, J. (2003) 'Tale of three (media) cities', *Australian Studies in Journalism* 12: 129–149

Rantisi, N. M. (2014) 'Exploring the role of industry intermediaries in the construction of "local pipelines": the case of the Montreal Fur Garment cluster and the rise of fur-fashion connections', *Journal of Economic Geography* 14 (5): 955–971.

Rantisi, N.M., and Leslie, D. (2006) 'Branding the design metropole: the case of Montréal, Canada', *Area* 38(4): 364–376.

Rantisi, N.M. Leslie, D. and Christopherson, S. (2006) 'Placing the creative economy: scale, politics and the material', *Environment and Planning A* 38 (10): 1789–1797.

Raunig, G., Ray, G. and Wuggenig, U. (eds) (2011) *Critique of Creativity: Precarity, Subjectivity and Resistance in the Creative Industries*, London: Mayfly Books.

Rendell, J. (2010) *Site Writing. The Architecture of Art Criticism*, London: I.B. Tauris.

Rendell, J. (2006) *Art and Architecture: A Space Between*, London: I.B. Tauris.

Revill, G. (2004) 'Performing French folk music: dance, authenticity and nonrepresentational theory', *Cultural Geographies* 11: 199–209.

Roberts, J. (2009a) 'Introduction: Art, Enclave Theory and the Communist Imaginary', *Third Text* 23 (4): 353–367.

Roberts, J. (2009b) 'Preface', *Third Text* 23 (4): 351.

Robinson, J. (2002) 'Global and World Cities: A View from off the Map', *International Journal of Urban and Regional Research* 26(3): 531–554.

Roelvink, G., St Martin, K. and Gibson-Graham, J.K., (2015) *Making Other Worlds Possible: Performing Diverse Economies*, Minneapolis: University of Minnesota Press.

Rogers, A. (2014) *Performing Asian Transnationalisms: Theatre, Identity and the Geographies of Performance*, New York: Routledge.

Rogers, A. (2011) 'Butterfly takes Flight: The translocal circulation of creative practice', *Social and Cultural Geography* 12 (7): 663–683.

Rogoff, I. (2000) 'In Conversation', in Dogherty, C. (ed.) *From Studio to Situation*. London: Black Dog Publishing, 104–108.

Roodhouse, S. (2006) *Cultural Quarters: Principles and Practice*, Bristol: Intellect.

Roodhouse, S. (2001) 'The Creative Industres: Definitions, Quantification and Practice', in Eisneberg, C., Gerlach, R. and Handke, C. (eds) *Cultural Industries the British Experience in International Perspective* [Online] Available from: *http://edoc.hu-berlin.de* [Accessed 29/7/2015].

Rose, G. (2007) *Visual Methodologies*, London: Sage.

Rose, P. (2002) *On Whitehead*, Belmont: Wadsworth.

Rosenbaum, S. (2011) *Curation Nation: How to Win in a World Where Consumers are Creators*, London: McGraw-Hill.

Rosenberg, B.C. (2011) 'Home Improvement: Domestic Taste, DIY, and the Property Market', *Home Cultures* 8 (1) 5–23.

Ross, A. (2008) 'The New Geography of Work Power to the Precarious?' *Theory, Culture & Society* 25(7–8): 31–49.

Ross, A. (2007) 'Nice work if you can get it. The mecurial career of creative industries policy', *Work Organisation, Labour and Globalisation* 1 (1): 13–30.

Rothman, J. (2014) 'Creativity Creep', *The New York Times*, September 2nd [Online] Accessed from: http://www.newyorker.com/books/joshua-rothman/creativity-creep [Accessed 26/7/15].

Routledge, P. (2010) 'Sensuous Solidarities: Emotion, Politics and Performance in the Clandestine Insurgent Rebel Clown Army', *Antipode* 44(2): 428–452.

Rugg, J. (2010) *Exploring Site-Specific Art. Issues of Space and Internationalism*, London, Oxford: I.B. Tauris.

Rugg, J. and Martin, C. (2012) *Spatialities: The Geographies of Art and Architecture*, London: Intellect.

Ryan, J. (2013) *Photography and Exploration*, London: Reaktion.

Rycroft, S. (2011) *Swinging City: A Cultural Geography of London. 1950–1974*. London: Ashgate.

Rycroft, S. (2007) 'Towards an Historical Geography of Nonrepresentation: Making the Countercultural Subject in the 1960s', *Social & Cultural Geography* 8 (4): 615–33.

Rycroft, S. (2005) 'The Nature of Op Art: Bridget Riley and the Art of Nonrepresentation', *Environment and Planning D: Society and Space* 23 (3): 351–71.

Sadasivam, B. (2000) 'Community Justice: West Bengal's Women Draw on Village Tradition to Stop Domestic Violence', *Ford Foundation Report Winter 2000: A Special Issue on Women: Now it's a Global Movement*, Ford Foundation, 6–9.

Sadler, S. (1998) *The Situationist City*, Boston: MIT Press.

Said, E. (1978) *Orientalism*, Penguin: London

Sava, O. (2010) 'The collective turkish home in Vienna: aesthetic narratives of migration and belonging', *Home Cultures* 7 (3): 313–340.

Scott, A. (2014) 'Beyond the Creative City: Cognitive–Cultural Capitalism and the New Urbanism', *Regional Studies* 48 (4): 565–578.

Scott, A. (2005) *On Hollywood: The Place, the Industry*, Redwood: Princeton University Press.

Scott, A. (2004) 'The other Hollywood: The organizational and Geographic bases of television-program production', *Media, Culture, Society* 26 (2): 183–205.

Scott, A. (2002) 'A New Map of Hollywood: the Production and Distribution of American Motion Pictures', *Regional Studies* 36 (9): 957–975

Scott, A. (2000) *The Cultural Economy of Cities: The Geography of the Image Producing Industries*, London: Sage.

Scott, A. (1999) The cultural economy: geography and the creative field, *Media, Culture and Society*, 21 (6): 807–817.

Seib, P. (ed) (2007) *New Media and the New Middle East*, London: Palgrave Macmillan.

Self, W. (2014) 'Give the freedom of our city to the urban explorers', *The Evening Standard*, 25/4/15 [Online] Accessed from: http://www.standard.co.uk/comment/will-self-give-the-freedom-of-the-city-to-our-urban-explorers-9286780.html [Accessed 27/7/2015].

Self, W. (2007) *Psychogeography*, London: Bloomsbury.

Sennett, R. (2009) *The Craftsman*, London: Penguin.

Senplades (2013) *Plan Nacional de Desarrollo/Plan Nacional para el Buen Vivir 2013-2017*, Quito: SENPLADES (Secretaría Nacional de Planificación y Desarrollo).

Seymour, S., Daniels, S. and Watkins, C. (1998) 'Estate and Empire: Sir George Cornewall's Management of Moccas, Herefordshire and La Taste, Grenada, 1771–1819', *Journal of Historical Geography* 24 (3): 313–51.

Seymour, S., Daniels, S. and Watkins, C. (1995) 'Picturesque Views of the British West Indies', *The Picturesque* 10: 22 28.

Sharp, J. (2007) 'The Life and Death of Five Spaces: Public Art and Community Regeneration in Glasgow', *Cultural Geographies* 14 (2): 274–92.

Shaw, I.G.R. and Sharp, J.P. (2013) 'Playing with the future: social irrealism and the politics of aesthetics, *Social and Cultural Geography* 14(3): 341–359.

Shaw, I. (2010) 'Playing war', *Social and Cultural Geography* 11(8): 789–803.

Shaw, I., and Warf, B. (2009) 'Worlds of affect: virtual geographies of video games', *Environment and Planning A* 41(6): 1332–1343.

Shaw, K. (2013) 'Independent creative subcultures and why they matter', *International Journal of Cultural Policy* 19(3): 333–352.

Shaw, W.S. (2006) 'Sydney's SoHo Syndrome? Loft Living in the Urbane City', *Cultural Geographies* 13 (2): 182–206.

Simpson, P. (2012) 'Apprehending everyday rhythms: Rhythmanalysis, time-lapse photography, and the space-times of street performance', *Cultural Geographies* 19(4): 423–445.

Simpson, P. (2011a) 'Street Performance and the City: Public Space, Sociality, and Intervening in the Everyday', *Space and Culture* 14(4): 415–430.

Simpson, P. (2011b) 'So, as you can see . . .': some reflections on the utility video methodologies in the study of embodied practices', *Area*, 43(3): 343–352.

Sinclair, I. (2003) *London Orbital*, Penguin: London.

Singer, I. (2010) *Modes of Creativity*, Boston: MIT Press.

Sjöholm, J. (2014) 'The art studio as archive: tracing the geography of artistic potentiality, progess and production', *Cultural Geographies* 21 (3): 505–514.

Sjöholm, J. (2012) *Geographies of the Artist's Studio*, London: Squid and Tabernacle.

Slater, A. (2011) *The Dress of Working Class Women in Bolton and Oldham, Lancashire 1939–1945*. Ph.D., Manchester: Manchester Metropolitan University.

Slater, T. (2011) 'Gentrification of the City', in Bridge, G. and Watson, S. (eds) *The New Blackwell Companion to the City*, London: Wiley, 571–585.

Smith, A. and von Krogh Strand, I. (2010) 'Oslo's new Opera House: Cultural flagship, regeneration tool or destination icon?', *European Urban and Regional Studies* 18(1): 93–110.

Smith, B. (1992) *Imagining the Pacific in the Wake of the Cook Voyages*, Melbourne: Melbourne University Press.

Smith, B. (1988) *European Vision and the South Pacific 1768–1850: A Study in the History of Art and Ideas*, New Haven: Yale University Press.

Smith, N. (2006) 'The Endgame of Globalization', *Political Geography* 25 (1): 1–14.

Smith, N. (1996) *The New Urban Frontier: Gentrification and the Revanchist City*, New York: Psychology Press.

Smith, N. (1987) 'Gentrification and the rent gap', *Annals of the Association of American Geographers* 77(3): 462–465.

Smith, N. and Katz, C. (1993) 'Grounding Metaphor: Towards a Spatialized Politics,' in Keith, M. and Pile, S. (eds) *Place and the Politics of Identity*, London: Routledge.

Solnit, R. (2000) *Hollow City: The Siege of San Francisco and the Crisis of American Urbanism*, London: Verso.

Sorensen, T. (2009) 'Creativity in rural development: an Australian response to Florida (or a view from the fringe)', *International Journal of Foresight and Innovation Policy* 5 (1): 24–43.

Sternberg, R.J. (eds) (1988) *The Nature of Creativity*, Cambridge: Cambridge University Press.

Storper, M. and Christopherson, S. (1987) 'Flexible Specialization and Regional Industrial Agglomerations: the Case of the US Motion Picture Industry', *Annals of the Association of American Geographers*, 77 (1): 104–117.

Storper, M. and Scott, A.J. (2009) 'Rethinking human capital, creativity and urban growth', *Journal of Economic Geography* 9: 147–167.

Storper, M. and Venables, A.J. (2004) 'Buzz: face to face contact and the urban economy', *Journal of Economic Geography* 4 (4): 351–370.

Strasser, S. (2000) *Waste and Want: A Social History of Trash*, New York: Owl Books.

Straughan, E. (2015) 'Crafts of Taxidermy: ethics of skin', *Geohumanities*, online first view.

Stupples, P. (2015) 'Beyond the predicted: expanding our understanding of creative agency in international development through practice and policy', *International Journal of Cultural Policy*, online first view.

Stupples, P. (2014) 'Creative Contributors: The role of the arts and the cultural sector in development', *Progress in Development Studies* 14 (2): 115–130.

Sunderberg, E. (2000) *Space, Site, Intervention. Situating Installation Art*, Minneaopolis: University of Minnesota Press.

Swyngedouw, E. (2002) 'The strange respectability of the situationist city in the society of the spectacle', *International Journal of Urban and Regional Research* 26(1): 153–165.

Tachibana, S. and Watkins, C. (2010) 'Botanical trans-culturation: Japanese and British knowledge and understanding of Aucuba japonica and Larix leptolepis 1700–1920', *Environment and History* 16 (1): 43–71.

Tatarkiewicz, W. (2012) *A History of Six Ideas: An Essay in Aesthetics*, London, New York: Springer.

Taylor, P. (2004) *World City Network*, London: Routledge.

Terranova, T. (2004) *Network Culture: Politics for the Information Age*, London: Pluto Press.

Terranova, T. (2000) 'Free labour: producing culture for the digital economy', *Social Text* 2 (18): 35–58.

Thomas, N.J., Hawkins, H. Harvey, D.C. (2010) 'The geographies of the creative industries: scale, clusters and connectivity', *Geography* 95: 14–19.

Thompson, G. (2000) 'Logos and Psychoanalysis: The role of truth and creativity in Heidegger's conceptual of language', *Psychologist Psychoanalyst* 20 (4): 1–12.

Thompson, J. (2009) *Performance Affects: Applied Theatre and the End of Effect*, Basingstoke: Palgrave Macmillan.

Thompson, N. (2009) *Experimental Geographies: Radical Approaches to Landscape, Cartography, and Urbanism*, London: Melville House Publishing.

Thrift, N. (2007) *Non-Representational Theory: Space, Politics, Affect*, London: Routledge.

Till, K. (2012) 'Wounded Cities: Memory Work and Place-Based Ethics of Care', *Political Geography* 31 (1): 3–14.

Till, K. (ed) (2010) *Mapping Spectral Traces*, Virginia: Virginia Tech.

Till, K. (2008) 'Artist and Activist Memory Work: Approaching Place Based Practice', *Memory Studies* 1 (1): 99–113.

Till, K. (2005) *The New Berlin: Memory, Politics, Place*, Minneapolis: University of Minnesota Press.

Tokumitsu, M. (2014) In the Name of Love. *Slate,* January 16th 2014. Last accessed 20//7/2-15, http://www.slate.com/articles/technology/technology/2014/01/do_what_you_love_love_what_you_do_an_omnipresent_mantra_that_s_bad_for_work.single.html

Tolia-Kelly, D.P. (2012) 'The Geographies of Cultural Geography II: Visual Culture', *Progress in Human Geography* 36 (1): 135–142.

Tolia-Kelly, D.P. (2006) 'Mobility/Stability: British Asian Cultures Of "Landscape and Englishness" ', *Environment and Planning A* 38 (2): 341–58.

Tolia-Kelly, D. (2004) 'Locating processes of identification: studying the precipitates of re-memory through artifacts in the British Asian home', *Transactions of the Institute of British Geographers* 29 (3): 314–329.

Toscano, A. (2009) 'The sensuous religion of the multitude: Art and abstraction in Negri', *Third Text* 23 (4): 369–382.

Townley, B., Beech, N. and McKinlay, A. (2009) 'Managing in the creative industries: Managing the motley crew', *Human Relations* 62 (7): 939–962.

Tuan, Y.F. (1977) *Space and Place: The Perspective of Experience*, Minneapolis, MN: University of Minnesota Press.

Turok, I. (2003) 'Cities, clusters and creative industries: the case of film and television in Scotland', *European Planning Studies* 11 (5): 549–565.

Umney, C. and Kretsos, L. (2015) 'That's the Experience: Passion, Work Precarity and Life Transitions Among London Jazz Musicians', *Work and Occupations* 42 (3): 313–334.

Vanolo, A. (2008) 'The image of the creative city: some reflections on urban branding in Turin', *Cities* 25 (6): 370–382.

Veblen, T. (1994) *The Theory of the Leisure Class*, New York: Dover.

Waitt, G. and Gibson, C. (2009) 'Creative Small Cities: Rethinking the Creative Economy in Place', *Urban Studies* 46 (5–6): 1223–1246.

Wallace, J. (2013) 'Yarn Bombing, Knit Graffiti and Underground Brigades: A Study of Craftivism and Mobility', *Journal of Mobile Media: Sound Moves,* 7(1), [Online] Accessed from: http://wi.mobilities.ca/yarn-bombing-knit-graffiti-and-underground-brigades-a-study-of-craftivism-and-mobility/ [Accessed 15/8/2016].

Wang, J. (2004) 'The global reach of a new discourse: How Far can "creative industries" travel?', *International Journal of Cultural Studies* 7 (1): 9–19.

Warhurst, C. and Nickson, D. (2001) *Looking Good, Sounding Right*, London: Industrial Society.

Warner, M. (1985) *Monuments and Maidens: The Allegory of the Female Form*, Berkeley: University of California Press.

Warren, A. and Evitt, R. (2012) 'Indigenous hip hop: overcoming marginality, encountering constraints', in Gibson, C. (ed) *Creativity in Peripheral Places: Redefining the Creative Industries*, London: Routledge, 142–159.

Warren, A. and Gibson, C. (2014) *Surfing Places, Surfboard Makers*, Honolulu: University of Hawai'i Press.

Warren, A. and Gibson, C. (2011) 'Blue collar creativity: reframing custom-car culture in the imperiled industrial city', *Environment and Planning A* 14 (11): 2705–2722.

Warren, S. (2013) 'Audiencing James Turrell's Deershelter Skyspace: Encounters Between Art and Audience at Yorkshire Sculpture Park', *Cultural Geographies* (20) 83–102.

Watson, J. (1983) 'The Soul of Geography,' *Transactions of the Institute of British Geographers* (8): 385–399.

Watson, M. and Shove, E. (2008) 'Product, Competence, Project and Practice: DIY and the dynamics of craft consumption', *Journal of Consumer Culture* 8 (1): 69–89.

Watson, S. (2009) 'Performing religion: Migrants, the church and belonging in Marrickville, Sydney', *Culture and Religion* 10 (3): 317–338.

Wei Li, L. (2006) *From Urban Enclave to Ethnic Suburb: New Asian Communities in the Pacific Rim,* Honolulu: University of Hawai'i Press.

Weismantel, M. (2003) 'Mothers of the Patria: La Chola Cuencana and La Mama Negra', in Whitten, N. (ed) *Millennial Ecuador: Critical Essays on Cultural Transformations and Social Dynamics*, Iowa City: University of Iowa Press, 325–354.

Weizmann, E. (2012) *Hollow Land. Israel's Architecture of Occupation*, London: Verso.

Weizmann, E. (2010) 'Interview on the Green Ling', [Online] Available from: http://francisalys.com/greenline/weizman.html [Accessed 29/7/2015].

Whatmore, S. (2002) *Hybrid Geographies: Natures, Cultures, Spaces*, London: Sage.

Wheeler, K. (1983) *The Creative Mind in Coleridge's Poetry*, Mass: Harvard University Press.

Williams, R. (1989) 'Politics and Policies: The Case of the Arts Council', in Williams, R. (1996 ed.). *The Politics of Modernism*. Verso: London.

Williams, R. (1958) 'Culture is Ordinary', in Williams, R. (1989a) *Resources of Hope: Culture, Democracy, Socialism*, London: Verso, 3–14.

Willis, P. (1998) 'Notes on Common Culture: Towards a Grounded Aesthetics', *European Journal of Culture* 1 (2): 163–176.

Willis, P. (1990) *Common Culture: Symbolic Work at Play in the Everyday Cultures of the Young*, London: Open University Press.

Wilson, D. and Keil, R. (2008) 'The real creative class', *Social and Cultural Geography* 9 (8): 841–847.

Woman's Weekly (1947). Make What You Have Look New!, 23 August, p. 233.

Wong, W. (2014) *Van Gogh on Demand*, Chicago: University of Chicago Press.

Wong, W. (2010) '*After the Copy: Creativity, Originality and the Labor of Appropriation: Dafen Village, Shenzhen, China, 1989–2010*'. Ph.D. Thesis [Online] Available from: http://dspace.mit.edu/handle/1721.1/61556 [Accessed: 25/7/2015].

Woodward, K. and Lea, J. (2010) 'Geographies of Affect', in Smith, S., Pain, R., Marston, S.A. and Jones III, J.P. (eds) *The Sage Handbook of Social Geographies*, London: Sage, 154–175.

Woodward, K., Vigdor, L., Jones, J.P. III, Hawkins, H., Marston, S.A. and Dixon, D.P. (2015) 'One Sinister Hurricane: Rethinking Collaborative Visualization', *Annals of the Association of American Geographers* 105: 496–511.

Wright, C. (2014) *Curation is Creation*. London: Asymmetrical Press.

Wright, J.K. (1947) 'Terrae Incognitae: The Place of Imagination in Geography', *Annals of the Association of American Geographers* 37: 1–15.

Wylie, J. (2005) 'A single day's walking: narrating self and landscape on the South West Coast Path', *Transactions of the Institute of British Geographers*, 30 (2) 234–247.

Yarwood, R. and Shaw, J. (2010) '"N-Guaging" Geographies: craft consumption, indoor leisure and model railways', *Area* 42 (4): 425–433.

Yorgason, E. and della Dora, V. (2009) 'Geography, religion and emerging paradigms: problematizing the dialogue', *Social & Cultural Geography* 10 (6): 629–637.

Yusoff, K. (ed) (2008) *Bipolar*, London: Arts Catalyst.

Yusoff, K. (2007) 'Antarctic Exposure: Archives of the Feeling Body', *Cultural Geographies* 14: 211–233.

Yusoff, K. and Gabrys, J. (2006) 'Time Lapses: Robert Smithson's Mobile Landscapes', *Cultural Geographies* 13 (3): 444–50.

Zaring, J. (1977) 'Romantic Face of Wales', *Annals of the Association of American Geographers* 67(3): 397–418.

Zebracki, M. (2013) 'Beyond public artopia: Public art as perceived by its publics', *GeoJournal* 78(2): 303–317.

Zerefos, C.S. Gerogiannis, V.T. and Balis, D. (2007) 'Atmospheric Effects of Volcanic Eruptions as Seen by Famous Artists and Depicted in Their Paintings', *Atmospheric Chemistry and Physics* 7 (15): 4027–4042.

Zimmerman, J. (2008) 'From brew town to cool town: Neoliberalism and the creative city development strategy in Milwaukee', *Cities* 25(4): 230–242.

Zukin, S. (2010) *Naked City: The Death and Life of Authentic Urban Places*, Oxford: Oxford University Press.

Zukin, S. (1989) *Loft Living: Culture and Capital in Urban Change*, New Brunswick: Rutgers University Press.

Zukin, S. and Maguire, J.S. (2004) 'Consumers and Consumption', *Annual Review of Sociology*, 30: 173–191.

Index